International Conference on Agricultural and Biological Sciences (ABS 2016)

IOP Conference Series: Earth and Environmental Science
Volume 41

Shanghai, China
23 – 26 July 2016

Editor:

Mohammad H. Golabi

ISBN: 978-1-5108-3051-6
ISSN: 1755-1307

Printed from e-media with permission by:

Curran Associates, Inc.
57 Morehouse Lane
Red Hook, NY 12571

Printed by Curran Associates, Inc. (2016)

For permission requests, please contact the Institute of Physics
at the address below.

Institute of Physics
Dirac House, Temple Back
Bristol BS1 6BE UK

Phone: 44 1 17 929 7481
Fax: 44 1 17 920 0979

techtracking@iop.org

Additional copies of this publication are available from:

Curran Associates, Inc.
57 Morehouse Lane
Red Hook, NY 12571 USA
Phone: 845-758-0400
Fax: 845-758-2633
Email: curran@proceedings.com
Web: www.proceedings.com

Table of contents

Volume 41

2nd International Conference on Agricultural and Biological Sciences (ABS 2016)
23–26 July 2016, Shanghai, China

Accepted papers received: 11 August 2016
Published online: 27 September 2016

Preface

Papers

Preface

The 2nd International Conference on Agricultural and Biological Sciences (ABS 2016) was successfully held from July 23rd to July 26th in Shanghai, China, which were one of the largest urban areas in the world and the showpiece of the booming economy of mainland China.

Agriculture is the cultivation and production of animals, plants, fungi, and other life forms for food, fiber, and biofuel, medicinal and other products used to sustain and enhance human life. Biological Sciences deals with the study of the many living organisms, which is closely related to agricultural sciences. Agriculture was the key development in the rise of sedentary human civilization. Promoting scientific and technological progress in the field of agricultural and biological sciences has become the consensus of all the countries in the world. Further challenges come from the climate change caused by global warming, environment pollutions, shortage of water resources and various animal and plant diseases pose new requirements for sustainable agricultural development. ABS conference series was held accordingly to provide a platform for scientists and researchers from all over the world to exchange their researches and new findings in those fields. Based on the success of ABS2015, ABS2016 attracts more than 100 participants from all over the world. With the theme of plant and animal sciences, the key focus of ABS2016 is on the areas of "Botanical Sciences", "Animal Sciences", "Soil and Environmental Science" and "Other fields in Agricultural Sciences". A special session named "Soil and environmental Science" also enriched the content of the Conference.

More than 300 papers were submitted to ABS2016, and 82 submitted manuscripts have met the scope of IOP Conference Series on Earth and Environmental Sciences to form this special book. After pre-review on the originality and language, peer review process was arranged by the editorial committee. The editorial board was led by the Guest Editor Professor Mohammad Golabi and Associated Professor Peiyue Li, with support from following editors: Dr. Dionisios Gasparatos, Dr. Zisheng Xing, Dr. Ademir de Oliveira Ferreira, Dr. Elaheh Moussavi Kiassari, Dr. Khaled Ibrahimi, Dr. B. C. Dhananjaya, Hamed Kashi, Maria Regina Gusmao Valerio Menino and Dr. Ankit Garg. And 33 best manuscripts were selected after the pre-review and peer review, and those manuscripts provide some insights in the field of soil and environmental sciences.

On behalf of the conference organizers, we would like to take this opportunity to express our sincere thanks to all the reviewers for their tremendous efforts and dedication to the Conference, to all the authors for their trust and contributions to the Conference, as well as all the colleagues from IOP publisher for their support and their endless efforts towards the publication of the proceedings for the Conference.

The International Conference on Agricultural and Biological Sciences is convened annually where experts and scholars gather to exchange their newest ideas and experience in this field, and we believe that with their earnest support and contributions, future ABS Conferences would scale new heights.

Guest Editor

Prof. Mohammad H. Golabi

University of Guam, Guam-USA

IOP Conf. Series: Earth and Environmental Science **41** (2016) 011002 doi:10.1088/1755-1315/41/1/011002

Peer review statement

All papers published in this volume of *IOP Conference Series: Earth and Environmental Science* have been peer reviewed through processes administered by the proceedings Editors. Reviews were conducted by expert referees to the professional and scientific standards expected of a proceedings journal published by IOP Publishing.

IOP Conf. Series: Earth and Environmental Science **41** (2016) 012001 doi:10.1088/1755-1315/41/1/012001

Analysis of the relationship between rusty root incidences and soil properties in *Panax ginseng*

Q X Wang[1], C L Xu[1], H Sun[1], L Ma[1], L Li[1], D D Zhang[1] and Y Y Zhang[1,2]

[1]Institute of Special Wild Economic Animal and Plant Sciences, Chinese Academy of Agricultural Sciences, No. 4899 Juye Street, Jingyue District, Changchun, Jilin Province 130112, China.

E-mail address: zyy1966999@sina.com

Abstract. Rusty root is a serious problem in ginseng cultivation that limits the production and quality of ginseng worldwide. The Changbai Mountains are the most famous area for ginseng cultivation in China. To clarify the relationship between rusty root and soil characteristics, physico-chemical properties and enzymatic activities of soil collected from five different fields in the Changbai Mountains were analyzed and a controlled experiment carried out by increasing the concentration of Fe (II). Soil bulk density, moisture, total iron (Fe) and total manganese (Mn) concentrations and polyphenol oxidase (PPO) activity were significantly higher in rusty root than healthy root groups (two-sample test, $P<0.05$ or $P<0.01$), respectively. Pearson test showed that there was a significant positive correlation between rusty root index and pH, N, Fe, Mn, Al, Zn and Ca of soil samples collected from fields ($P<0.05$ or $P<0.01$), and a significant positive correlation also occurred between rusty root index and Fe (II) added to soil in Fe (II) inducing rusty root ($P<0.01$). Physiological factors may be very important roles giving rise to ginseng rusty root. Fe (III) reduction and Fe (II) oxidation could be important in increasing the incidence of rusty root. Soil moisture and bulk density of non-rhizosphere soil not attached to the root surface, and pH, N and PPO content of rhizosphere soils attached to the root surface were heavily involved in the reduction, oxidation and sequestration of metal ions.

1. Introduction

Panax ginseng C.A.Mey is a valuable herbal root with multifunctional properties, including anti-cancer, anti-emetic, anti-oxidative, anti-angiogenic, anti-proliferation and apoptosis properties [1-4]. During cultivation, pharmacological/bioactive constituents of root in *P. ginseng* tend to increase with plant age [5]. But a widespread problem with ginseng roots such as rusty root is caused by continuous cultivation over years [6,7]. Ginseng roots are commercially graded primarily on their size, shape and overall appearance. Roots with an off white to beige color and smooth surface without blemishes are highly valued [8,9]. It is the development of reddish-brown superficial discolored areas that eventually become dry and slough-off [10]. When roots are dried, the affected areas are dark and sunken; a root shape that is distorted limits the production and quality of ginseng worldwide [11].

The pathogenesis of rusty root has always been controversial, it having been suggested that it arises due to physiological stresses [12-14]. It has inferred that the production of phenolic compounds and their transportation to the outside skin of ginseng is the cause, and iron compounds deposited on the root surface may be another rusty root inducing factor [8,15,16]. Most studies have focused on element content and the production of phenolic compounds in ginseng roots, whereas only a few have explored soil characteristics. The rhizosphere is immensely important to terrestrial life on Earth, being very closely related to plant growth. Thus, the rhizosphere, the component of soil adherent to the root, is where most of the important soil-plant relationships occur [17]. There may be a very close relationship between the incidence of rusty root and ginseng rhizosphere soil properties. Ferrous ions in the soil might be the leading factor related to rusty root. Li et al. [18] and Wang et al. [19] suggested that active reducing organic substances formed in soil under high moisture conditions could promote the activation of iron oxides and enhanced the accumulation of its divalent iron, and rusty root could be attributed to the oxidizing and depositing of ferrous iron (Fe (II)) in epidermis of ginseng root.

The object of this study was characterization and comparison of the physico-chemical properties and enzymatic activities of soil collected from Changbai Mountains to identify any possible relationship between rusty root of ginseng and soil characteristics. A controlled experiment was carried out by increasing the concentration of Fe (II) to see if this induced rusty root.

2. Materials and methods

2.1 Soil sampling and analysis

The sampling areas were located in Fusong country, which is one of the most well-known areas for ginseng cultivation in the Changbai Mountains. Rhizosphere soils of ginseng roots were sampled from five ginseng cultivation fields (A to E in Table 1) suggested by FuSong ginseng research institute. The 20 × 20 m plots were sampled in autumn 2014 during the ginseng harvesting. The middle of each field was divided into 4-6 plots (1×1m) to avoid border undesired effects on experiment accuracy. Ten ginseng roots were harvested randomly from each plot, and 40-60 ginseng roots in each field were used for rusty root index statistics. Four to six soil samples were collected in each field consisting of soils from 10 ginseng roots of the same rusty-root grade. The rusty root index statistics were calculated according to Liu et al. [14] and Li et al. [18] as follow:

$$\text{Rusty root index} = \frac{\sum(n \times \text{rusty root grade})}{\text{N} \times \text{the highest rusty root grade}}$$

where n equals the number of ginsengs with different rusty root grade, N equals the number of total ginsengs. The rusty-root grades ranged from 0 to 4, which were defined by healthy ginseng root without rust, and a rust area <10%, 10-25%, 25-50%, and >50% (Table 1).

Non-rhizosphere soils not attached to the root surface for measuring bulk density and moisture were collected from a depth of 5-10 cm using a 100 cm^3 stainless steel cylinder, and dried at 105°C. Rhizosphere soil samples for chemical and enzymatic activity analysis of adhering soil were collected by gently shaking roots into sterile bags, which were sealed and placed in ice [20]. Samples were air-dried and passed through a 0.15 mm sieve for metal and organic matter content analysis, and 2 mm sieve for other properties, respectively. Soil pH was determined with a glass electrode pH-meter (PHS-3C; Shanghai, China) in a suspension of 1:2.5 soil/water ratio (w/v) [21]. Organic matter content was determined by the Walkley and Black method [22]. Available nitrogen was detected by alkaline hydrolysis; available phosphorous was determined by the sodium hydroxide-molybdenum stibium anti-color method, and available potassium was measured by the ammonium acetate-flame photometer method [23].

For analysis of metal content (total Ca, Mg, Al, Fe, Mn, Cu, Ni), samples were digested with 15 mL HNO_3 and 20 mL $HClO_4$ (70%), and analyzed by ICP-AES [24]. And the measured metals in our study were selected according to Rahman and Punja [11].

Table 1. Samples from five fields.

Field name	Cultivation years	Latitude (N)	Longitude (E)	Soil sample	Rusty-root grade	Rusty root index
A	5	41°58'40"	127°31'38"	A-1	4(>50%)	
				A-2	4(>50%)	
				A-3	4(>50%)	
				A-4	1(<10%)	
				A-5	1(<10%)	
				A-6	1(<10%)	
B	9	41°58'54"	127°31'37"	B-1	4(>50%)	0.50
				B-2	4(>50%)	
				B-3	4(>50%)	
				B-4	0	
				B-5	0	
				B-6	0	
C	5	42°14'31"	127°45'22"	C-1	2(10%-25%)	0.69
				C-2	4(>50%)	
				C-3	3(25%-50%)	
				C-4	2(10%-25%)	
D	5	41°18'15"	127°24'8"	D-1	3(25%-50%)	0.33
				D-2	0	
				D-3	0	
				D-4	1(<10%)	
E	2	42°25'27"	127°20'56"	E-1	4(>50%)	0.80
				E-2	3(25%-50%)	
				E-3	4(>50%)	
				E-4	4(>50%)	

E-5	1(<10%)

Samples for enzymatic activity were stored at 4°C and analyzed within a few days after collection [23]. Soil enzymatic activities (urease, acid phosphatase, catalase and polyphenol oxidase) were assayed according to Guan et al. [25]. Each assay was repeated independently at least three times and averaged.

2.2 Fe (II) induction of rusty root

Rusty root could be attributed to the oxidation and deposition of ferrous iron (Fe (II)) in epidermis of ginseng root [18, 19]. Rusty root was induced through a controlled experiment by increasing the concentration of Fe (II) at a time to elucidate the cause in our study. Four-year-old fresh ginseng roots were collected in April from Fusong country. They were placed in sand at 23°C. Six days after germination, the roots of uniform seedlings were washed with deionized water, and transferred in batches of 5 to separate pots containing 2.5 kg soil. To induce rusty root in ginseng by Fe (II), different amounts of Fe(II)EDTA were added to the soils (pH 6.36, 3.60g/kg iron, and 0.245 g/kg manganese) to reach the Fe (II) concentration as follows: 0.67, 1.34, 2.68, 5.36, and 10.72 g/kg. No Fe (II) was added to the control treatment, and the different treatments were referred to as Fe0 (0), Fe1 (0.67 g/kg), Fe2 (1.34 g/kg), Fe3 (2.68 g/kg), Fe4 (5.36 g/kg) and Fe5 (10.72 g/kg) based on the amount of Fe (II). No external nutrients were supplied during the experiment but the plants were watered. After 28-day, plants were harvested and visual inspection of root symptoms were recorded for statistical analysis.

2.3 Statistical analysis

The soil samples from the five fields were divided into 2 groups: healthy root group (the corresponding rusty-root grades were 0 and 1) and rusty root group (the corresponding grades were from 2 to 4). Differences between the means in the 2 groups and relationship between rusty root index and soil properties were analyzed by the two-sample t-test and Pearson correlation test according to SPASS 20.0 software, with significant being $P<0.05$ or less.

3. Results

3.1 Analysis of rhizosphere soil properties

Soil physico-chemical properties including bulk density (0.98 vs 0.90 $g \cdot cm^{-3}$), moisture (24.27 vs 21.01%), total iron (Fe) (7.37 vs 5.62 g/kg) and total manganese (Mn) (0.58 vs 0.35 g/kg) concentrations were significantly higher in rusty root groups than healthy root groups ($P<0.05$ or $P<0.01$) (Table 2 and 3). Polyphenol oxidase (PPO) activity of rusty root groups was 67.17% higher than the healthy groups ($P<0.01$; Table 4). No significant differences in other soil properties between rusty root groups and healthy groups were observed ($p>0.05$).

Soil samples were acidic with pH_{H2O} ranging from 4.29 to 6.56. The soil samples from fields B (4.55) and E (6.26) had lowest and highest pH_{H2O} values in the five fields (Table 2). Pearson test showed a significant positive correlation between rusty root index and pH, Al, Zn, Ca and Ni ($P<0.05$), and a very significant positive correlation between rusty root index and N, Fe and Mn

($P<0.01$; Tables 2 and 3). There was no significant correlation between rusty root index and other environment factors and cultivation years ($p>0.05$).

3.2 Fe-inducing rusty root

After 28-days, reddish-brown spots appeared at the root surface, and the ginseng root showed symptoms of rust (Figure 1). There was a very significant positive correlation between rusty root index and the amount of Fe added to the soil ($P<0.01$; Table 5). The extensive of rusty root was Fe(II) concentration-dependent.

Figure 1. Rusty roots of *P. ginseng* inducing by the Fe (II) concentrations of 5.36 and 10.72 g/kg.

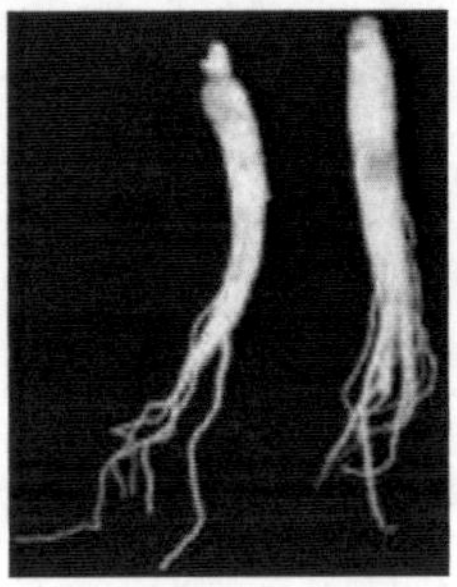

Table 2. Soil moisture, bulk density, organic matter, alkaline solved N, effective P, effective K and pH of soil samples from five fields.

	non-rhizosphere soil		rhizosphere soil				
Soil sample	Soil moisture (%)	Bulk density ($g.cm^{-3}$)	Organic matter ($g.kg^{-1}$)	Alkaline solved N ($mg.kg^{-1}$)	Effective P ($mg.kg^{-1}$)	Effective K ($mg.kg^{-1}$)	pH
A-1	23.82±0.09	1.0492±0.0074	166.21±4.46	541.56±4.91	43.96±0.50	294.68±2.41	5.13±0.0033
A-2	22.98±0.39	1.0234±0.0052	159.24±7.58	536.94±40.0	42.08±0.54	254.45±5.72	4.86±0.0057
A-3	23.46±0.46	1.0721±0.0186	110.70±8.30	383.83±91.4	37.15±0.66	294.67±8.23	5.05±0.0088
A-4	17.36±0.20	0.7518±0.0141	136.02±3.12	482.21±34.6	32.61±0.48	316.14±3.10	5.68±0.0067
A-5	18.06±0.19	0.8641±0.0059	85.48±21.97	401.36±5.46	34.17±2.67	261.62±4.25	5.31±0.0100
A-6	17.64±0.08	0.8567±0.0072	146.87±1.67	536.42±21.9	58.61±0.50	377.47±4.81	5.31±0.0088
B-1	26.36±0.38	0.9048±0.0040	105.98±1.15	318.73±36.2	15.48±0.16	247.94±1.08	4.65±0.0133
B-2	27.14±0.17	0.9751±0.0020	104.52±5.49	352.75±31.9	14.17±0.17	361.85±3.62	4.29±0.0100
B-3	28.03±0.28	0.9963±0.0050	127.96±16.4	375.59±42.1	6.89±0.30	291.87±6.63	4.33±0.0033
B-4	21.27±0.15	0.9110±0.0239	79.40±2.69	234.35±21.7	10.33±0.47	280.06±2.83	4.65±0.0033
B-5	20.87±0.22	0.9012±0.0036	85.91±8.06	238.41±13.3	16.89±6.96	350.98±2.73	4.72±0.0033
B-6	22.01±0.37	0.8964±0.0078	77.50±4.02	238.49±20.7	12.18±0.42	433.77±3.02	4.68±0.0057
C-1	21.23±0.16	0.8986±0.0044	108.01±3.02	414.80±13.9	35.83±3.13	254.54±2.89	5.35±0.0088
C-2	22.31±0.25	0.8769±0.0017	104.18±3.53	384.49±7.2	46.83±4.66	211.21±2.79	5.33±0.0120
C-3	21.98±0.23	0.9012±0.0036	133.17±25.5	369.70±18.3	60.21±2.40	254.54±2.89	5.31±0.0145
C-4	18.62±0.19	0.8276±0.0053	120.70±6.24	479.25±24.6	68.94±1.97	340.90±2.10	5.67±0.0115
D-1	28.58±0.11	1.0232±0.0031	50.88±0.99	240.76±11.2	33.09±1.81	132.67±3.34	5.47±0.0348
D-2	24.37±0.23	0.9557±0.0029	129.63±1.81	254.45±26.3	64.33±1.54	207.94±6.05	5.68±0.0233

IOP Conf. Series: Earth and Environmental Science **41** (2016) 012001 doi:10.1088/1755-1315/41/1/012001

D-3	23.84±0.26	0.9814±0.0039	108.69±1.90	235.17±11.8	59.86±1.91	191.75±3.67	5.61±0.0321
D-4	25.01±0.31	0.9631±0.0025	123.78±1.16	258.50±7.4	32.47±0.79	145.45±3.10	5.75±0.0120
E-1	24.03±0.37	1.0429±0.0024	37.97±14.22	646.40±48.4	10.78±0.21	142.13±2.85	6.52±0.0252
E-2	24.97±0.17	0.9875±0.0036	35.08±8.22	507.27±21.2	20.74±1.07	215.22±2.68	6.14±0.1645
E-3	25.31±0.09	1.0332±0.0074	36.67±2.05	512.51±25.5	39.54±0.29	439.68±4.04	6.50±0.0440
E-4	25.19±0.11	1.0194±0.0215	37.17±1.99	479.30±42.2	58.40±1.62	567.65±2.06	5.59±0.0305
E-5	19.71±0.23	0.9635±0.0946	24.94±0.90	683.70±16.8	34.18±1.35	260.30±3.53	6.56±0.0240

Values are means±SE (n = 3)

Table 3. The metal element content of soil samples from five fields.

Soil sample	rhizosphere soil						
	Total Ca (g/kg)	Total Mg (g/kg)	Total Al (g/kg)	Total Fe (g/kg)	Total Mn (g/kg)	Total Cu (mg/kg)	Total Ni (mg/kg)
A-1	1.855±0.011	1.572±0.006	12.971±0.282	7.454±0.056	0.429±0.003	13.599±0.140	15.358±0.238
A-2	1.772±0.031	1.546±0.018	12.821±0.507	7.511±0.191	0.450±0.018	14.058±0.429	15.938±0.718
A-3	1.553±0.010	1.472±0.011	12.349±0.279	7.195±0.132	0.456±0.008	12.421±0.348	14.513±0.384
A-4	1.600±0.017	1.411±0.004	12.483±0.608	6.931±0.184	0.433±0.031	13.153±0.376	13.955±0.384
A-5	1.535±0.009	1.369±0.001	12.464±0.416	6.739±0.080	0.418±0.015	12.578±0.273	13.718±0.155
A-6	1.764±0.010	1.286±0.010	11.983±0.218	6.484±0.046	0.429±0.010	14.078±0.222	13.598±0.314
B-1	1.232±0.028	1.248±0.014	11.294±0.359	6.494±0.099	0.306±0.007	10.209±0.174	15.035±0.353
B-2	1.169±0.013	1.197±0.004	10.729±0.094	6.228±0.027	0.306±0.001	9.756±0.104	14.009±0.276
B-3	1.182±0.004	1.199±0.003	10.989±0.067	6.270±0.026	0.312±0.015	10.366±0.106	14.507±0.195
B-4	1.084±0.013	1.188±0.004	10.474±0.262	5.201±0.073	0.264±0.013	8.201±0.128	14.431±0.232
B-5	1.271±0.009	1.195±0.008	11.758±0.129	5.420±0.064	0.326±0.007	10.729±0.308	14.740±0.303
B-6	1.242±0.015	1.212±0.007	12.140±0.178	5.479±0.060	0.306±0.007	9.545±0.256	15.118±0.248
C-1	1.779±0.011	1.147±0.008	13.738±0.383	6.998±0.102	0.722±0.034	18.058±0.403	13.730±0.210
C-2	1.456±0.001	0.749±0.01	18.172±0.42	5.711±0.088	0.459±0.01	15.298±0.76	11.309±0.30

IOP Conf. Series: Earth and Environmental Science **41** (2016) 012001 doi:10.1088/1755-1315/41/1/012001

		2	8		4	2	9
C-3	1.480±0.021	0.748±0.013	18.857±0.539	5.680±0.105	0.446±0.021	18.151±0.347	11.960±0.507
C-4	1.481±0.007	0.721±0.006	17.280±0.298	4.332±0.043	0.412±0.013	18.274±0.849	11.672±0.326
D-1	1.394±0.038	0.672±0.019	15.367±0.823	4.705±0.136	0.251±0.008	19.151±0.319	12.945±0.725
D-2	1.398±0.018	0.661±0.013	14.560±0.346	3.518±0.094	0.192±0.006	19.948±0.451	12.192±0.511
D-3	1.358±0.005	0.655±0.006	14.746±0.230	3.483±0.028	0.197±0.005	19.107±0.112	12.423±0.211
D-4	1.439±0.017	0.618±0.013	14.678±0.182	3.669±0.099	0.300±0.029	22.901±0.249	12.820±0.363
E-1	1.237±0.002	0.589±0.003	14.722±0.150	10.701±0.079	0.898±0.038	15.141±0.585	11.976±0.151
E-2	1.210±0.023	0.642±0.012	11.989±1.164	10.027±0.031	0.915±0.031	14.964±0.921	13.105±0.171
E-3	1.892±0.003	1.663±0.016	18.187±0.351	11.095±0.038	1.174±0.046	17.111±0.448	24.451±0.364
E-4	1.855±0.009	1.590±0.007	18.084±0.209	10.188±0.010	1.086±0.006	17.176±0.448	24.217±0.421
E-5	1.453±0.008	1.629±0.023	20.624±0.273	9.221±0.112	0.655±0.024	17.891±0.104	22.031±0.230

Values are means±SE (n = 3)

Table 4. Polyphenol oxidase, catalase, urease and acid phosphatase activity of soil samples from five fields.

Soil sample	rhizosphere soil			
	Polyphenol oxidase m(purpurogallin)/ (mg.g^{-1}.soil.3h^{-1})	Catalase (mL of 0.1 mol/L $KMnO_4$ g^{-1}.soil)	Acid phosphatase m(phenol)/ (mg.g^{-1}.soil.24h^{-1})	Urease m(NH_3-N)/ (mg.g^{-1}.soil.24h^{-1})
A-1	1.24±0.022	0.209±0.0042	28.15±7.15	0.683±0.0729
A-2	1.22±0.018	0.242±0.0058	27.71±1.09	0.448±0.1702
A-3	1.35±0.036	0.232±0.0142	31.74±3.64	0.549±0.0151
A-4	1.08±0.035	0.231±0.0053	31.19±1.55	0.530±0.0296
A-5	1.15±0.019	0.266±0.0156	33.24±2.64	0.504±0.1105
A-6	1.10±0.014	0.219±0.0046	31.33±3.84	0.436±0.0245
B-1	2.45±0.075	0.152±0.0054	32.16±4.63	0.077±0.0521
B-2	2.50±0.028	0.153±0.0032	26.98±0.48	0.069±0.0260

B-3	2.45±0.113	0.128±0.0002	36.09±3.65	0.078±0.0048
B-4	1.34±0.031	0.153±0.0004	32.88±1.62	0.019±0.0332
B-5	1.17±0.043	0.147±0.0057	35.98±3.29	0.087±0.0297
B-6	1.47±0.031	0.140±0.0031	34.22±2.95	0.057±0.0211
C-1	2.11±0.004	0.133±0.0184	29.25±1.41	0.478±0.0324
C-2	2.45±0.014	0.142±0.0108	36.56±1.56	0.518±0.0355
C-3	2.43±0.011	0.153±0.0008	39.60±2.97	0.485±0.0093
C-4	1.71±0.038	0.135±0.0195	40.03±2.81	0.384±0.0355
D-1	2.44±0.070	0.133±0.0092	28.16±2.83	0.285±0.0173
D-2	1.22±0.014	0.094±0.0022	28.25±1.52	0.217±0.0097
D-3	1.28±0.037	0.084±0.0153	43.39±6.51	0.113±0.0513
D-4	1.46±0.088	0.122±0.0030	36.73±6.14	0.278±0.0042
E-1	2.46±0.024	0.123±0.0105	31.53±1.58	0.232±0.0941
E-2	2.08±0.052	0.131±0.0059	33.28±1.23	0.228±0.0081
E-3	2.62±0.025	0.109±0.0052	47.97±3.53	0.173±0.0462
E-4	2.41±0.032	0.080±0.0061	40.09±2.06	0.213±0.0190
E-5	1.49±0.061	0.120±0.0006	20.74±1.29	0.125±0.0125

Values are means±SE (n = 3)

Table 5. Rusty root index of Fe (II) inducing rusty root experiment.

No.	The concentrations of Fe(II)	Rusty-root grade	Number of ginsengs	Rusty root index
Fe0	0	0	8	0.27
		1(<10%)	3	
Fe1	0.67g/kg	0	5	0.28
		1(<10%)	3	
		2(10%-25%)	1	
Fe2	1.34g/kg	0	3	0.38
		1(<10%)	3	
		2(10%-25%)	1	
		3(25%-50%)	2	
		4(>50%)	1	
Fe3	2.68g/kg	0	1	0.44
		1(<10%)	4	
		2(10%-25%)	1	
		3(25%-50%)	2	
		4(>50%)	1	
Fe4	5.36g/kg	0	1	0.44
		1(<10%)	5	
		3(25%-50%)	1	
		4(>50%)	2	

Fe5	10.72g/kg	1(<10%)	1	0.67
		2(10%-25%)	1	
		3(25%-50%)	3	
		4(>50%)	1	

4. Discussion

Non-rhizosphere and rhizosphere soils of ginseng roots sampled from five ginseng cultivation fields in Changbai Mountains were used to identify the relationship between rusty root and soil characteristics in our study. Soil bulk density, soil moisture, total iron (Fe), total manganese (Mn) concentrations and polyphenol oxidase (PPO) activity were significantly higher in rusty root groups than healthy root groups by two-sample test, respectively. High soil bulk density and moisture of non-rhizosphere soils in rusty root groups may lead to poor soil permeability, ventilation and low oxygen content, which could promote the reduction of insoluble metal ions into soluble ions, especially Fe (III) into Fe (II) which also provides the basis for Fe (II) oxidation [26]. Rahman and Punja [11] suggested that excessive Fe (II) in rhizosphere soil can be toxic to plant roots, and ginseng could induce host defense responses, producing phenolic compounds. An increase in PPO activity of the rhizosphere soil is closely related to its efficiency in promoting oxidation of polyphenols to quinones in the soil [27]. Quinones can bind with amino acids, metal ions and peptides, etc., to form initially humic acid molecules and pigment under suitable conditions [28,29]. Phenolic compounds secreted by ginseng roots are oxidized by PPO in rhizosphere soils and sequestrated metal ions consequently form pigments molecules. These adhere to the ginseng root surface, making it red-brown. This was also suggested by Rahman and Punja [11], based on quantification of phenolic and metals in tissues from rusty and healthy root.

Pearson correlation showed that there was a significant positive correlation between rusty root index and pH, N, Fe, Mn, Al, Zn, Ca and Ni. Plant root can influence their environment by secreting oxidizing substances into the rhizosphere [30]. Ginseng root under iron toxicity stress can produce oxidizing substances, resulting in oxidation and deposition of metal ions (especially iron), which can be largely responsible for the symptoms in epidermal and underlying cortical cells of ginseng root. Thus rusty root index becomes higher with increase of metal ion content (especially Fe) in the rhizosphere. pH and nitrate also are involved in controlling Fe-cycling [26]. The soils were acidic with pH_{H2O} ranging from 4.29 to 6.56. The neutral pH conditions and nitrogen content are conducive to Fe (II) oxidation [26]. Higher pH and N content in rhizosphere may drive the oxidation of metal ions. Ginseng rusty root may be a physiological disorder that is closely related to soil environmental factors. Fe (III) reduction and Fe (II) oxidation could be an important influencing factor giving rise to rusty root; soil moisture and bulk density of non-rhizosphere soil can also promote the reduction of Fe (III) to Fe (II), and higher pH, N and PPO content of rhizosphere soils are important during oxidation and sequestration of metal ions. Increasing rusty root index with increasing addition of Fe (II) indicated that Fe (II) is a main factor inducing rusty root of *P. ginseng*.

5. Conclusions

Fe (III) reduction and Fe (II) oxidation could be very important in increasing the incidence of rusty root. Soil moisture and bulk density of non-rhizosphere soil, and pH, N and PPO content of rhizosphere soils were heavily involved in the reduction, oxidation and sequestration of metal ions (especially Fe). Physiological factors may be very important roles giving rise to ginseng rusty root.

Acknowledgements
Financial support provided by Ministry of Science and Technology of the People's Republic of China (No. 20140520157JH, No. 20140GB2B100025, No. 20150204053NY) is gratefully appreciated. We are very grateful to BioMedES Ltd (Inverurie, Aberdeenshire, AB51 0LX, UK) for improving the article linguistically.

References

[1] Kang K S, Yokozawa T, Kim H Y and Park J H 2006 Study on the nitric oxide scavenging effects of ginseng and its compounds *Journal of Agricultural and Food Chemistry* **54** 2558-62

[2] Sagar S M, Yance D and Wong R K 2006 Natural health products that inhibit angiogenesis: a potential source for investigational new agents to treat cancer *Current Oncology* **13** 1426

[3] Koo H N, Jeong H J, Choi I Y, An H J, Moon P D, Kim S J, Jee S Y, Um JY, Hong S H and Shin S S 2007 Mountain grown ginseng induces apoptosis in HL-60 cells and its mechanism have little relation with TNF-α production *The American Journal of Chinese Medicine* **35** 169-82

[4] Shin B K, Kwon S W, Park J H 2015 Chemical diversity of ginseng saponins from Panax ginseng *Journal of Ginseng Research* **16** 287-98

[5] Shi W, Wang Y, Li J, Zhang H and Ding L 2007 Investigation of ginsenosides in different parts and ages of Panax ginseng *Food Chemistry* **102** 664-8

[6] Reeleder R D, Roy R and Capell B 2002 Seed and root rots of ginseng (panax quinquefolius l) caused by cylindrocarpon destructans and fusarium spp *Journal of Ginseng Research* **26** 151-8

[7] Seifert K A, McMullen C R, Yee D, Reeleder R D and Dobinson K F 2003 Molecular differentiation and detection of ginseng-adapted isolates of the root rot fungus Cylindrocarpon destructans *Phytopathology* **93** 1533-42

[8] Yang D C, Kim Y H, Yun K Y, Kwon JN, Kang H M and Lee S S 1997 Red-colored phenomena of ginseng (Panax ginseng CA Meyer) root and soil environment *Korean Journal of Ginseng Science* **21** 91-7

[9] Campeau C, Proctor J T A, Jackson C J C and Rupasinghe H P V 2003 Rust-spotted north american ginseng roots: phenolic, antioxidant, ginsenoside, and mineral nutrient content *Hortscience* **38** 179-82

[10] Punja Z K, Wan A, Goswami R S, Verma N, Rahman M, Barasubiye T, Seifert K A and Lévesque C A 2007 Diversity of fusarium species associated with discolored ginseng roots in british columbia *Canadian Journal of Plant Pathology* **29** 340-53

[11] Rahman M and Punja Z K 2005 Biochemistry of ginseng root tissues affected by rusty root symptoms *Plant Physiology and Biochemistry* **43** 1103-14

[12] Campeau C, Proctor J T A, Murr D P and Schooley J 2003 Characterization of north american ginseng rust-spot and the effects of ethephon *Journal of Ginseng Research* **27** 188-94

[13] Punja Z K, Wan A M and Rahman M 2006 Role of fusarium species in rusty root development on ginseng roots *Phytopathology* **96** S170-1

[14] Liu X, Yang Z, Gao L, Xiang W, Zhang B, Xie Z and You J 2014 Comparison of the characteristics of artificial ginseng bed soils in relation to the incidence of ginseng red skin disease *Experimental Agriculture* **50** 59-71

[15] Lee S S, Lee M G and Choi K T 1999 Rusty-root tolerance and chemical components in 4-year old ginseng superior lines *Journal of Ginseng Research* **23** 61-6

[16] Choi J E, Lee J S, Yoon S M and Cha S K 2002 Comparison of inorganic elements and epidermis structures in healthy and rusty ginseng *Korean Journal of Crop Science* **47** 161-6

[17] Zhang X, Cao Y, Tian Y and Li J 2014 Short-term compost application increases rhizosphere soil carbon mineralization and stimulates root growth in long-term continuously cropped cucumber *Scientia Horticulturae* **175** 269-77

[18] Li Z, Guo S, Tian S, Liu Z and Long B 1997 Study on the causes for ginseng red skin sickness occurred in albic bed soil *Acta Pedologica Sinica* **34** 328-35

[19] Wang Y, Li Z, Sun Y, Guo S, Tian S and Liu Z 1997 Studies on the genesis of ginseng rust spots *Journal of Ginseng Research* **21** 69-77

[20] Rumberger A, Yao S, Merwin I A, Nelson E B and Thies J E 2004 Rootstock genotype and orchard replant position rather than soil fumigation or compost amendment determine tree growth and rhizosphere bacterial community composition in an apple replant soil *Plant Soil* **264** 247-60

[21] Moodie C D, Smith H W and McCreery R A 1951 Laboratory manual for soil fertility *Soil Science* **71** 400

[22] Walkley A and Black I A 1934 An examination of the degtjareff method for determining soil organic matter, and a proposed modification of the chromic acid titration method *Soil Science* **37** 29-38

[23] Adamczyk B, Adamczyk S, Kukkola M, Tamminen P and Smolander A 2015 Logging residue harvest may decrease enzymatic activity of boreal forest soils *Soil Biology and Biochemistry* **82** 74-80

[24] Rattan R K, Datta S P, Chhonkar P K, Suribabu K and Singh A K 2005 Long-term impact of irrigation with sewage effluents on heavy metal content in soils, crops and groundwater-a case study *Agriculture, Ecosystems and Environment* **109** 310-22

[25] Guan S Y 1986 *Soil Enzymes and Research Methods* (China: Agricultural Science Press) pp 274-97

[26] Hu M and Li F 2014 Soil microbe mediated iron cycling and its environmental implication *Acta Pedologica Sinica* **51** 683-98

[27] Liu R, Dai Y and Sun L 2015 Effect of rhizosphere enzymes on phytoremediation in PAH-contaminated soil using five plant species *Plos One* **10** e0120369

[28] Duran N and Esposito E 2000 Potential applications of oxidative enzymes and phenoloxidase-like compounds in wastewater and soil treatment: a review *Applied Catalysis B: Environmental* **28** 83-99

[29] Trasar-Cepeda C, Leiros M C, Seoane S and Gil-Sotres F 2000 Limitations of soil enzymes as indicators of soil pollution *Soil Biology and Biochemistry* **32** 1867-75

[30] Fu Y, Yu Z, Cai K and Shen H 2010 Mechanisms of iron plaque form ation on root surface of rice plants and their ecological and environmental effects :A review *Plant Nutrition and Fertilizer Science* **16** 1527-34

Effects of composite soil with feldspathic sandstone and sand on soil aggregates and organic carbon

J Li[1,2,3], J C Han[1,3,4], Y Zhang[1,3], G Y Lei[1,3], H Y Wang[1,3] and D W Zhu[1,3]

[1] Shaanxi Province Land Engineering Construction Group, Xi'an 710075, China

[2] College of Forestry, Northwest A&F University, Yangling 712100, China

[3] Key Laboratory of Degraded and Unused Land Consolidation Engineering, the Ministry of Land and Resources of China, Xi'an 710075, China

E-mail：kele502514@163.com

Abstract. The case was to study the effects of soils with different proportions of feldspathic sandstone and sand on soil stability and organic carbon at 0-30 cm soil depth with four different ratios(C1、C2、C3 and C4), They were used to prepare the composite soil in Fu Ping, Shaanxi Province of China, then the soil aggregates distribution, WASR, MWD, GMD, D valueand and organic carbon content were measured and analysed.The results showed : the soil stability of C1, C2 and C3 was better than C4, i.e., the composition could improve the soil stability. With the increasing of the planting years, the contents of soil aggregates with the size >0.25 mm and MWD, GMD and SOC increased for each treatment at 0- 30 cm soil depth, which was contrary to D values. WASR of C2 was significantly higher than others ($p<0.05$) after 3-year planting. The significant logarithmic relationships were found between the D values and the ratios in C1, C2 and C3. Besides C1 and C2 could increase the stability and content of large soil aggregates to improve soil structure; C2 could significantly increase the SOC than others at 0- 30 cm soil depth.

1. Introduction

Mu Us Sandland is located between the southeast of Ordos Plateau and the north of Loess Plateau, China. Due to the improper agricultural reclamation and overgrazing, the region has encountered severe grassland degradation, desertification and productivity decrease [1,2,3]. Feldspathic sandstone and sand both are important compositionin Mu Us Sandland. The bare feldspathic sandstone that weathered is easy to carry away with the wind and rain, which has been called the "cancer of environment" [4]. Chinese scientists and the local people studied the management of the Mu Us Sandland in a long time, and theymainlyfocused on the distribution, qualities and types of feldspathic sandstone, the characteristics of water loss and soil erosion, control methods and ecological remediation. Han Liwen (2005)and ZhengYuanrun (2006) reported that air seeding, afforestation and grass breeding in the sand area, and other integrated methods related to eco-system restoration and developing new models, which were all limited by the local natural conditions, resulting innumerous measures with whatever physical or biological means merely had significant effects to reverse the desertification[5,6]. The ultimate reason could be the unstable soil structure, meaning that only by improving soil structure, can we rehabilitate the damage of desertification [7,8].

As the most basic soil structural element, soil aggregate has tremendous effects on soil physical and chemical characteristics, and thus its related parameters are very important to reflectthe soil structure and stabilities [9,10]. Soil organic carbon is very important factor during the formation of soil aggregates. The close relationship between soil organic carbon and soil aggregates and maintaining soil organic carbon has been thought as a very vital strategy in releasing soil erosion, holding soil structure and improving soil grains condition[11,12]. The quantitative and dimensional distributions of soil aggregates in various sizes determine the distributions of soil pores and their continuity [13,14]. Among many different kinds of soil aggregates, water stable aggregate (WSA) is commonly used to evaluate the stability and anti-erosion ability of soil structure [15]. The characteristics of the distributions and stability of soil aggregates in a quantitative way was study in earlier time, but the study only focused on the soil grains' proportions different sizes and the related parameters of soil aggregates in size over 0.25 mm ($R_{0.25}$) .Then Van Bavel calculated the weighed sum of the data of soil aggregates in different sizes, and came up with the concept "mean weight diameter (MWD)"[16].The higher values of MWD means the larger proportion of soil aggregates and more stable of soil.Soil particles and their aggregates also have considerable characteristics of distribution, which could be analyzed by fractal dimension (D) method. Basically, higher D value means better soil aggregate structure and stability[17,18,19].

Generally, the structure of soil matrixcouldbe improved by coving healthy soil on the local area.But Mu Us Sandland lacks loess soil, and the cost of long distance transportation is considerably high. From the perspective of resources, feldspathic sandstone and sand both are very important local natural resources.Feldspathic sandstone is in the form between soil and parent materials, but apt to rigidify during drying. So it affects the permeability of gas and water, and root extension.But feldspathic sandstone swells rapidly in water, becoming very good material to hold moisture.The sand, without solid structure, has very good permeability in both dry and wet conditions but is poor at water retention. It would be great if we could use their complementarity in texture and permeability to improve the local soil quality. Related study has been focusing on soil texture, parameters related to the variation of water and fertility under compositesoilwith different proportions of feldspathic sandstone and sand. The object of this study was to explore the contents of soil aggregates in different sizes, the contents of $R_{0.25}$, the stability of soil aggregates, the mean weight diameter (MWD), fractal dimension value, the contents of soil organic carbon and their variations under different composite artificial soil with different proportions of feldspathic sandstone and sand.Then the winter wheat was planted in the composite artificial soil,and the effects of different artificial soils on the wheat were compared.

2. Materials and methods

2.1 Site description

The experiment was carried out at Fu Ping Research Station at Zhe Yuan Village of Du Cong, Fu Ping County, City of Weinan, Shaanxi Province, China (109°11′N，34°42′E, 492m above sea level). The area belongs to temperate semi - humid climate, the average annual rainfall capacity is 472.97 mm and 49% of it happens during July and September. The annual evaporation ranged from 1000 to 1300 mm, and frost-free period, average temperature, highest temperature in the summer, lowest temperature in the winter were 225 d, 13.4℃, 41.8℃, -2℃respectively and annual total energy from optical radiation were 123.9~127.8kca/cm^2. All the parameters could satisfy the basic demand for crop growth.

2.2 Experimental design

The experiment was performed during September 2008 to June 2011. To simulate the soil composited by feldspathic sandstone and sand in situ in Mu Us Sandland the experimental plots were filled with 70 cm depth of parent materials using pure sand, and 30 cm of artificial soil, which were consisted of feldspathic sandstone and sand in different ratio.in top layer. The feldspathic sandstone and sand was from Da Jihan Village, Xiaojihan Town, Yu Yang District, the City of Yulin, Shaanxi Province, which belongs to the typical Mu Us Sandland region. According to the study that the turning point of the decreasing rate of the hydraulic conductivity of composite soils appeared when 1:2 volume ratio of feldspathic sandstone to sand was used. Hence the volume ratio gradients of feldspathic sandstone to sand for the 4 treatments were 1:1(C1), 1:2(C2), 1:5(C3), and 1:2 (C3) for the C4, consisted of sand. The experiment measured the differences of soil aggregates and organic carbon between the each other treatment. Each plot had an area of 4 m^2 (ie., 2m ×2m), with three replicates. The whole experimental area laid out from the north to the south. The composition of composite soil in different treatments was showed in table 1 before planting.

Table 1. Different mixed ratio of feldspathic sandstone and sand on soil texture befor planting.

Treatments	Diameter of particles content (%)			Soil texture
	Sand	Silt	Clay	
C1	39.57b	49.38a	8.24a	Loamy soil
C2	52.74ab	38.88b	8.38a	Sandy loam
C3	70.79a	24.08b	5.13a	Sandy loam
C4	94.07b	3.20b	2.73b	Sand

Note: Multiple comparisons were between the treatments in the same line. Letters indicate significant differences among different treatments

2.3 Materials

Winter wheat: Xiaoyan 22, thousand grain weight: 38g, germination percentage: 90.1%, rate of seeding: 150 kg/hm^2, row spacing: 20 cm. sowing by drilling.

2.4 Fertilizing treatment

Granulated fertilizer was added in the form of phosphorus (P) as $(NH_4)_2HPO_4$ (180 kg/hm^2), potassium (K) as KCl (90 kg/hm^2), and nitrogen (N) as $CO(NH_2)_2$ (255 kg/hm^2,) in each plot.

The fertilizers were applied on 18th September 2008, 20th September 2009 and 24th September 2010, when the winter wheat was planted. To make sure thorough blending of the top soil and fertilizers, seeding and tillage were used in this experiment. No fertilizers were used during the growing season. And winter wheat were irrigated once prior to the tillering stage with the same amount of water in 4 treatment plots.

2.5 Sampling and Measurement

2.5.1 Soil aggregates measurement. At harvest (June 6th 2009, June 10th 2010 and June 8th 2011), we collected the surface soil at 0- 30 cm soil depth were collected, and the residuals and large rocks were removed. At last samples were dried by natural air. Quartering method was used to get the soil samples (total weight not less than 400 g). Subsequently, each soil sample was separated into three types by sieves of 5 mm and 2 mm, ie., larger than 5 mm, 5 –2 mm and less than 2 mm. Then samples were weighed separately as a_1, a_2, a_3 (g) and the total weight a (g). Finally, the separated soil was mixed into 4 plastic bags in total, and each weight was 50 g.

The soil aggregates sizes distributions and stabilities were measured using both dry sieving and wet sieving methods. The dry sieving method: soil were sieved through5, 2, 1, 0.5 and 0.25 mm mesh vertically from top to bottom. The piled sieves were shaken at 200 round/min for 2 min. then the soil was collected from each sieve with different grain sizes such as larger than 5 mm, 5 – 2 mm, 2 – 1 mm, 1 – 0.5 mm, 0.5 – 0.25 mm and less than 0.25 mm. Then samples were weighted as Wdi. The wet sieving method: four separated samples were put into different plastic cups, and then were added into water to stabilize for 10 min. Then the dispersed samples were sieved through 5 sieves (the hole diameters were 5mm, 2mm, 1mm, 0.5mm and 0.25 mm, respectively), and shaken up and down within 3 cm vertically, and finally were washed into aluminum boxes and then dried at 105°C for 12 hours. At last, samples were weighed as W_{wi}.

2.5.2 Soil organic carbon measurement. The top soil samples were collected at 0- 30 cm soil depth at harvest and the crop residuals and small rocks were removed. Then samples were dried by natural air. Quartering method was used to get the soil samples through the 0.149 mm sieves for measuring the carbon. using Potassium dichromate volume method were used (by using $K_2Cr_2O_7$—H_2SO_4 solution to oxidize the carbon in soil organic matter under oil-bath heating condition, $Cr_2O^{-2}{}_7$turned to Cr^{3+}, and the remained $K_2Cr_2O_7$ was titrated by standard $FeSO_4$ solution, finally, the content of organic carbon could be calculated based on the consumed $K_2Cr_2O_7$.).

2.6 Statistical analyses

To calculate the content of water stable aggregates, with wet sieving method,

$$W_i = \frac{W_{di} or W_{wi}}{50} \times 100\% \tag{1}$$

Where W_i is the mass proportion of i size soil aggregates, e.g.

To calculate the mass proportion of the aggregates with size over 0.25 mm ($R_{0.25}$), mean weight diameter (MWD) and geometric mean diameter(GMD) [20].

$$R_{0.25} = \sum_{i=1}^{n}(W_i) \quad (2)$$

$$MWD = \sum_{i=1}^{n}(\overline{X_i} \times W_i) \quad (3)$$

$$GWD = \sum_{i=1}^{n}(\overline{X_i} \times W_i) \quad (4)$$

To calculate the water stable aggregates rate:

$$WSAR = WSA/MSA \times 100\% \quad (5)$$

Where WSA is the mass of water stable aggregates in size over 0.25 mm, MSA is the mass of stable aggregates in size over 0.25 mm.

Fractal dimension values were calculated using the equation below by Yang Peiling et al [17].

$$\frac{M(r \le \overline{X_i})}{M_t} = \left(\frac{\overline{X_i}}{X_{\max}}\right)^{3-D} \quad (6)$$

Take logarithm for both sides,

$$\lg\left[\frac{M(r \le \overline{X_i})}{M_t}\right] = (3-D)\lg\left(\frac{\overline{X_i}}{X_{\max}}\right) \quad (7)$$

Where $\overline{X_i}$ the mean diameter of the aggregates is in a specific size, $M(r \le \overline{X_i})$ is the mass of aggregates in size less than $\overline{X_i}$, M_i is the total mass of aggregates (50 g), $X_{max}$is the maximum diameter of aggregates. D value can be obtained by linear regression using equation (6) and (7) through least square fitting method.

Using Origin 8 to process the data and figures, and then SPSS (PASW Statistics 18) was used to do the data statistical analysis and the multiple comparisons were done using duncan's new multiple range method (SSR).

3. Results and analysis

3.1 Water-stable aggregatesdistribution characteristicsamong three planting seasons under Different mixed ratio of feldspathic sandstone and sand

Dry sieving and wet sieving were commonly used to *study* the distribution of soil aggregates Stable aggregates that are able to resist mechanical force, could be obtained by dry sieving which includes both aggregates of resisting water and unresisting water. Water resisting and stable soil aggregates could be acquired by wet sieving. For the composited artificial soils, quantities and distributions of aggregates could be obtain by wet sieving that demonstrate more characteristics of stability, water retention, permeability and ability of resisting erosion of soil structure[21]. Hence wet sieving method was used in this study to determine the proportions of soil aggregates by their grain sizes at >5 mm, 5 - 2 mm, 2 - 1 mm, 0.5 – 1 mm, 0.25 - 0.5 mm and >0.25 mm (table 1). The aggregates that were larger than 0.25 mm were called $R_{0.25}$ in this study and were very important for the soil structure, and their proportions have positive correlation with soil fertility[22,23]. This study found that the amount of $R_{0.25}$ increased with the years of planting but decreased with the increase of the proportion of the sand in the treatments. Treatment C4 had the highest amount of $R_{0.25}$ with an average of 18.93%; the treatments C1, C2, C3 and C4 all had the least amount of aggregates , with their average sizes during 1 - 0.25 mm in the three years; In 2009 and 2010, treatments C1, C2 and C3 had the most amount of aggregates, and their sizes were larger than 5 mm, with percentage of 24.84% - 37.36% of $R_{0.25}$. In 2011, the aggregates' sizes between 1 and 0.5 mm were 32.91% - 37.19% of $R_{0.25}$, indicating soil aggregates with their size larger than0.25 mm were enriched during 1 - 0.5 mm, and soil aggregates in treatment C1 were significantly higher than that in others by 0.44~2.85 percentages ($p<0.05$). Treatments C1, C2, and C3 all had significantly more the amount of $R_{0.25}$of treatments C1, C2, and C3 in 2011were significantly higher than that in 2009 by 28.99%, 48.67%, 41.45% and 48.32 % ($p<0.05$) respectively. With the increasing of planting years, treatments C1, C2 and C3 all had higher the amount of average $R_{0.25}$ in treatments C1, C2 and C3 were significantly higher than C4.

Table 2. Soil aggregate size fraction content by wet sieving at 0-30 cm depth under different percentage of mixed sand on soil texture from 2009to 2011.

Year	Size fraction (mm)	Treatments /WSA (%)			
		C1	C2	C3	C4
2009	>5	6.63a	6.88a	5.61ab	1.62b
	2-5mm	6.78a	4.79ab	4.12bc	2.44c
	1-2mm	2.43a	1.65a	1.02b	1.58b
	0.5-1mm	4.65a	2.44b	2.48b	2.51c
	0.25-0.5mm	5.80ab	5.33ab	4.79b	7.35a
	>0.25mm	26.69b	21.08ab	18.02a	15.50a
2010	>5	7.15a	7.77a	7.78a	3.39b
	2-5mm	6.44a	6.54b	4.75b	8.61a
	1-2mm	3.20a	2.23ab	1.99b	1.69b
	0.5-1mm	4.64a	4.80b	2.68bc	1.78c
	0.25-0.5mm	5.89a	4.48a	3.62b	3.45b
	>0.25mm	27.32b	25.82a	20.82a	18.92a
2011	>5	8.11a	6.42a	5.61a	3.28a
	2-5mm	6.86a	6.01ab	4.34bc	6.39c
	1-2mm	2.89a	2.06a	1.41b	1.25b
	0.5-1mm	11.33a	10.89b	9.48bc	8.48c
	0.25-0.5mm	6.01a	5.96a	4.66ab	3.59b
	>0.25mm	34.43b	31.34ab	25.49a	22.99a

Multiple comparisons were between the treatments in the same line. Letters indicate significant differences among different treatments at 5% level of probability (SSR).

3.2 Water-stable aggregates rateamong three planting seasons

Equation (5) calculated the water aggregate stability rate (WASR) of each treatment in 2009, 2010 and 2011. In 2009 and 2010, at 0-30 cm soil depth, the WSARs of different treatments decreased with the increase of sand proportions in the composite soils, but increased the planting years. The WASRs in each treatment ranked as C1> C2> C3> C4.There were significant differences between the C4 and C1 or C2, but there were no significant differences compared with C3 ($p<0.05$). In 2011, for C2 the highest WASR was 22.76%, which was higher than that in other treatments ($p<0.05$). According to Figure 1, the average WASR in C1 treatment was higher than that in C2, C3 and C4 by 2.97%, 21.88% and 31.90%, respectively, but no significant difference with C2 ($p<0.05$).

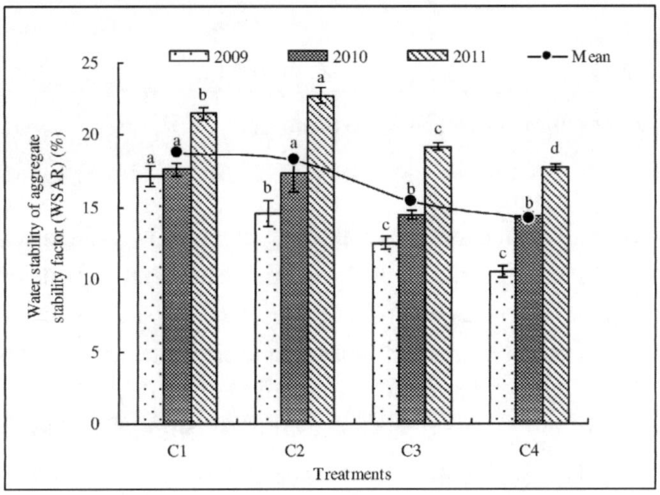

Figure 1. WASR of wet sieving at 0-30 cm depth under different percentage of mixed sand on soil texture form 2009 to 2011.Letters indicate significant differences among different treatmentsat the 0.05 levels ($p < 0.05$).

3.3 Mean weight diameter (MWD) and Geometric mean diameter(GMD)

The mean weight diameter (MWD) and geometric mean diameter (GMD) of soil aggregates indicated their size distributions. And the higher MWD and GMD demonstrated the higher degree of concentration of aggregates and stabilities. Different sizes of soil aggregates have different abilities to coordinate, maintain, supply soil nutrients, and to improve soil porous networks, hydraulic and biological properties. So that is reason that the sizes of distributions are very important to soil qualities. This study used the soil aggregates with sizes over 0.25 mm were used in this study and their MWD and GMD were calculated to evaluate the soil aggregates stabilities (Table 3).

During the same year, negative correlation was found between the proportions of sand and MWD or GMD of each treatment. Both the MWD and GMD decreased with the increasing proportion of sand. The MWD and GMD were positively correlated with the planting year in the same treatment, ie., they both increased in the year after. In 2009, there were no significant differences between the MWD from each treatment, but the reverse was true for the GMD ($p<0.05$). MWD and GMD in treatments C1, C2 and C3 were significantly higher than that in C4 ($p<0.05$)in 2010 and 2011.The three-year average MWD and GMD of C1 were significantly higher than that of C2, C3 and C4 by 9.88%, 29.61% and 39.06% for MWD, and 18.84%, 26.15% and 43.85% for GMD, respectively ($p<0.05$).

Table 3. The mean weight diameter (MWD) and geometric mean diameter (GMD) at 0-30 cm depth under different percentage of mixed sand on soil texture from 2009 to 2011.

Years	Parameters	C1	C2	C3	C4
2009	MWD	0.83a	0.72a	0.64a	0.63a
	GWD	0.26a	0.22ab	0.20b	0.15c
2010	MWD	0.85a	0.74b	0.69b	0.64c
	GWD	0.27a	0.23b	0.22bc	0.21c
2011	MWD	0.99a	0.97a	0.79a	0.65ab
	GWD	0.29a	0.24a	0.23a	0.21ab

MWD : mean weight diameter, GWD : mean geometric diameter.

3.4 Fractal dimensioncharacteristics

The fractal dimension of soil aggregates under each treatments was fitted using equation (7) (the values of R^2 were all higher than 0.92).Then the values of fractal dimension (D value) for each treatment at 0 – 30 cm depth could be calculated (Figure 2) as 2.89 – 2.96. The figure showed that the D values for each treatment had similar trend of curves, and all went downward with the passing of the planting years. The D values in 2011 were lower than that in 2010 by 1.21% - 3.20%.The average D values of C1 and C2 were lower than that of C3 and C4 by 13.75% and 16.76%, respectively, and 0.79% and 10.99%, respectively. The D values were mainly related to the proportions of water stable aggregates for each treatment, ie., with the increasing of the planning years, more interactions of crop roots and soil particles were involved, which may increase the contents of larger soil aggregates, thus resulting in the decreasing trend of D values.

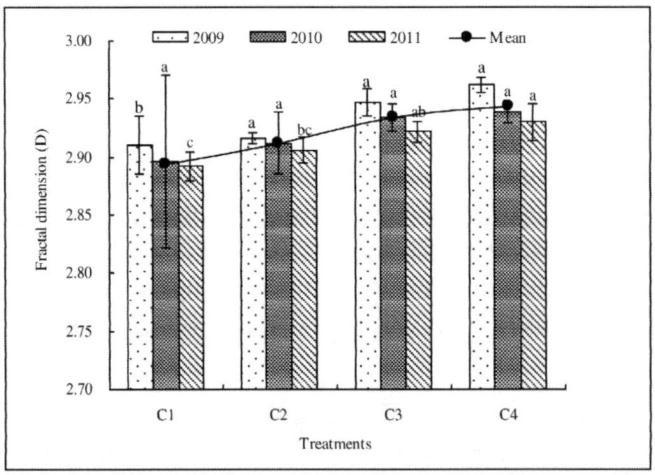

Figure 2. Fractal dimension of soil aggregates at 0 - 30 cm depth under different percentage of mixed sand on soil texture from2009 to 2011.

3.5 Soil organic carbon

This study found that soil organic carbon mainly contributed to the formation of soil water stable aggregates, which was an important element between soil aggregates and B. The decrease of soil organic carbon would result in the loss of larger soil aggregates and deteriorated soil quality[24]. Crops could significantly increase the amount of soil organic carbon, showing positive effect of carbon sequestration. On the other hand, with the increasing of the proportions of sand in the artificial soils, soil stability decreased, which could result in decomposition of soil organic carbon and emission of CO_2.

The study showed that soil organic carbon (SOC) contents significantly differed from each other treatment ($p<0.05$) (table 4).Significant trends that with the year increase the SOC also increased at 0 – 30 cm soil depth, were found. There were no significant differences among the contents of the SOC of C1 in the three years. There were significant differences between the SOC from C2, C3 and C4 in 2009 and 2010, and that from 2011 ($p<0.05$). In 2011, the SOC contents of C1(1.18g/kg), C2(1.22g/kg), C3(1.11g/kg) and C4(0.41g/kg)in 2011were higher than that in 2009 by 52.54% - 77.29%.

Table 4. Effects of different percentage of mixed sand on soil texture on soil organic carbon in 0-30 cm from 2009 to 2011. /g/kg.

Treatments	C1	C2	C3	C4
2009	0.42±0.03aAB	0.57±0.13bA	0.45±0.03bB	0.28±0.04bC
2010	0.98±0.84aB	1.24±0.11bA	1.06±0.28bA	0.38±0.11bC
2011	1.85±0.48aAB	1.98±0.35aA	1.83±0.87aB	0.59±0..09aC

Note: values followed by the different letters are significantly different at P<0.01 and P<0.05, the data: mean ± standard deviation.

4. Discussion

Soil aggregates, as a very important component in soils, are the key factors to maintain and coordinate water and nutrition in soil, and to affecte the types and activities of enzymes, and to stabilize the loose slaking soil layers that directly influence the productivity of crops[25]. The size of aggregates over 0.25 mm ($R_{0.25}$) are called soil structural aggregates, which are the best structural units in soil, whose percentages are very important to indicate structure related to soil qualities, eg., the $R_{0.25}$ contents had positive correlation with soil fertility. The size of soil aggregates less than 0.25 mm are called miro-soil aggregates[14]. Many researchers had confirmed that different tillage regimes would affect the transition and redistribution between the soil structural aggregates and miro-soil aggregates[26,27] 0), resulting in changing the structural stabilities and anti-erosion abilities. The wet sieving method is used to determine the characteristics of water stable aggregates, since water stable aggregates also has great contribution to maintaining soil structural stability[28]. In this study, the 3-year experiment with different ratios of feldspathic sandstoneto sand in different composited soil treatments showed thatthe soil aggregates over 0.25 mm from each treatment increased with the increase of the planting years.Result may indicate that the mixture of artificial soils could enhance the capability to maintain water and nutrition. And the SOC increasing could lead to forming more water stable aggregates and improving the incompact structure of sand, also could alleviate the hardening phenomena of feldspathic sandstones. In the same planting year, the soil aggregates over 0.25 mm decreased with the increase of the proportions of sand, and the average size of soil aggregates over 0.25 mm in treatments C1 and C2 were 29.48% and 26.08% over the three years, respectively. which were higher than that in C3 and C4, meaning proper mixture of the composited soils could improve the structural quality of soil.

The mean weight diameter of soil aggregates (MWD) and geometrical mean diameter (GMD) at some degree could reflect the distribution characteristics of soil aggregates[29]. This study found that there were negative correlations between the values of MWD and GMD and the percentages of the sand in the same year,, and there were positive correlations between the values of MWD and GMD and the planting years. The MWD and GWD values in 2011 were higher than that in 2009 by 3.17% - 23.43% and 7.41% - 40.1%, respectively.

The values of fractal dimension (D values) also could be used to indicate the change of soil structure[30]., The higher proportions of clay particles, finer texture and weaker permeability of soil indicated higher D values[3,18]. It was reported that the D values of soil had significant negative correlations with the contents of large soil aggregates[31,32]. Liu Yunpeng et al studied the D values of 4 types of soils in Shaanxi, China, indicating that D values were ranged from 2.63to 2.87, and the ideal value should be around 2.75[33]. This study discovered that the D values of each treatment all decreased with the passing planting year, but increased with the increase of the sand proportions in soil. And the 3-year average D values were between 2.89 and 2.94, which were higher than the ideal value. Result may need related practices,eg., increasing the planting years and reducing tillage activities to reduce D values.

The soil organic carbon contents would significantly affect the proportions and distributions of soil aggregates[34]. Higher amount of SOC contents could improve soil structure and enhance its stability[35].The sequestration of SOC could reduce the scattering of large aggregates, which in turn reduces the decomposition rate of SOC[36]. This study showed that the SOC contents at 0 – 30 cm soil depth increased with the increase of planting years. Puget et al. found that SOC in soil miro-aggregates was more stable than that in large soil aggregates, but the amount of SOC was more in the larger sizes of soil aggregates[37], which was consistent to this study. The increase of SOC of treatment C1 in 2011 were higher than that in 2010 by 77.92%. The amount of average SOC was the highest in the 3 years. Total organic carbon content was lower than normal organic carbon content, but through the reasonable fertilizer and farming methods, the soil organic carbon content could increase, and thus the reasonable fertilizer and farming methods could contribute to improve the large aggregate content.

This study only tested the effects of the composite soils under a 3-year experiment with limited treatments, because longer term of experiment are need to further study the effects of other mixture ratios of feldspathic sandstone to sand on soil structure. Only by synthesizing the economic benefits, technique difficulties and soil structure, water movement and crop growth status affected by specific combination of feldspathic sandstone and sand, could we evaluate its application more accurately[38,39].

5. Conclusion

The 3-year experiment showed that the soil structure was improved with the increase of the planting years, and the amount of the soil aggregates over 0.25 mm were significantly increased, and the stability of soil structure also improves. After 3-year experiment, there were significant differences among the contents of soil aggregates in composite artificial soil. With the increase of the planting years, the soil aggregates stabilities, mean weight diameter, geometric mean diameter and fractal dimension were all affected. Treatment C1 and C2 all had better comprehensive indices than C3 and C4. There were high correlations between the ratios of sand in treatments C1, C2 and C3 and the values of fractal dimension (R^2=0.9701). Significant improvements were found in treatments C1 and C2, and their composite ratios are recommended.

After 3-year experiment, there were increasing trends of soil organic carbon at 0 – 30 soil depth, and the highest content of SOC was 1.22g/kg, observed in treatment C2 at 0- 30cm soil depth.

Acknowledgments

This study was supported by the China public welfare scientific research of the ministry of land and resources(201411008). We are grateful to Mr Han Jichang for providing laboratory,so we could managing the field experiments and professional English editor Zhu Daiwen， and kind help with the language of this manuscript.

References

[1] Kong D Z 1996 The ecological economic principle for comprehensive development and mangement of Maowusu sandland *Prataculturał Science* **13**(3) 13-5

[2] Zheng Y R and Zhang X S 1998 The diagnosis and optimal design of high efficient ecological economy system in Mowusu sand land *Act Phytoecologica Sinica* **22**(3) 262-8

[3] Wu C Z and Hong W 1999 Study on fractal features of soil aggregate structureunder different management patterns *Acta Pedologica Sinica* **36**(2) 162-7

[4] Wang Y C, Wu Y H, Kou Q, Min D A, Chang Y Z and Zhang R J 2007 Definition of arsenic rock zone borderline and its classification *Science of Soil and Water Conservation* **5**(1) 14-8

[5] Han L W, Li Z H, Shan X P and Zhang X S 2005 Research on strategy for land desertification control *Research of Soil and Water Conservation* **12**(5) 210-3

[6] Zheng Y R 2006 Trends of desertification in China and the countermeasures *Science & Technology Review* **11**(24) 67-70

[7] Wu W 2001 Desertification process and its controlling countermeasures in Maowusu sandy land in China *Chinese Journal of Eco-Agriculture* **9**(3) 15-8

[8] Hu B H, Yuan Q, Hai J B and Liao Y C 2009 Research on optimization models of agricultural ecosystem in Mu Us sandland *Agricultural Research in the Arid Areas* **27**(01) 212-8

[9] Tisdall J M and Oades J M 1982 Orgnic matter and water-stable aggregate *Journal of Soil Science* **33**(2) 141-63

[10] Lu J W and Li Z B 2002 Advance in soil aggregate study *Research of Soil and Water Conservation* **9**(1) 85-91

[11] Mikha M M and Rice C W 2004 Tillage and manure effects on soil and aggregate-associated carbon and nitrogen *Soil Science Society of America Journal* **68**(3) 809–16

[12] Bronick C and Lal R 2005 Soil structure and management: a review *Geothermal* **124**(1/2) 3-22

[13] Dexter A R 1988 Advances in characterization of soil structure *Soil&Tillage Research* **11** 199-238

[14] Huo L 2007 Effects of long-term fertilization on soil aggregate size distribution and stability in calcic kastanozems of loess plateau (Lanzhou: Gansu Agricultural University) pp16-17

[15] Su J and Zhao S W 2009 Comparison of the analysis methods for soil aggregate stability *Bulletin of Soil and Water Conservation* **29**(5) 114-7

[16] Bavel C H M V 1950 Mean weight-diameter of soil aggregates as astatistical index of aggregation *Soil Science Society of America Journal* **14**(C) 20-3

[17] Young I M and Crawford J W 1991 The fractal structure of soil aggregates: itsmeasurement and interpretation *Journal of Soil Science* **42** 187-92

[18] Yang P L, Luo Y P and Shi Y C 1993 Use the weight-size distribution tocharacterize the soil fractal features *Chinese Science Bulletin* **38**(20) 1896-9

[19] Erfect E and Kay B D 1995 Applications of fractals in soil and tillage research:a review *Soil & Tillage Research* **36** 1-20

[20] Qiu L P, Zhang X C and Zhang J A 2006 Distribution of nutrients and enzymes in Loess Plateau soil aggregates after long-term fertilization *Acta Ecological Sinica* **26**(2) 364-72

[21] Yoder R E 1936 A direct method of aggregate analysis of soils and a study of the physical nature of erosion losses *Journal of the American Society of Agronomy* **28** 337-51

[22] Six J, Elliott E T, Paustian K and Doran J W 1998 Aggregation and soil organic matter accumulation in cultivated and native grassland soils *Soil Science* **62**(5) 1367-77

[23] Six J, Elliot E T and Paustian K 2000 Soil structure and soil organic mater: II.A normalized stability index and the effect of mineralogy *Soil Science Society of America Journal* **64**(3) 1042-9

[24] Wang L, Li J and Bai W X 2014 Effects of tillage rotation and fertilization on soil aggregates and organic carbon content in corn field in Weibei Highland *Chinese Journal of Applied Ecology* **25**(3) 759-68

[25] Cheng K, Li J and Mao H L 2013 Effects of different rotational tillage patterns on soil physical properties in rainfed wheat fields of the Loess Plateau *Scientia Agricultura Sinica* **46**(18) 3800-8

[26] Hou X Q, Jia Z K and Han Q F 2012 Effects of different rotational tillage patterns on soil structure, infiltration andwater storage characteristics in dryland *Transactions of the CSAE* **28**(5) 85－94

[27] Puget P Chenu C and Balesdent J 2000 Dynamic of soil organicmatter associated with particle-size fractions of water-stableaggregate *European Journal of Soil Science* **51**(4) 595-605

[28] Li A Z, Zhang R Z and Wang J 2008 Effect of tillagemethods on the formation of water-stable aggregates in loesssoil *Chinese Journal of Soil Science* **39**(3) 481-4

[29] Zhou H, Lu Y Z, Yang Z C and Li B G 2007 Effects of conservation tillage on soil aggregates in huabei plain, China *Scientia Agricultura Sinica* **40**(9) 1973-9

[30] Perfect E and Kay B D 1995 Applications of fractals in soil and tillage research:a review *Soil*

and Tillage Research **36** 1-20

[31] Zhou Z F, Hu Y J, Ma C, Xue T Z, Guo Y and Wei G X 2012 Effects of long-term conventional cultivation on stability and distributions of organic carbon in soil aggregates *Agricultural research in the Arid areas* **30**(6) 112-8

[32] Gao H W, Li W Y and Li H W 2003 Conservation tillage technology with Chinese characteristics *Transactions of the CSAE* **19**(3) 1–4

[33] Liu Y P 2002 A fractal study on soil structure and soil water movement model (Shaanxi: Northwest A&F University) pp 9-12

[34] Unger P W 1997 Aggregate and organic carbon concentration interrelationships of a Torrertic Pale Stoll *Soil and Till-age Research* **42** 95—113

[35] Peng X H, Zhang B and Zhao Q G 2004 A review on relationship between soil organ-ic carbon pools and soil structure stability *Acta Pedologica Sinica* **41**(4) 618—23

[36] Lal R and Shukila M K 2004 Principles of soil physics New York *European Journal of Soil Science* **56**(5) 683-4

[37] Lu R K 1999 *Method of analysis in soil and agro chemistry* (Beijing: Agricultural Press)

[38] Wang R D and Wu X X 2009 New pattern to control Maowusu sandland *Research of Soil and Water Conservation* **16**(5) 176-80

[39] Wang R, Liu W Z and Li Z 2008 Physical properties of soils along a 10m deep soil profile in loess tableland *Acta Pedologica Sinica* **45**(3) 550-4

IOP Conf. Series: Earth and Environmental Science **41** (2016) 012003 doi:10.1088/1755-1315/41/1/012003

Antioxidant activity of seedling growth in selected soybean genotypes (*Glycine max (L.)* Merrill) responses of submergence

R I Damanik[1, 3], P Marbun[1] and L Sihombing[2]

[1]Agroecotechnology Department, Faculty of Agriculture, University of North Sumatera, 20155 Medan, Indonesia

[2]Agribusiness Department, Faculty of Agriculture, University of North Sumatera, 20155 Medan, Indonesia

E-mail: d_revandy@hotmail.com

Abstract. In order to better understand the physiological and biochemical responses relating to direct seeding establishment in soybeans, the plant growth rate and antioxidative defense responses of seedlings in seven Indonesian soybean genotypes (Anjasmoro, Detam-1, Detam-2, Dieng, Grobogan, Tanggamus, and Willis) at different submergence periods (4, and 8 days) were examined. Twelve-day old seedlings were hydroponically grown in limited oxygen conditions. The results showed that the chlorophyll content in soybean seedlings was reduced beginning as early as 4 d under submerged condition, except for Detam-1, Detam-2, and Grobogan genotypes. The dry weight and protein concentration of seedlings were significantly higher at control condition (0 d) than those in submerged condition. The activities of superoxide dismutase (SOD) increased linearly until 8 d submerged for all genotypes. On the other hand, our results showed that catalase (CAT) and ascorbate peroxidase (APX) activities did not work together, meaning that CAT is activated and APX deactivated, or vice versa, in response to submergence conditions, except for Grobogan and Tanggamus genotypes which had an effect on both CAT and APX activities. Submergence stress led to a significant increase in glutathione reductase (GR) together with APX activity for Detam-2 and Dieng genotypes at 8 d submerged.

1. Introduction

The soybean is one of the most important food commodities in Indonesia after rice and corn. About 60% of the national soybean production comes from the marginal lands and rice fields, therefore the harvest has fluctuated. Soybean planting at rice fields comes with many problems, especially flooding, which causes premature senescence that results in leaf chlorosis, necrosis, defoliation, cessation of growth, and reduction of yield.

In higher plants, the adverse effects of abiotic restrictions (submergence, salinity, drought, heat/cold, light) are dependable under oxidative stress which happens by an antioxidant defense system in support of the plants [1-2]. In fact, a common response to the stress causing all these adverse conditions is an oxidative burst, which means there is a temporary overproduction of the reactive oxygen species (ROS) within the diverse cellular stall of the plant cell [3]. ROS have divided into two main classes consisting of non-radical species (H_2O_2) or free radical forms O^{-2}; OH^-; OH^{2-}. Accumulation of high concentrations of ROS is potentially detrimental to plants cells causing damage to valuable biomolecules like DNA, proteins, lipids, chlorophyll, and membranes [4].

Plants are strictly affected by environmental tension because under such stress the production and quenching of the reactive oxygen species (ROS) in plants will stop working in order to maintain the impartial state. One of the major biological consequences of soil flooding is oxygen deficiency. Roots suffer from periodic or prolonged deprivation of oxygen, which interferes with respiration at the level of electron transport. The lack of a suitable electron acceptor leads to saturated redox chains, accumulation of NAD (P)H, and a decline in the generation of ATP. The toxic radicals can be removed through the mobilization of antioxidant reserves, which react both enzymatically and chemically with the toxic molecular species and their products. Compound constituents have been recognized to scavenge free radicals and consequently defend active plant cells adjacent to oxygen toxicity. To counteract the hazardous properties of ROS while under stress, plants have evolved a multifarious antioxidative defense system composed of antioxidant enzymes and metabolites such as ascorbate peroxidase (APX), catalase (CAT), superoxide dismutase (SOD), glutathione reductase (GR), ascorbic acid (AsA), glutathione (GSH), oxidized glutathione (GSSG), and Vitamin E [4-5].

The construction and scavenging of activated oxygen species are correlative, which maintains the plants in a reasonably steady state. Plants are modified to minimize radical damage using their strategy protection mechanisms. Thus, the sense of balance between the arrangement and detoxification of activated oxygen species is significant to cell endurance during flooding stress. [4].

2. Materials and methods

1.1. Plant materials and stress condition

The study was conducted using seven soybean genotypes to examine the differences in their tolerance of submergence at the early seedling stage. All those genotypes were provided by Indonesian Agricultural Research Centre, Malang, Indonesia. The experiments were conducted in laboratory conditions with temperature maintained at 25 ± 2°C. The healthy seeds were surface sterilized in an aqueous elucidation of 25 mM sodium hypochlorite for 25 min, rinsed four times with distilled water, and germinated on two sheets of Whatman filter paper (No 1; Macherey-Nagel, Germany) in 300 ml bottles moistened with 5 ml distilled water in dark conditions for 4 days. Three replicates of 20 seeds in each bottle were used for germination. The germinated seeds were maintained and continued to grow in the bottles and watered for 1 week before submergence treatment. It was designed for the experiment that 12-day-old seedlings were completely submerged in tap water for 4, 8, and 12 d in a concrete glass tank (dimension 50 cm x 20 cm x 20 cm) under 12 h light period/day and 70% relative humidity, with water levels maintained at 10 cm above the seedlings. Non-stressed seedlings were cultured in a similar concrete glass tank, but without the filled with water. A sufficient number of bottles were used within each replicate to provide adequate plant material for daily sampling during the first 4 d following the beginning of the experiment. During the submerged period, germinated seeds were kept submerged until measurements were completed.

1.2. Determination of plant biomass and chlorophyll analysis

Plants were harvested prior to stress imposition 4, 8, and 12 d after the start of submergence treatment. For each submergence duration, 10 plants (shoot + root) were harvested, and rinsed for 1 min with tap water. Two randomly chosen groups of five plants were used for determination of growth in terms of plant biomass and chlorophyll analysis. Each measurement was done independently. For plant biomass response, the whole plants were used for dry mass basis after drying in an oven for 72 h at 80°C. The data obtained was expressed in gram per plant. The total chlorophyll content estimation was measured following the experiment [6]. Fresh leaves (0.1 g) were added to 80% acetone and centrifuged at 5000 g for 10 minutes, and supernatants were measured at absorbance A_{645} and A_{663}. Calculation data used the equation: total chlorophyll (µg/ml) = 20.2 (A_{645}) + 8.02 (A_{663}).

1.3. Preparation of crude enzyme extracts
Three repititions of clean plants were collected from the bottles. The collected plants were treated with liquid nitrogen in a mortar and were homogenized with ice frozen 50 mmol L^{-1} sodium phosphate buffer (pH 7.8) containing 0.1 mM ethylenediaminetetraacetic acid (EDTA).Na_2 and 2% (w/v) insoluble polyvinyl polypyrrolidone (PVPP). Then, the homogenate was centrifuged in an eppendorf tube for 25 min at 10 000 *g* and the resultant supernatant was used for enzymes activity measurements. The soluble protein content of the soybean seedlings was determined by the method of Bradford [7].

1.4. ROS scavenging enzyme assays
The assay for superoxide dismutase (SOD, EC 1.15.1.1) activity was estimated according to the method of Steward and Bewley (1980) using nitroblue-tetrazolium (NBT) [8]. The reaction mixture contained enzyme, 390 mM L-Methionine, 2.25 mM NBT, 1 mM EDTA, 7% Na_2CO_3, 0.05 mol L^{-1} Na-Phosphate buffer (pH 7.8), and 60 µM riboflavin. The riboflavin was added last to start the reaction. The reaction mixtures were placed 30 cm from the light source (about 15W fluorescent lamps) for 10 min, and the absorbance decrease was recorded at 560 nm. The non-irradiated reaction mixture without enzymes, which gave the maximal color, served as the control and was deducted from A_{560}. One unit of SOD activity was defined as the amount of enzyme that caused 50% inhibition of the initial rate of reaction in the absence of enzyme, expressed in units min^{-1} mg protein $^{-1}$.

Catalase (CAT, EC 1.11.1.6) activity was assayed by measuring the initial rate of disappearance of H_2O_2 [9]. The extraction mixture contained 300 µL of enzyme extract, 0.5 mL of 10 mM H_2O_2, and 600 µL of 30 mM potassium phosphate buffer (pH 7.0), and the decrease in H_2O_2 following the decline in absorbance was recorded at 240 nm for 30 s. The enzyme activity was calculated as µmol H_2O_2 oxidized mg^{-1} protein min^{-1} by using an extinction coefficient (€ = 36 μM^{-1} cm^{-1}).

Ascorbate peroxidase (APX, EC 1.11.1.11) activity was measured following the method of Nakano and Asada [10]. The assay mixture contained 100 µl enzyme extract, 600 µl 0.1 mM EDTA in 0.05 M Na-phosphate (pH 7.0), and 400 µl 0.5 mM Ascorbic acid. The reaction started with the addition of 400 µl of 3 % H_2O_2 and the absorbance decrease was recorded after 1 minute at 290 nm. The activity of APX was calculated using an extinction coefficient (€ = 2.8 mM^{-1} cm^{-1}). .

Glutathione reductase (GR, EC 1.6.4.2) activity was estimated following Goldberg and Spooner [11]. The assay mixture contained 100 µl enzyme, 2.5 ml of 0.1 mM EDTA in 0.05 M Na-phosphate (pH 7.0), and 100 µl 0.5 mM oxidized glutathione (GSSG). After 5 minutes 50 µl of 9.6 mM NADH was added and mixed thoroughly. The absorbance decrease was recorded at 290 nm spectrophotometrically at an interval of 1 minute. The expression of 1 unit of GR activity is nmol glutathione reduced per minute, calculated using an extinction coefficient (€ = 6.22 mM^{-1} cm^{-1}).

1.5. Statistical analysis
All the data obtained was subjected to a one-way analysis of variance (ANOVA), and the mean differences were compared by a lowest standard deviations (LSD) test. Comparisons with $P < 0.05$ were considered statistically significant.

2. Results

2.1. Plant biomass
The plant biomass of the seedlings was observed, and the dry weight of plants was determined as shown in Figure 1. Although the seven soybean genotypes were able to grow under stress conditions, we observed a significant difference in the biomass response of the seedlings.

The biomass of the genotypes significantly decreased during submergence periods both for the tolerant (Dieng) and intolerant genotypes compared with their controls, with a large decline in the intolerant genotypes. The longest period of submergence 8 d led to decreased biomass of the seedlings. The accumulation of biomass for intolerant genotypes ranged from 42% to 47%, whereas the accumulation of biomass in the tolerant genotype Dieng was 54% and almost the same as intolerant genotypes Detam-1 (56%) and Detam-2 (57%) compared to their respective control plants. Although the intolerant genotypes of soybean seedlings seemed to have higher contents of dry weight compared with the tolerant cultivar, submergence stress caused differential reductions that were greater in intolerant genotypes than in the tolerant Dieng genotype. The lowest depletion of biomass at 8 d submergence was at the Willis genotype, compared with the control.

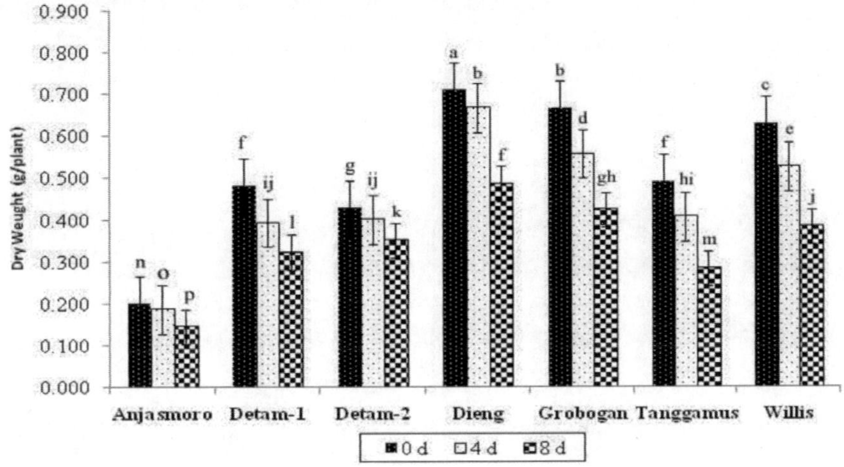

Figure 1. Changes in dry weight of soybean seedlings genotype at different days of submergence.

2.2. Effect of submergence on chlorophyll analysis

To investigate the physiological responses to submergence during the growth of the early seedlings of soybeans, the total chlorophyll substance was determined. The total chlorophyll content showed diverse profiles depending on the genotypes (Figure 2). Relative to the control, a marked decrease in chlorophyll was observed in submerged plants at 4 d except for Detam-1 and Grobogan genotypes, which maintained higher total chlorophyll content compared to the control. Accumulation of chlorophyll progressively declined when extending the period of submergence treatment, and remained significantly lower compared to the non-submerged seedlings in the tolerant genotype.

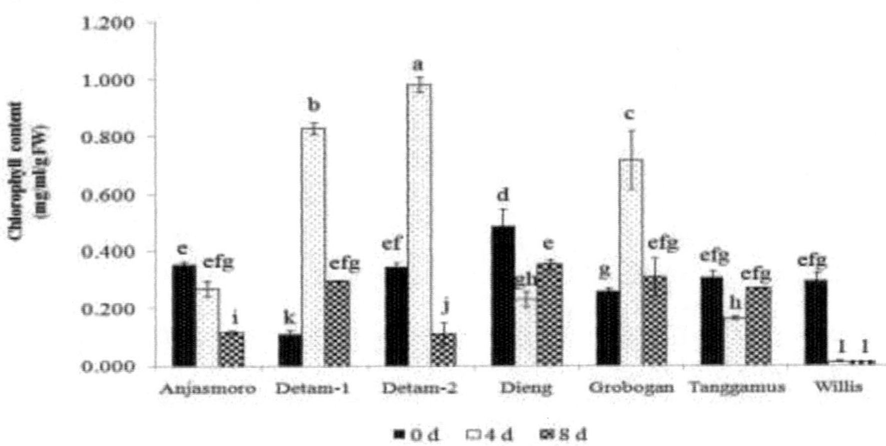

Figure 2. Changes of chlorophyll content of soybean seedlings genotype at different days of submergence.

2.3. Effect of submergence on antioxidant enzymes activities

A considerable discrepancy change was detected in the actions of the antioxidant enzymes within responses to submergence treatment in all genotypes. Under submerged conditions, the superoxide dismutase activity in all genotypes was higher than the control. The activity of SOD was greatly increased at 8 d for all genotypes except for the Willis genotype (Figure 3). For the soybean-tolerant Dieng genotype, SOD activity under 4 d and 8 d submerged were 14 times and 5.7 times higher than the control, respectively. Detam-1 and Detam-2 genotypes recorded a continuous increase in SOD activity up until 8 d submerged higher than the control, respectively. The establishment of superoxide dismutase through submergence stress is probably related to increased production of hydrogen peroxide, which is scavenged by peroxidase and/or catalase. Peroxidase and catalase activity was considered for soybean plants in different submergence periods.

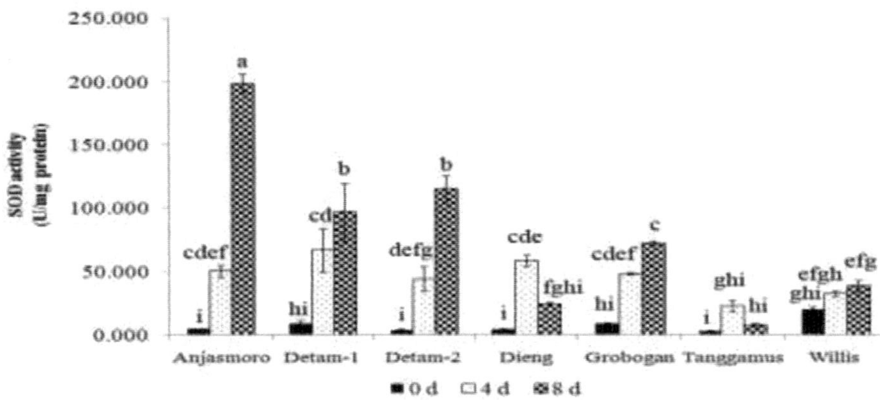

Figure 3. Superoxide dismutase activity of soybean seedlings genotype at different days of submergence.

The response of CAT action to submerged stress varied with the genotype and submerged periods. There was a considerable increase in catalase activity detected for Detam-1 and Detam-2 genotypes, but not for the flood-tolerant Dieng soybean genotype compared to control at 4 d submerged (Figure 4).

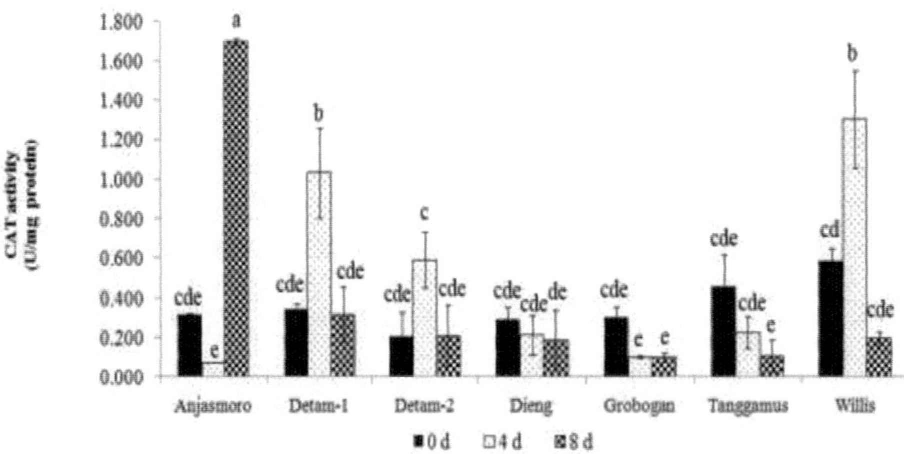

Figure 4. Catalase activity of soybean seedlings genotype at different days of submergence.

The ascorbate peroxidase (APX) activity of the Anjasmoro, Grobogan, and Willis genotypes showed a significant decreased under 4 d submerged seedlings, although APX activity still decreased until 8 d submergence compared with the control seedlings (Figure 5). The consequences also showed that APX activity significantly increased for Detam-1, Detam-2 and Dieng genotypes compared with the control seedlings starting at 4 d and 8 d submerged.

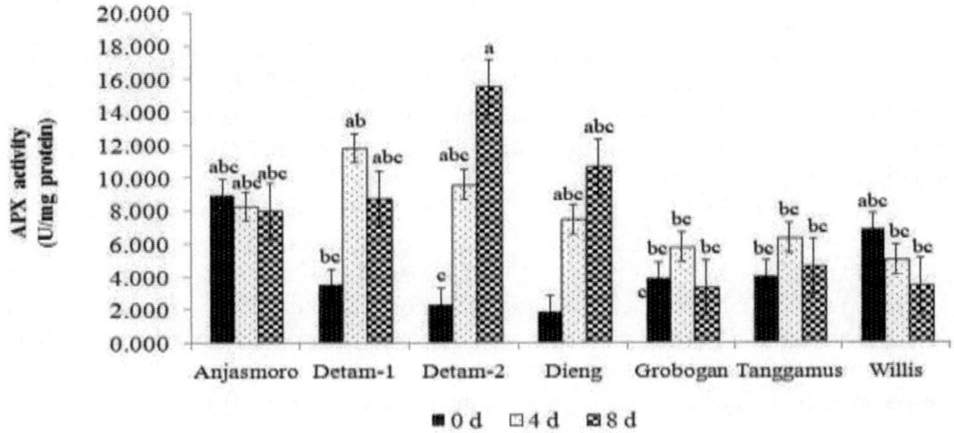

Figure 5. Ascorbate peroxidase activity of soybean seedlings genotype at different days of submergence.

The glutathione reductase action in submerged seedlings showed a considerable difference between submergence treatments, for all genotypes. The Detam-2 genotype had increased GR activity after 8 d submerged by almost twenty-folds, relative to the non-submerged control (Figure 6).

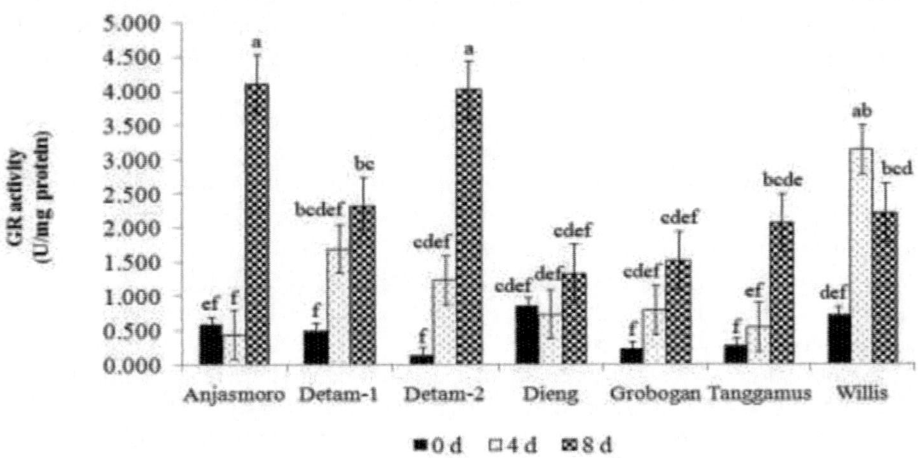

Figure 6. Glutathione reductase activity of soybean seedlings genotypes at different days of submergence.

3. Discussion

The capability to continue to exist under particularly elevated water for flood-tolerant soybean genotypes such as Dieng (Figure 1) was related to the sufficient biomass and to quick regeneration growth (shoots and roots) before and after the water stage decreased [12]. Moreover, the genotype anticipated the properties of shoot elongation competing with maintenance processes for energy. For that reason, energy running appears to be crucial for submergence acceptance. For example, the quantity of starch and total carbohydrates before and after submergence seemed to have a considerable impact on favored survival compared with the regeneration of the plant [13].

In this study, we have investigated the characteristics of soybean plants treated under non-submergence conditions, which appeared progressively green, and the level of chloroplast pigment was as expected. The seedlings began to display evident symptoms during a 4-day period of submergence, such as some leaves developing chlorosis and turning yellow, which was followed by defoliation at the bottom of the stem segments. Some of the genotypes maintained their membrane integrity during 4 d submerged (Figure 2), which then decreased during longer submergence periods. This was possibly due to the fact that competence for submerged photosynthesis was decreased between intolerant genotypes, which depend on energy consumption to survive prolonged submergence. Carbohydrate starvation has been shown to be one of the possible reasons for hypoxia/anoxia-induced injuries [14].

Aerobic respiration was inhibited due to oxygen deficiency, and consequently, the carbohydrate and energy status dropped to harmful levels in susceptible submerged plants [15]. Also, under submergence conditions, the hasty drop in the CO_2 photosynthetic assimilation rate was probably due to severely limited CO_2 because of the small absolute concentrations involved compared with O_2 and the structural damage suffered by the photosynthetic apparatus as evident from the fall in the values of Fo, Fm and Fv/Fm ratio. In addition, an irreversible loss in Rubisco activity and plant pigment content might be responsible for a poor CO_2 photosynthetic assimilation rate [12,16].

There is no doubt that submersion can cause anaerobiosis and make plants suffer from anoxia. The enhancement of the production of antioxidant enzymes may play an important role in metabolic stress tolerance. Plants have evolved an enzymatic mechanisms to scavenge the rapidly evolving ROS under stress. The effects of submergence stress on soybean plant genotypes were monitored by tracking changes in antioxidative enzyme activities, SOD, CAT, APX and GR which work in concert to detoxify ROS on the various submergence days of treatment. The changes in CAT may vary according to the intensity and time of stress. The activity of the enzyme ascorbate peroxidase showed a significant reduction in the Willis and Anjasmoro genotypes compared to control plants (Figure 5). The level of antioxidative reaction depends on the species, the development and the metabolic state of the plant, as well as the duration and intensity of the stress [17]. CAT scavenges H_2O_2 to water, but it has a lower affinity for H_2O_2 than APX [18]. The correlation action between SOD and CAT activity under a pressure of submersion showed a match for Anjasmoro and Willis genotypes (Figs. 3, 4), while the combination of SOD and APX was found in Dieng, Grobogan and Tanggamus genotypes (Figs. 3, 5). CAT is known as an enzyme that works to eliminate the negative effects of hydrogen peroxide, and aligns with APX to accomplish the same tasks as neutralizing hydrogen peroxide [17].

In our experiments, we found a gradual increase of GR among the submerged seedlings in the 8 days of submergence for all genotypes compared with the control. The improvement of GR enzyme activity also plays an important role in abiotic stress. Glutathione is a crucial antioxidant associated with regenerating AsA in the AsA-GSH cycle, and thus GSH is also involved in the regulation of H_2O_2 concentrations and control of the redox state in plant cells [19]. The relationship between GR functions and APX activity under stress conditions (in the AsA-GSH cycle) has been well recognized [4, 15, 19]. Hence, the increase in GR activity alone is not sufficient to put off stress induced H_2O_2 production. In the present study, APX activity increased at 8 d submerged for Detam-2 and Dieng genotypes, which correlated with the improved in GR activity at the same days of submerged (Figs. 5, 6).

Attention to the Detam-2 genotype can be used for tolerance to flooding, in addition to the Dieng genotype, which is known to be tolerant to flooding. This also shows that the activity of enzyme GR can be used as a criteria to evaluate tolerance flooding soybean genotypes. However, the evaluation of biochemical aspects needs to be further explored in order to analyze the enzyme activity of GR in cooperation with other antioxidants associated with cycles of AsA-GSH in the regulation of stress signals on soybean plants.

Acknowledgments
This work was supported by project Fundamental Research grant Ministry of Research, Technology and Higher Education Republic of Indonesia (Grant No.4806/UN5.IR/KEU/2014). Moreover, R.I.Damanik acknowledges the University of North Sumatera, Medan, Indonesia for the support of scientific labs.

References

[1] Sreenivasulu N, Sopory S K and Kavi K P B 2007 Deciphering the regulatory mechanisms of abiotic stress tolerance in plants by genomic approaches *Gene* **388** 1-13

[2] Farooq M, Basra S M A, Wahid A, Cheema Z A, Cheema M A and Khaliq A 2008 Physiological Role of Exogenously Applied Glycinebetaine to Improve Drought Tolerance in Fine Grain Aromatic Rice (*Oryza sativa* L.) *J. Agr. Crop Sci.* **194** 325-33

[3] Alscher R G, Erturk N and Heath L S 2002 Role of superoxide dismutases (SODs) in controlling oxidative stress in plants *J. Exp. Bot.* **372** 1331-41

[4] Blokhina O, Eija V and Kurt V F 2003 Antioxidants, oxidative damage and oxygen deprivation stress: a review *Annals of Bot.* **91** 179-94

[5] Dewir Y H, Chakrabarty D, Ali M B, Hahn E J and Paek K Y 2006 Lipid peroxidation and antioxidant enzyme activities of *Euphorbia millii* hyperhydric shoots *Environ. and Exp. Bot.* **58** 93-99

[6] Arnon D J 1949 Copper enzymes in isolated chloroplasts. Polyphenoloxidase in *Beta vulgaris Plant Physiol.* **24** 1-15

[7] Bradford M 1976 A rapid and sensitive method for the quantities of microgram quantities of protein utilizing the principle of protein-dye binding *Anal. Biochem.* **72** 248-54

[8] Stewart R C and Bewley J D 1980 Lipid peroxidation associated with accelerated aging of soybean axes *Plant Physiol.* **65** 245-8

[9] Bergmeyer H U 1970 *Methoden der enzymatischen, analyse* vol. I (Berlin: Akademie Verlag) pp 636-47

[10] Nakano Y and Asada K 1981 Hydrogen peroxide is scavenged by ascorbate-spesific peroxidase in spinach chloroplast *Plant Cell Physiol.* **22** 67-80

[11] Goldberg D M and Spooner R J 1983 Glutathione reductase vol III, 3rd ed. ed Bergmeyer H U *Methods of Enzymatic analysis* (Basel Weinheim) pp 258-65

[12] Macek P, Rejmankova E and Houdkova K 2006 The effect of long-term submergence on functional properties of *Eleocharis cellulosa* Torr. *Aquatic Bot.* **84** 251-8

[13] Panda D, Sharma S G and Sarkar R K 2008 Chlorophyll fluorescence parameters, CO_2 photosynthetic rate and regeneration capacity as a result of complete submergence and subsequent re-emergence in rice (*Oryza sativa* L.) *Aquatic Bot.* **88** 127-33

[14] Schluter U and Crawford R M M 2001 Long term anoxia tolerance in leaves of *Acorus calamus* L. and *Iris pseudacorus* L. *J. Exp. Bot.* **52** 2213-25

[15] Sarkar R K, Das S and Ravi I 2001 Changes in certain antioxidative enzymes and growth parameters as a result of complete submergence and subsequent re-aeration of rice cultivars differing in submergence tolerance *J. Agr. Crop Sci.* **187** 69-74

[16] Jackson M B and Ram P C 2003 Physiological and molecular basis of susceptibility and tolerance of rice plants to complete submergence *Annals of Bot.* **91** 227-41

[17] Reddy A R, Chaitanya K V and Vivekanandan M 2004 Drought-induced responses of photosynthesis and antioxidant metabolism in higher plants *J. of Plant Physiol.* **161** 1189-202

[18] Graham D and Patterson B D 1982 Responses of plants to low, non-freezing temperatures: Proteins, metabolism, and acclimation *Annu. Rev. Plant Physiol.* **33** 347-72

[19] Kocsy G, Galiba G and Brunold C 2001 Role of glutathione in adaptation and signaling during chilling and cold acclimation in plants *Physiol. Plant* **113** 158-64

Application of *Glomus sp.* and *Pseudomonas diminuta* Reduce the Use of Chemical Fertilizers in Production of Potato Grown on Different Soil Types

A Nurbaity[1,2], E T Sofyan[1] and J S Hamdani[1]

[1]Faculty of Agriculture, Universitas Padjadjaran, Jl. Raya Bandung Sumedang km.21 West Java, 45363, Indonesia

E-mail: a.nurbaity@unpad.ac.id

Abstract. The use of high chemical fertilizer rates in potato production has been applied on the farm in Indonesia. Application of biofertilizer consists of arbuscular mycorrhizal fungi has been tested to reduce the use of NPK rates in production of potato and to determine whether different soil types will have different response to this biofertilizer. A greenhouse experiment was conducted using mixtures of spores of *Glomus* sp. and inoculant of mycorrhizal helper bacteria *Pseudomonas diminuta*, applied at different rates of NPK fertilizer (0, 25, 50, 75 and 100% of recommended rates) and different soil types (Andisols and Inceptisols). Results of experiment showed that application of *Glomus sp.* and *P. diminuta* reduced the use of NPK up to 50%, where the growth (plant height and tuber number), N,P,K uptake and tuber yields of potato had similar effect to the highest recommendation rate of NPK fertilizer. Inceptisols in general had better response to the biofertiliser compared to Andisols. Findings from this experiment confirmed the evidences that biofertilizer could reduce the use of chemical fertilizer, and the widely distributed soil in Indonesia such as Inceptisols, is potential to be used as a medium for potato production.

1. Introduction

Potato is one of the most important food plants after rice, hence, some strategies need to be applied to improve the production of potato yield. One of the growing problems in potato production is high demands for fertilizer inputs, as its N, P and K requirements are higher than other vegetable crops [2]. The chemical practice used in intensive agriculture, has brought attention to alternative methods such as applying beneficial microbes for the establishment of a sustainable potatoes cultivation system [12].

Arbuscular mycorrhizal fungus is one of the beneficial microbes that play a vital role in maintaining soil and plant health [7, 11]. Some bacterias, known as Mycorrhizal Helper Bacteria (MHB), can directly stimulate mycorrhizal fungal activities. Enhancement of productivity of potatoes due to application of AMF [16] and MHB has also been reported [6]. Increased plant growth in the presence of mycorrhizal infection has been attributed mainly to the enhanced uptake of P. Previous study has shown that AMF occurred in potato production area in Andisols West Java Indonesia, and several AM fungi isolated from this area were effective as microbial inoculants for potato production system [12].

Andisols in highland areas is the most common soil type used for potato plantation in Indonesia. However, it has important characteristics due to its high capacity to immobilize phosphorus [13] and unstable particle sizes [15]. High rates of agro-chemical application are common to Andisols [13].

Furthermore, due to the limited high land areas for the potato plantation, it is now essential to find other areas such as medium land-Inceptisols. Inceptisols is widely distributed across the nation, covers about 37.5% of land in Indonesia that need to be maximized for agriculture production. Nevertheless, Inceptisols is known to have poor fertility especially the availability of phosphorus [15], hence improvement of the soil condition is vital prior to the cultivation of potato. Improvement in the productivity of Inceptisols will have significant contribution to the improvement of agricultural productivity in Indonesia.

Production of mycorrhizal propagules, as conducted in this experiment, permits the inoculation of these organisms in plants growing in soils where AMF inoculum levels are reduced [14]. Different type of soils will have different responses to the AMF inoculum and soil type influenced spore density as well as the percentage of mycorrhizal colonization of roots during cultivation period [9].

The purpose of this experiment was to determine the beneficial effects of AMF mixed with MHB on soil nutrients availability, nutrients uptake, growth and yield of potato grown in different soil types at different NPK levels. It was expected that application of this biofertilizer could reduce the use of NPK rates in the production of potato crops cultivated on different soil types.

2. Materials and methods

The experiment was conducted at a greenhouse of Assessment Institute for Agricultural Technology (AIAT) West Java, Indonesia (1.312 m above sea level). Two types of soil, Andisols from Lembang and Inceptisols from Jatinangor, were collected from 0-20 cm depth, air-dried and used as growing medium. The basic properties of these soils are presented in Table 1.

Table 1. Properties of the soils used in experiment.

	Andisols	Inceptisols
pH H_2O	6.3	6.0
Organic Carbon (%)	2.7	2.6
Total N (%)	0.5	0.4
Available P (ppm)	4.3	2.5
Available K (mg 100 g^{-})	18.7	18.1
CEC ($cmol.kg^{-1}$)	51.5	25.4
Base Saturation	24.4	65.1
Sand (%)	22	6
Silt (%)	64	37
Clay (%)	14	57
Soil Texture	Clay loam	Clay

Microbial inoculum consisted of AM Fungi and Mycorrhizal Helper Bacteria was used in this experiment. AM fungus species *Glomus sp.* in this study was isolated from potato plantation area in a farm in West Java Indonesia [12]. The spores were mass produced using sorghum as trap plant grown on sterilised zeolite as substrate for 3 months in 20 cm (diameter) x 10 cm (height) pot in a greenhouse. The mycorrhizal inoculum consisted of soil, spores (an average of 50 spores per g), mycelium and infected root fragments [8]. Ten gram of inoculum was incorporated in each pot. The MHB used was *Pseudomonas diminuta* [6]. Twenty mL of microbial suspension with the density of 10^7 colony forming units mL^{-1} was mixed with AMF inoculum in each pot at the day of planting.

Certified potato seed-tubers of Atlantic cultivars from Potato Center in Pangalengan West Java were placed after microbial inoculants covered with a layer of soil in the pots containing 10 kg of soil according to the treatment (Andisols or Inceptisols at different rates of NPK). Prior to sowing, the soils were steam sterilized at 121°C to kill indigenous microbes. Seedlings were grown under greenhouse condition with natural light interception, and watered every day to maintain 80 % of water holding capacity.

A factorial design was used with two factors: 1. Combination of fertilizer (FMA + MHB inoculum) combined with NPK fertilizers at 0, 25%, 50%, 75% and 100% recommended rate (0, 250, 500, 750,

100 kg ha^{-1}), and 2. Two type of soils (Andisols and Inceptisols). The experiment, laid out in randomized block design with three replications.

Plants were evaluated for plant height every week, number of total tuber and tuber weight at 12 weeks after planting (WAP). To quantify AMF infection at the vegetative phase (7 WAP), roots were cut into 1-cm lengths and were stored in lactic acid glycerol until further processing by clearing and staining. Percent infection was assessed using the gridline intersect method [3]. Nutrients concentrations were analyzed in soil and plant shoot at maximum vegetative phase (7 WAP) using Kjeldahl apparatus for N, while phosphorus and potassium were determined by acetic acid extraction and measured with spectrophotometer for P and flame photometer K. Chlorophyll content measured by CCM 200 plus (Opti-science.inc®). All data collected were then analyzed by two-way ANOVA using SPSS software (version 12). Differences at the 5% level of significance were tested using Fisher's Least Significance Difference (LSD) Test.

3. Results

There was a significant difference on root length colonized by AMF when grown on different soil types. However, no significant differences between fertilizer and soil type treatments on the percentage of mycorrhizal colonization on root of potato (Table 2). Mycorrhizal root length colonized of potato grown on Inceptisols was higher than those grown on Andisols.

Table 2. Effect of different rates of NPK and soil types on percentage of root colonization and root length colonized by AMF

Treatments	Mycorrhizal Colonization (%)	Root Length Colonised (m)
NPK levels (kg ha^{-1})		
0	42	28
250	51	33
500	60	40
750	46	58
1000	47	41
LSD 0.5	17	26
Soil Type		
Andisols	47	28 a
Inceptisols	52	52 b
LSD 0.5	11	16

Measurements on soil parameters showed that there were significant differences between NPK levels and type of soils on available P and exchangeable K, but not to pH and total nitrogen (Table 3). There was a notable increased in soil P availability on Inceptisols compared to Andisols. Inceptisols contain clay more than Andisols.

Table 3. Soil pH and nutrient availability (Total N, Available P, Exchangeable K) as the effect of different rates of NPK and soil types.

Treatments	pH	Total N (%)	Available-P (ppm)	Exchangeable-K (cmol/kg)
NPK levels (kg ha^{-1})				
0	6.4	0.41	60.96	2.50 a

250	6.2	0.49	75.88	3.60 b
500	6.0	0.36	93.62	2.80 a
750	6.2	0.35	71.23	4.90 c
1000	6.5	0.39	97.33	4.30 bc
LSD 0.5	0.6	0.12	37.86	0.82
Soil Type				
Andisols	6.2	0.42	9.60 a	3.90 b
Inceptisols	6.3	0.38	150.0 b	3.20 a
LSD 0.5	0.4	0.08	23.95	0.52

Increase shoot nutrients content and chlorophyll content index were observed with application of NPK fertiliser. No differences found in shoot nutrient as the effect of two soil types, but the CCI was higher in plant grown on Inceptisols (Table 4). There was positive correlation between content of N in shoot and CCI in this experiment ($r^2 = 0.741$) (data not shown).

Table 4. Nutrient uptake in plants (N, P, K) and chlorophyll content index (CCI) as the effect of different rates of NPK and soil types.

Treatments	N (%)	P (%)	K (%)	CCI
NPK levels (kg ha^{-1})				
0	3.79	0.35	2.30	18.20
250	4.00	0.39	2.62	19.90
500	3.99	0.44	2.54	20.90
750	4.24	0.47	3.17	22.10
1000	4.18	0.41	3.02	22.60
LSD 0.5	0.39	0.09	0.96	4.29
Soil Type				
Andisols	4.05	0.43	2.95	20.50
Inceptisols	4.03	0.40	2.50	23.80
LSD 0.5	0.25	0.05	0.61	2.71

Plant height, number of tuber and tuber yield were significantly higher in Inceptisols than Andisols (Table 5). No significant difference was observed on tuber weight per plant between NPK applications. This indicates that plants inoculated with mycorrhizal received benefit of its occurrence which results in similar yield of potato compared to the 100% of recommended rate of NPK.

Table 5. Growth and yield of potato as the effect different rates of NPK and soil types.

Treatments	Plant Height (cm)	Number of tuber per plant	Tuber weight per plant (g)
NPK levels (kg ha^{-1})			
0	62.5	7.5	74.0
250	56.6	8.8	72.2
500	59.0	6.7	69.2
750	58.7	8.0	76.2
1000	65.3	8.3	93.2
LSD 0.5	13.2	1.6	42.0
Soil Type			
Andisols	53.8 a	6.5 a	55.0 a

Inceptisols	67.0 b	8.4 b	90.9 b
LSD 0.5	20.9	2.5	26.5

4. Discussion

In general, finding from this experiment showed that potato crops grown on soil inoculated with biofertiliser consisted of AM fungi (*Glomus sp.*) and mycorrhizal helper bacteria (*Pseudomonas diminuta*) combined with reduced NPK levels (500-750 kg ha^{-1}) gave similar response to the crops grown with NPK recommendation rate (1000 kg ha^{-1}). Interesting results also found that different soil types (Andisols and Inceptisols) gave significant different responses on potato growth. Plants grown on Inceptisols which known have lower fertility levels gave better response compared to Andisols when biofertiliser applied.

In term of mycorrhizal parameters, difference responses between mycorrhizal colonisation and root length colonised in different soil types (Andisols and Inceptisols) were found. This could be due to the diferrent fertilitity levels and textures of both soils. Soil texture was one of the most important factors influencing colonization percentage in several plants such as maize, sorghum and peanuts by AMF. Soils with low fertility limit plant development and increase the dependence of plants on mycorrhizal association [4].

Inceptisols as used in this experiment, had more clay than Andisols (Table 1). Although clayey soils are rarely represent the best growth medium for plants, high level of clay can increase cationic exchange capacity, and soil pH. High acidity enchances the potential for cationic percolation, which will immobilize phosphate. The lower the soil pH, the higher the amount of phosphate fixed in the soil, and therefore the lower its availability to the plant [5].

Nutrient uptakes, growth and yield of potato in this experiment were also showed similar response when grown at different levels of NPK (lowest to the highest). Several experiments carried out in P-deficient soils (such as Inceptisols) have shown improved nitrogen-fixation on inoculation of the plants with mycorrhiza [1]. Other works also revealed the increasing of shoot nutrient contents of mycorrhizal plants, ie. mineral nutrient such as potassium is also assimilated more quickly and in greater amounts by mycorrhizal plants [10]. Andisols was generally has higher soil fertility level compared to Inceptisols. The decrease of plant growth and yield in Andisols is probably related to high soil fertility levels, which reduced the dependence of the plants on mycorrhizae and restricted the development of these fungi to their root cortex [14]. Furthermore, this experiment showed that low fertility soils such as Inceptisols, could be more productive when biofertiliser (*Glomus sp* and *Pseudomonas diminuta*) applied and gave similar potential yield with Andisols and high level of chemical fertiliser (NPK) which is now widely used for potato production in Indonesia.

5. Conclusion

Application of *Glomus* sp. and *Pseudomonas* sp. reduced the use of NPK up to 50%, where the growth, nutrients uptake (N,P,K), and tubers of potato grown on different soil types had similar effect to the highest recommendation rate of NPK fertilizer that being applied by local farmers. Findings from this experiment confirmed the evidences that application of AMF and MHB could reduce the use of chemical fertilizer which support sustainable farming system.

Acknowledgements

Project supported by a project grant from The Ministry of Research, Technology and Higher Education, Indonesia (PUPT 2015). The technical help of Derisfha Sri Anggraeni, Windy Amorita, Dwi Harvani Sirait, and Siti Nisrina Hasna is gratefully acknowledged.

References

[1] Abbott L K and Robson A D 176 Growth Stimulation of Subterannean clover with Vesicular-Arbuscular Mycorrhizas *Aust J Agric Res* **28** 639-49
[2] Adavi Z and Tadayoun M R 2014 Effect of Mycorrhiza Application on Plant Growth and Yield

in Potato Production Under Field Condition *Iranian Journal of Plant Physiology* **4**(3) 1087-93

[3] Brundrett M, Bougher N, Bell R W and Hatton T J 1996 *Working with mycorrhizas in forestry and agriculture ACIAR* (Australia: Canberra)

[4] Carrenho R, Trufem S F B, Bononi V L R and Silva E S 2007 The Effect of Different Soil Properties on Arbuscular Mycorrhizal Colonization of Peanuts, Sorghum and maize *Act abo.bras.* **21**(3) 723-30

[5] Gosling P, Mead, A, Proctor M, Hammond, J P and Bending G D 2013 Contrasting Arbuscular Mycorrhizal Communities Colonizing Different Host Plants Show a Similar Response to a Soil Phosphorus Concentration Gradient *New Phytologist* **198** 546-56

[6] Hidayat C, Arief D H, Nurbaity A and Sauman J 2013 Rhizobacteria Selection to Enhance Spore Germination and Hyphal Length of Arbuscular Mycorrhizal Fungi In Vitro *Asian Journal of Agriculture and Rural Development* **3**(4) 199-204

[7] Jeffries P, Gianinazzi S, Perotto S, Turnau K and Barea J M 2003 The Contribution of Arbuscular Mycorrhizal Fungi in Sustainable Maintenance of Plant Health and Soil Fertility *Biol. Fertil. Soils* **37** 1-16

[8] Kartika Y, Nurbaity A, Fitriatin B N and Sofyan E T 2014 Effect of Sterilization and Composition Carrier of Arbuscular Mycorrhiza and Mycorrhizal Helper Bacteria (MHB) on Mycorrhizal Spores, MHB Population, and Shoot Root Ratio of Sorghum (*Sorghum bicolor*) *Agric. Sci. J.* **I**(4) 262-8

[9] Land S and Schönbeck F 1991 Influence of Different Soil Types on Abundance and Seasonal Dynamics of Vesicular Arbuscular Mycorrhizal Fungi in Arable Soils of North Germany *Mycorrhiza* **1**(1) 39-44

[10] Moore D, Robson G D and Trinci A P J 2011 *21st Century Guidebook to Fungi* (Cambridge: Cambridge University Press)

[11] Nurbaity A 2014 Application of AM Fungi in Remediation of Saline Soils ed Z M Solaiman et al. *Mycorrhizal Fungi: Use in Sustainable Agriculture and Land Restoration* Soil Biology **41** (Berlin Heidelberg: Springer) pp 313-21

[12] Nurbaity A, Hidayat C, Hudaya C and Sauman J 2013 Mycorrhizal Fungi and Organic Matter Affect Some Physical Properties of Andisols *Soil Water Journal* **2**(1) 639-44

[13] Prasetya B and Anderson C 2011 Assesment of the Effect of Long Term Tillange on The Arbuscular Mycorrhiza Colonization of Vegetable Crops Grown in Andisols *Agrivita* **33**(1) 85-92

[14] Smith S E and Read D J 1997 The Roles of Mycorrhizas in Ecosytems ed S E Smith and D J Read *Mycorrhizal Symbisosis* (London: Academic Press) pp 409-52

[15] Tan K H 2008 Soils in The Humid Tropics and Monsoon Region of Indonesia (Bocca Raton: CRC Press)

[16] Wu F, Wang W, Ma Y, Liu Y, Ma X, An L and Feng H 2013 Prospect of beneficial microorganisms applied in potato cultivation for sustainable agriculture *African Journal of Microbiology Research* **7**(20) 2150-8

IOP Conf. Series: Earth and Environmental Science **41** (2016) 012005 doi:10.1088/1755-1315/41/1/012005

Effects of soil amendment on soil characteristics and maize yield in Horqin Sandy Land

L Zhou[1], J H Liu[1, 4], B P Zhao[1], A Xue[2] and G C Hao[3]

[1] College of Agronomy, Inner Mongolia Agricultural University, Hohhot, Inner Mongolia 010019, China

[2] Ottawa Research and Development Centre, Agriculture and Agri-Food Canada, Ottawa, ON K1A 0C6, Canada

[3] Inner Mongolia Trirock co, Ltd, Naiman Banner, Tongliao, Inner Mongolia 028300, China

E-mail address: cauljh@163.com

Abstract. A 4-year experiment was conducted to investigate the inter-annual effects of sandy soil amendment on maize yield, soil water storage and soil enzymatic activities in sandy soil in Northeast China in 2010 to 2014. We applied the sandy soil amendment in different year, and investigated the different effects of sandy soil amendment in 2014. There were six treatments including: (1) no sandy soil amendment application (CK); (2) one year after applying sandy soil amendment (T1); (3) two years after applying sandy soil amendment(T2); (4) three years after applying sandy soil amendment(T3); (5)four years after applying sandy soil amendment(T4）; (6) five years after applying sandy soil amendment （T5）. T refers to treatment, and the number refers to the year after application of the sandy soil amendment. Comparing with CK, sandy soil amendments improved the soil water storage, soil urease, invertase, and catalase activity in different growth stages and soil layers, the order of soil water storage in all treatments roughly performed: T3 > T5 > T4 > T2 > T1 > CK. the order of soil urease, invertase, and catalase activity in all treatments roughly performed: T5 > T3 > T4 > T2 > T1 > CK. Soil application of sandy soil amendment significantly ($p \leqslant 0.05$) increased the grain yield and biomass yield by 22.75%-41.42% and 29.92%-45.45% respectively, and maize yield gradually increased with the years go by in the following five years. Sandy soil amendment used in poor sandy soil had a positive effect on soil water storage, soil enzymatic activities and maize yield, after five years applied sandy soil amendment (T5) showed the best effects among all the treatments, and deserves further research.

1. Introduction

The Horqin Sandy Land located in the southeast part of Inner Mongolia, becoming the most seriously desertified areas in northeast China [1]. Sandy soil is the major soil type in this area, and sandy soils are defined by poor physical properties which leading to low water-holding capacity, high evapotranspiration and excessive drainage of rain and irrigation water below the root zone [2]. Due to its poor physical structure, soil water content and soil nutrient status are unable to meet the plant demands, which restricts plant growth, soil productivity and local economic development.

In recent years, soil amendments (such as PAM, humic acid, Biochar, et al.,) have been widely used in agriculture development to improve the soil quality and increase crop yield [3-5].To date, the published results have reported that soil amendments play a vital role in improving soil water and fertilizer retention properties. Such as Cross-linked chained PAM and PAA have superior ability to absorb and retain ultra-high amounts (1000–100,000 %) of water in comparison to their own weight, so they can improving water holding capacity of soil, at the same time they can curb the elution of nutrients by reducing the percolation of water. Humic acid not only can substantial increase yield, but also found it related to biochemical activity [6]. Bentonite is a clay soil that usually contains at least 70% of the three-layered (2:1) clay mineral montmorillonite [7]. As a result of its special layered structure [8], it is characterized by considerable swelling behaviour and high water absorbing capacity. Therefore, sodium bentonite is a good material for reducing infiltration [9], and it was extensively used in a number of applications, such as coatings [10], cosmetics, medicines and drilling fluids [11]. However, limited study is reported for bentonite as a soil amendment.

Soil water storage plays a vital role in maize growth through impact on anatomical, morphological, physiological and biochemical processes [12]. Soil enzymatic activities plays a crucial role in nutrient cycling and elements transfer [13]. Soil enzymes is also regarded as the main regulator of soil biological processes and also regarded as sensors, to reveal the effects of soil physico-chemical changes of soil quality [14]. Some studies argued that crop rotation, amendments, tillage and agricultural management affect the activity of soil enzyme [15].

The objective of this study was to determine whether the amendment of sandy soil has a positive effects on soil water storage, the soil profile distribution, microbiological soil quality indicators, and maize yield; and also if these amendment have increasing trend as the years goes by after application.

2. Materials and Methods

2.1. Study area

The effects of sandy soil amendment of this field experiment was conducted at the experimental base of the Inner Mongolia Trirock Co. Ltd, Nanman banner, Inner Mongolia, China (42° 50′ 0.03″ N , 120° 41′ 46.14″ E), located in the South of the Horqin Sandy Land. The local climate is continental monsoon with an annual average rainfall of 364.6 mm, concentrated over the months of June-September. However, the annual evapotranspiration can reach up to 1934.4 mm, and the annual average temperature was 6.4 ℃, the frost-free period was around 151 days. The soil type was aeolian sandy soil and contained 27.15 mg · kg^{-1} available nitrogen, 4.68 mg · kg^{-1} available phosphorus, 65.31 mg · kg^{-1} available potassium.

2.2. Material

The soil amendment is a novel product named sandy soil amendment, produced by the Inner Mongolia Trirock Co. Ltd, and its ingredient constitutes: Na-bentonite (91%), Humic acid (6%), Na_2co_3 (2%), Plant cellulose (1%). The application rate is 30 t · hm^{-2}, which is the optimal quantity conducted by this company with multiple field trials. The treatment is usually applied by spreading the sandy soil amendment evenly in the field before planting in spring, and mixed with 0-25 cm soil via tillage by rotary tiller before maize was seeded.

2.3. Experimental design

The experimental base was established in 2010 by the Inner Mongolia Trirock Co. Ltd. The experiment commenced from spring 2014 when the maize seeded to autumn 2014 when the maize was harvested. The field was laid out in randomized complete block design with three replications. Each plot was 40 ㎡. In this study, all the amendment treatments (T1, T2, T3, T4, and T5) were applied to the sandy soil only in their first year, so we can observe all the effects in different years in the same year. We added a new treatment in the spring every year planting season ,whereby, from two treatments in 2010 up-till six treatments in 2014 (Table 1). There were six treatments: (1) no sandy soil amendment application (CK); (2) one year after applying sandy soil amendment (T1); (3) two years after applying sandy soil amendment(T2); (4) three years after applying sandy soil amendment(T3);(5) four years after applying sandy soil amendment(T4); (6) five years after applying sandy soil amendment （T5）.

We used the "zhengdan 958" as the maize variety in 2014, which is a common, promoted local cultivar with high yield among the farmers. A standard seed rate of 45 kg · hm^{-2}, were used and all the treatments were broadcast with diammonium phosphate at a rate of 270 kg · hm^{-2} while seeding, and urea used as a top-dressing at a rate of 300 kg · hm^{-2} and 150 kg · hm^{-2} in huge bellbottom and tasseling stage, respectively. Seeds were at placed 6 cm depth, between-row spacing of 50 cm and within-row spacing of 30 cm. During the growing period, irrigation was applied uniformly across all plots when required according to the local practice. Maize was seeded in April 28 and harvested in September 28.

Table 1. The experimental design from 2010 to 2014.

Year	Treatment					
2010	CK	T1				
2011	CK	T1	T2			
2012	CK	T1	T2	T3		
2013	CK	T1	T2	T3	T4	
2014	CK	T1	T2	T3	T4	T5

Treatments: CK, T1, T2, T3, T4 and T5, T refers to treatment, and the number refers to the year after the application of sandy soil amendment. CK— no sandy soil amendment application.

2.4. Soil sampling and analysis

During Maize jointing stage, huge bell-bottom period and tasseling stage, there was an interval of 5 days between each irrigation. Soil samples were collected using soil auger at the depth of 0-100 cm from each replicated plot. The topsoil was taken at 0-20 cm with a 10 cm intervals and 20 cm intervals in the depth of 20-100 cm. Each sample was divided into two portions, where, one portion was placed in an aluminum box and then immediately transported to the laboratory for drying at 105 ℃ The other portion was stored on ice in the field and then immediately sieved through a 2-mm mesh in the laboratory, and stored at 4 ℃ in preparation for enzyme analysis.

The soil bulk density was measured by the method of cutting ring. The water storage (W) was calculated as follows:

$W_i = h \times d \times b_i\% \times 10$

where, W (mm), the soil water storage; h (cm), soil depth; d ($g \cdot cm^{-3}$) soil bulk density; b%, the gravitational water content. Soil water storage was calculated at 0–100 cm soil profile; subscript i referred to the soil layer.

There were three soil enzymes activity observed including urease, invertase and catalase. The soil enzyme activities was assayed as described by Guan [16], and all the enzyme activities were determined from air-dried samples. The urease activity was determined with sodium phenate colorimetry, and expressed as $mg \cdot g^{-1} \cdot 24\ h^{-1}$. The invertase activity was determined with constant temperature incubation, and expressed as $mg \cdot g^{-1} \cdot 24\ h^{-1}$. The catalase activity was determined with potassium permanganate titration, and expressed as mg $\cdot g^{-1} \cdot 30\ min^{-1}$.

2.5. Statistical analysis

All the data were analyzed by SAS ver. 9.0 software package. Significant difference were determined by the Duncan's multiple range test at $P<0.05$.

3. Results

3.1 Soil water storage

The soil water storage of different treatments at different growth stages of maize in 0 - 100 cm profile is shown in Figure 1. The results showed that the soil water storage from 0 – 40 cm was significantly ($p<0.05$) increased by sandy soil amendment compared with CK in all the three critical growth stages. The trend of variation in soil water storage of amendment treatments in 0 – 40 cm profile performed: 20 – 40 > 0 – 10 > 10 – 20 cm, but the soil water storage of CK was increasingly raised in 0-40 cm profile. The order of soil water storage in all treatments roughly performed was : T3 > T5 > T4 > T2 > T1 > CK. Applying sandy soil amendment increased soil water storage at the jointing, huge bell-bottom, tasseling stages compared with CK. Soil water storage increased by 167.65 - 258.24%, 95.88 - 126.75 %, and 73.71 - 197.94% in 0 – 10 cm profile, respectively, by15.53 - 53.24%, -1.67 - 30.69%, and 8.25 - 40.71% in 10 - 20cm profile, respectively ,and by 16.81 - 52.59%, 16.79 - 33.21%, and 14.14 - 39.9% in 20 – 40 cm profile.

The soil water storage of CK was significantly higher than any other treatments below the soil depth of 40 cm. but there was not a certain trend among the amendment treatments. The soil water storage of all the treatments reached the peak in 20-40 and 60-80 cm profile, respectively.

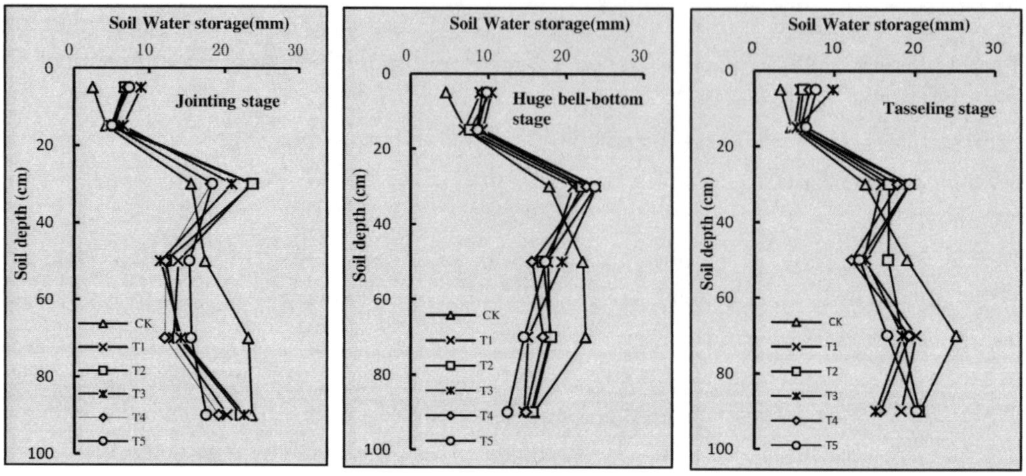

(a) (b) (c)

Figure 1. Soil water storage at different growth stages of maize in 0 – 100 cm profile. Treatments: CK, T1, T2, T3, T4 and T5, T refers to treatment, and the number refers to the year after the application of sandy soil amendment. CK— no sandy soil amendment application.

3.2 Soil enzymatic activities

The soil enzymatic activities of different treatments at different growth stages of maize in different soil layers are given in (Table 2, 3, 4). Soil urease, invertase and catalase activities in all treatments displayed the same trend in the three growth stages. Application of sandy soil amendment had a positive effect on soil enzymatic activities. All the treatments had a high activity in huge bell-bottom stage and in 10 – 20 cm soil profile, and all the amendment treatments were significantly higher than CK. All the treatments had a low activity in jointing stage and in 0 – 10 cm soil profile, and there were no significant differences between the all treatments.

After applying the sandy soil amendment, the soil enzymatic activities of amendment treatments showed an increasing trend with time. The T5 always had the highest soil enzymatic activities in different growth stages in 0 – 40 cm soil profile. The order of soil enzymatic activities performance in all treatments was: T5 > T3 > T4 > T2 > T1 > CK.

In the 0 - 10cm soil profile, T5 treatment significantly improved the soil urease, invertase, and catalase activity compared with CK by 4.88 - 7.05%, 8.99 - 17.72%, and 45.81 - 65.13% at the three critical stages, respectively. The soil enzymatic activities of T3 and T4 treatment were lower than T5, and they had no significant difference between them in soil urease and catalase activity, and in soil invertase activity of jointing and tasseling stage.

In the 10 – 20 cm soil profile, all the treatments significantly improved the soil enzymatic activities and had the highest values than in the other two soil profiles. The T5 treatment significantly improved the soil urease, invertase, and catalase activity compared with CK by 20.36 - 21.71%, 13.82 - 25.34%, and 39.89 - 48.65% at the three critical stages, respectively. The T3 and T4 treatment significantly and 23.29%, 17.12% and 15.82%, 10.09% and 10.56% at the three stages, respectively. Conversely, the differences between the two treatments were not significant at the huge bell-bottom and tasseling stage, and also significantly improved the soil invertase activity compared to CK by 35.31 - 45.27% and 35.88 - 43.80%, but there were no significant difference between the two treatments in all the three stages.

In the 20 – 40 cm soil profile, all the amendment treatments increased the soil enzymatic activities compared to CK by 0.21 - 5.92%, 0.66 - 15.38%,and 25.29 - 60.15% in soil urease, invertase and soil catalase activity (it is expected that the T1 treatment of soil urease was lower than CK at jointing stage), respectively

Table 2. Soil urease activity at different growth stages of maize in 0-40 cm profile.

Soil Layer	Treatments	Soil Urease activity（$mg.g^{-1}.24h^{-1}$)		
		Jointing stage	Huge bellbottom stage	Tasseling stage
0-10	CK	0.492bc	0.511dc	0.494c
	T1	0.485c	0.515c	0.480d
	T2	0.496bc	0.509d	0.502b
	T3	0.503b	0.528b	0.505b
	T4	0.493bc	0.525b	0.508b
	T5	0.516a	0.547a	0.520a
10-20	CK	0.496e	0.514d	0.502e
	T1	0.518d	0.604c	0.524d
	T2	0.521d	0.609bc	0.530d
	T3	0.574b	0.615ab	0.583c
	T4	0.560c	0.605bc	0.595b
	T5	0.597a	0.621a	0.611a
20-40	CK	0.473c	0.479d	0.483a
	T1	0.469c	0.485c	0.485a
	T2	0.474c	0.489c	0.492a
	T3	0.501a	0.506b	0.495a
	T4	0.485b	0.507b	0.500a
	T5	0.491b	0.518a	0.496a

Values in the same column followed by different letters indicate significant differences ($p<0.05$). Treatments: CK, T1, T2, T3, T4 and T5, T refers to treatment, and the number refers to the year after the application of sandy soil amendment. CK— no sandy soil amendment application

Table 3. Soil catalase activity at different growth stages of maize in 0-40 cm profile.

Soil Layer	Treatments	Soil Catalase activity（$mg.g^{-1}.30min^{-1}$)		
		Jointing stage	Huge bellbottom stage	Tasseling stage
0-10	CK	0.133c	0.170d	0.163d
	T1	0.178b	0.216c	0.207c
	T2	0.186b	0.221c	0.210c
	T3	0.213a	0.241b	0.219b
	T4	0.214a	0.239b	0.219b
	T5	0.219a	0.252a	0.238a
10-20	CK	0.148e	0.174d	0.165d

	T1	0.187d	0.224c	0.219bc
	T2	0.201c	0.228c	0.216c
	T3	0.215ab	0.246b	0.223b
	T4	0.213b	0.243b	0.224b
	T5	0.220a	0.254a	0.231a
20-40	CK	0.084c	0.167e	0.152d
	T1	0.126b	0.209d	0.196bc
	T2	0.129ab	0.216c	0.193c
	T3	0.131ab	0.227b	0.199ab
	T4	0.135a	0.226b	0.202a
	T5	0.136a	0.239a	0.201ab

Values in the same column followed by different letters indicate significant differences ($p<0.05$). Treatments: CK, T1, T2, T3, T4 and T5, T refers to treatment, and the number refers to the year after the application of sandy soil amendment. CK— no sandy soil amendment application

Table 4. Soil invertase activity at different growth stages of maize in 0-40 cm profile.

Soil Layer	Treatments	Soil Invertase activity（$mg.g^{-1}.24h^{-1}$)		
		Jointing stage	Huge bellbottom stage	Tasseling stage
0-10	CK	10.964e	12.861e	12.310e
	T1	10.650d	12.931e	12.413d
	T2	11.184c	13.270d	12.612c
	T3	11.573b	14.472b	13.162b
	T4	11.624b	14.351c	13.125b
	T5	11.939a	15.140a	14.260a
10-20	CK	10.357f	12.594d	12.180e
	T1	11.247e	13.426c	13.141c
	T2	12.036d	13.706c	12.976d
	T3	12.542c	14.75ab	13.409b
	T4	12.769b	14.586b	13.466b
	T5	12.981a	15.236a	13.863a
20-40	CK	9.866d	12.428f	11.540d
	T1	9.931d	12.564e	11.760c
	T2	10.150c	12.963d	11.583d
	T3	10.266b	13.640b	11.931b
	T4	10.513a	13.564c	12.140a
	T5	10.540a	14.340a	12.080a

Values in the same column followed by different letters indicate significant differences ($p<0.05$). Treatments: CK, T1, T2, T3, T4 and T5, T refers to treatment, and the number refers to the year after the application of sandy soil amendment. CK— no sandy soil amendment application

3.3 Maize grain and biomass yield

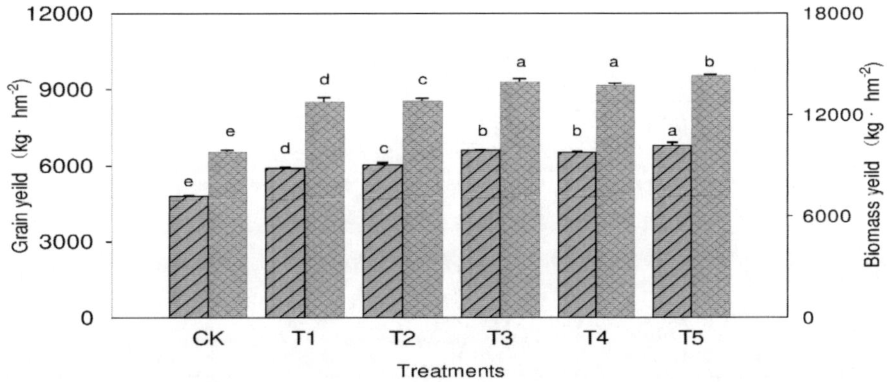

Figure 2. Maize grain yield and biomass yield in 2014.

Means followed by the same letter are not significantly different at (p<0.05). Treatments: CK, T1, T2, T3, T4 and T5, T refers to treatment, and the number refers to the year after the application of sandy soil amendment. CK— no sandy soil amendment application.

All the treatments amendment significantly improved the Maize grain yield and biomass yield (Figure 2.). The grain and biomass yield were gradually increased in the first three years. The T4 treatment a bit higher than the T3. However, there was no significant difference between them, and T3 and T4 treatments significantly increased grain and biomass yield by 41.67 and 39.71%, 37.71 and 35.57%, respectively, compared to CK. The T5 treatment had the highest grain and biomass yield, while, the T1 treatment performed least within the amendment treatments. The T5 and T1 significantly improved grain and biomass yield by 45.45 and 29.92%, 41.42 and 22.75%, respectively, compared to CK.

3.4 Relativity among water storage, grain yield and enzyme activities of soil

A correlation matrix (Table 5) shows that some significant relationships exist among the water storage, enzymatic activities and grain yield in three soil layers (0 - 10, 10 - 20, 20 – 40 cm). In 0 – 10 cm soil layer, the grain yield was strongly correlated ($r = 0.809$, $P < 0.001$; $r = 0.864$, $P < 0.001$), with the water storage and catalase activity of the soil, respectively. However, the soil invertase and urease activity have no significant effect ($p > 0.05$) on grain. In 10 – 20 cm soil layer, there was no significant effect between water storage and grain yield, however, the grain yield shared a strong correlation ($r = 0.774$, $P < 0.001$; $r = 0.863$, $P < 0.001$; $r = 0.616$, $P < 0.01$) with urease, catalase and invertase activities, respectively. In 20 – 40 cm soil layer, grain yield was correlated with urease activity ($r = 0.608$, $P < 0.001$), and ($r = 0.574$, $P < 0.01$) with water storage.

Table 5. Linear correlation coefficient among water storage , grain yield and enzyme activities of soil.

		WT	Ure	Inv	Cat	GY
	WT	1				
	Ure	0.610^{**}	1			
0-10 cm	Inv	0.574^{*}	0.837^{***}	1		
	Cat	0.860^{***}	0.688^{**}	0.748^{***}	1	
	GY	0.809^{***}	0.436	0.378	0.864^{***}	1

	WT	1				
	Ure	0.705^{**}	1			
10-20 cm	Inv	0.807^{***}	0.835^{***}	1		
	Cat	0.684^{**}	0.879^{***}	0.907^{***}	1	
	GY	0.368	0.774^{***}	0.616^{**}	0.863^{***}	1
	WT	1				
	Ure	0.319	1			
20-40 cm	Inv	0.321	0.716^{***}	1		
	Cat	0.329	0.717^{***}	0.931^{***}	1	
	GY	0.574^{*}	0.608^{**}	0.226	0.428	1

*, **, *** significant at 0.05, 0.01 and 0.001 probability levels. WT—water storage, Ure—urease, Inv—invertase, Cat—catalase, GR—grain yield.

4. Discussion

4.1 Water storage

Soil water storage is a vital soil quality indicator, and it plays a significant role in a series of hydrological and biological processes [17]. It has widely been reported that both synthetic and natural water absorbing soil amendments can increase water retention capacity of soils and regulate the plants available water supplies [18-19]. In our study, application of sandy soil amendment increased the soil water storage in the top-soil compared to CK, and had a little effect on deeper soil, and this might due to the application of sandy soil amendment in the deeper layers. However, it had also changed the distributions of soil water storage in 100 cm soil layer, and soil water storage of all the amendment treatments. The 0 – 100 cm soil profile had the highest values in 20 – 40 cm soil layer and higher than CK during 0 – 40 cm soil depth. However, soil water storage of CK reached its peak in 60 -100 cm soil layer and higher than any other treatments in 40 -100 cm soil depth. This might be attributed to the sandy soil which has a poor ability of retaining the water and reducing the infiltration [2, 20].

The long-term effects on soil water storage among the amendment treatments had increasing trend in the first three years. The order of soil water storage between T3, T4 and T5 was: T3 > T5 > T4. High level of water needed by maize [21], soybean [22] and Pueraria [23] at productive stage has been reported. The one reason why T5 had the highest yield but with the lower soil water storage compared T3 may be explained by much water demanded by the maize at productive stage. The other reason may be attributed to the soil bulk density as many studies have proved that soil amendments lower the soil bulk density [24-25].

4.2 Soil enzymatic activities

Soil enzymatic activities originated from active microorganisms [26], and plays a vital part in depolymerization of structurally diverse polymeric macromolecules [27]. Previous studies have indicated that soil amendments could increase the soil enzymatic activities [28]. In our study, we observed the three soil enzymatic activities during the three growth stages of maize in different soil layers. Evidence in this study showed that adding sandy soil amendment could improve the three soil enzymatic activities in all three soil layers, and as the applied years went on, the soil enzymatic activities had an increasing trend. This phenomenon maybe due to the improved soil physical properties (e g. soil bulk density, soil aggregates, and pH).

4.3 Grain and biomass yield

Soil amendments increased the yield, mainly due to the reduce irrigation water consumption and improve fertilizer retention in the soil [29]. In our study, Application of sandy soil amendment increased the grain and biomass yield compared to CK. Although the grain and biomass yield of T3 were higher than T4 in values, they have no significant difference between them. Hence, the long-term effects on yield are gradually increasing as the years goes by in 5 years. However, to ascertain the longer effects on sandy soil amendment further studies on continuing observation in the future is recommended.

5. Conclusions

1. The distribution of soil water storage in 0 – 100 cm soil layer was changed by adding sandy soil amendment, application of sandy soil amendment improved the water storage in 0 – 40 cm soil layer, and reduced the water storage below the 40 cm compared with CK. The T3 treatment performed best.
2. Sandy soil amendment on sandy soil can enhance enzymatic activities in the three crucial stages of maize and three soil layers (0 - 10, 10 - 20, 20 – 40 cm).and all the enzymatic activities had a similar trend. The T5 treatment possessed the highest values (soil urease and invertase were 0.621 and 5.23 mg · g^{-1} · 24 h^{-1}, soil catalase activity was 0.254 mg · g^{-1} · 30 min^{-1}).
3. We observed that sandy soil amendment can increase maize biomass and grain yield, and improved with the year increases. The order of maize yield in all treatments roughly performed: T5 > T3 > T4 > T2 > T1 > CK.

Acknowledgements

We are grateful for the financial support from the National Special Fund for Agro-scientific Research in the Public Interest of China (201303126) and Agricultural science and technology achievements transformation 'Demonstration of production and application technology and popularization of sandy soil amendment'（sq2012eca400008）.We also thank the Inner Mongolia Trirock co, Ltd for providing us the experimental base, field assistance and accommodation.

References

[1] Zhao H L, Zhou R L, Zhang T H and Zhao X Y 2006 Effects of desertificationon soil and cropgrowth properties in Horqin sandy cropland of Inner Mongolia,NorthChina *Soil Till. Res.* **87** 175-85

[2] Andry H, Yamamoto T, Irie T, Moritani S, Inoue M and Fujiyama H 2009 Water retention, hydraulic conductivity of hydrophilic polymers in sandy soil as affected by temperature and water quality *J.Hydrol.* **373** 177-83

[3] Chan K Y, Van Z L, Meszaros I, Downie A and Joseph S 2007 Agronomic values of greenwaste biochar as a soil amendment *Aust. J. Soil. Res.* **45** 629-34

[4] Belyaeva O N and Haynes R J 2012 Comparison of the effects of conventional organic amendments and biochar on the chemical, physical and microbial properties of coal fly ash as a plant growth medium *Environ. Earth. Sci.* **66** 1987-97

[5] Dorraji S.S, Golchin A.and Ahmadi S 2010 The effects of hydrophilic polymer and soil salinity on corn growth in sandy and loamy soils *Clean-Soil. Ai.r Water.* **38** 584-91

[6] Fong S S, Seng L and Mat H B 2007 Reuse of nitric acid in the oxidative pretreatment step for preparation of humic acids from low rank coal of Mukah, Sarawak *J. Brazil. Chem. Soc.* **18** 41-6

[7] Puma S, Dominijanni A, Manassero M and Zaninetta L 2015 The role of physical pretreatments on the hydraulic conductivity of natural sodium bentonites *Geotext. Geomembranes.* **43** 263-71

[8] Yesilyurt Z, Boylu F, Cinku K, Esenli F and Celik M S 2014 Simultaneous purification and modification process for organobentonite production *Appl. Clay. Sci.* **95** 176-81

[9] Duan Y L 2011 The Modification of Acrylamide Superabsorbent Resins and Adsorption Performance (Qinhuangdao: Yan Shan University) pp: 1-9

[10] Choudalakis G and Gotsis A D 2014 Morphology and gas transport properties of acrylic resin/bentonite nanocomposite coatings *Prog. Org.Coat.* **77** 845-52

[11] Murray H H 2000 Traditional and new applications for kaolin,smectite,and palygorskite:a general overview *Appl. Clay. Sci.* **17** 207-21

[12] Khalili M, Naghavi M R, Aboughadareh A P and Rad H N 2013 Effects of drought stress on yield and yield components in Maize cultivars (Zea mays L) *Int J Agron. Plant. Prod.* **4** 809-12

[13] Burns R G 1982 Enzyme activity in soil: location and a possible role in microbial ecology *Soil. Biol. Biochem.* **14** 423–7

[14] Baum C, Leinweber P and Schlichting A 2003 Effects of chemical conditions in re-wetted peats on temporal variation in microbial biomass and acid phosphatase activity within the growing season *Soil. Biol. Biochem.* **22** 167-74

[15] Eivazi F, Bayan M R and Schmidt K 2003 Select soil enzyme activities in the historic Sanborn Field as affected by long-term cropping systems *Commun. Soil. Sci. Plan.* **34** 2259-75

[16] Guan S Y 1986 *Soil enzyme and its research method* (Beijing: China Agriculture Press) pp 274-323

[17] She D L, Liu D D, Liu Y Y, Liu Y, Xu C L, Qu X and Chen F 2014 Profile characteristics of temporal stability of soil water storage in two land uses *Arab. J. Geosci.* **7** 21-34

[18] Xu S, Zhang L, McLaughlin N B, Mi J, Chen Q and Liu J 2015 Effect of synthetic and natural water absorbing soil amendment soil physical properties under potato production in a semi-arid region *Soil. Till. Res.* **148** 31-9

[19] Yu Y, Peng R, Yang C and Tang Y 2015 Eco-friendly and cost-effective superabsorbent sodium polyacrylate composites for environmental remediation *J. Mater. Sci.* **50** 5799-808

[20] Atkinson C J, Fitzgerald J D and Hipps N A 2010 Potential mechanisms for achieving agricultural benefits from biochar application to temperate soils: a review *Plant. Soil* **337** 1-18

[21] Plaut Z 1995 Sensitivity of crop plants to water stress at specific developmental stages: re-evaluation of experimental findings *Israel J. Plant Sci.* **43** 99-111

[22] Choi K, Lee H, Kim S and Hong E 1996 Dry matter and grain yield production of

soyabean cultivars as affected by excessive water stress at vegetative growth stage and flowering stage *Agric. Sci.* **38** 117-22

[23] Tijani F O, Oyedele D J and Aina P O 2008 Soil moisture storage and water-use efficiency of maize planted in succession to different fallow treatments *Int. Agrophys.* **22** 81-7

[24] Oguntunde P G, Abiodun B J, Ajayi A E and van de Giesen N 2008 Effects of charcoal production on soil physical properties in Ghana *J. Plant. Nutr. Soil. Sc.* **171** 591-6

[25] Glaser B, Lehmann J and Zech W 2002 Ameliorating physical and chemical properties of highly weathered soils in the tropics with charcoal - a review *Biol. Fert. Soils* **35** 219-30

[26] Bowles T M, Acosta-Martinez V, Calderon F and Jackson L E 2014 Soil enzyme activities, microbial communities, and carbon and nitrogen availability in organic agroecosystems across an intensively-managed agricultural landscape *Soil. Biol. Biochem.* **68** 252-62

[27] Schimel J P and Bennett J 2004 Nitrogen mineralization: challenges of a changing paradigm *Ecology.* **85** 591-602

[28] Wang X B, Zhou W. Liang G Q, Song D L and Zhang X Y 2015 Characteristics of maize biochar with different pyrolysis temperatures and its effects on organic carbon, nitrogen and enzymatic activities after addition to fluvo-aquic soil *Sci. Total. Environ.* **538** 137-44

[29] Guo M Y, Liu M Z, Zhan F L and Wu L 2005 Preparation and properties of a slow-release membrane-encapsulated urea fertilizer with superabsorbent and moisture preservation *Ind. Eng. Chem. Res.* **44** 4206-11

Evaluation of the Morpho-physiology characteristics of maize inbred lines introduced from CIMMYT to identify the best candidates for planting in acidic soil in Jasinga, Indonesia

K Lubis[1, 3], S H Sutjahjo[2] , M Syukur[2] and Trikoesoemaningtyas [2]

[1] Department of Agroecotechnology, Faculty of Agriculture, University of North Sumatra, Medan-Indonesia

[2] Department of Agronomy and Horticulture, Bogor Agricultural University, Indonesia

Email : nisalbz@gmail.com

Abstract. Technological developments and climate change have affected crop planting strategies. For example, maize production has expanded to sub-optimal lands, including acidic soil common in areas like Indonesia. Breeding programs have created inbred lines of maize introduced from CIMMYT; they were tested locally in acidic soils to determine their adaptability and tolerance mechanisms. Breeds CLA 46 and NEI 9008 were found to be excellent candidates for acidic soil due to their ASI, high number of grains per year, and suitable dry seed weight.

Keywords: Maize, morpho-physiology, acid soil Indonesia

1. Introduction

High rainfall in some parts of Indonesia causes soil to become acidic. Sub-optimum land in Indonesia includes 91.9 million ha; of this, 62.6 million hectares (68.1%) is acidic dry land in Sumatra, Kalimantan, and parts of Java [10].

The main constraints of acid soil are Aluminium (Al) toxicity and Phosphor (P) deficiency, which decrease plant productivity [1]. Aluminum can affect plants morphologicaly and physiologicaly. In crop studies, morphological and physiological characteristics are used to identify crop tolerate to abiotic stress because these characteristics are easy to observe [5]; [13]; [14]. These selectors include plant height, ASI (Anthesis to Silking Interval), panicle length, and the weight of 100 seeds have been used to test maize tolerance to sub-optimum conditions [7]; [3].

To expand maize production, it is important to develop a crop variety tolerant to acidic soil. Introductions is one step in the breeding program used to increase the diversity of materials selection (citation). Crosses of maize inbred lines introduced from CIMMYT have been developed for acidic soils. Similarly, local inbred lines of maize have been developed for optimum soil, which increased the genetic diversity of materials from where selection for breeding plants tolerant aluminum can be done. Evaluating combining ability can be performed to evaluate and select prospective candidates of tolerant inbred line [11].

Evaluating and selecting characteristics associated with aluminium tolerance can be performed to identify prospective maize lines. This research evaluated the morpho-physiological characteristics of maize inbred lines from CIMMYT in acidic soil.

2. Materials and methods

The research was conducted on acid soil, with criteria Al-dd 1.87 me. 100 g -1 , pH 4.4 in Jasinga-Bogor from January 2013 - April 2013. The experiment used the Randomized Complete Block Design format [6], with three replications consisting of 20 treatments (genotype). Treatments were the result of crossbreeding maize inbred lines of CIMMYT (CLA 84, 46, 106) and local inbred lines obtained from Cereal Crops Research Institute in Maros, Indonesia (1042-71, NEI 9008), namely P1 - P20.

Table 1. F1 population.

F1 Code	Crosses Parent	F1 Code	Crosses Parent
P1	CLA 84 X CLA 46	P11	CLA 46 X 1042-71
P2	CLA 46 X CLA 84	P12	1042-71 X CLA 46
P3	CLA 84 X CLA106	P13	CLA 46 X NEI 9008
P4	CLA106 X CLA 84	P14	NEI 9008 X CLA 46
P5	CLA 84 X 1042-71	P15	CLA 106 X 1042-71
P6	1042-71 X CLA 84	P16	1042-71 X CLA 106
P7	CLA 84 X NEI 9008	P17	CLA 106 X NEI 9008
P8	NEI9008 X CLA 84	P18	NEI 9008 X CLA 106
P9	CLA46 X CLA 106	P19	1042-71 X NEI 9008
P10	CLA 106 X CLA46	P20	NEI 9008 X 104271

3. Data analysis

The observed variables are:

1. Analysis of variance

Analysis of variance was performed using SAS version 9.1 software. Analysis used a randomized block design with three replications using a statistical model:

$$Y_{ijkl} = m + T_{ij} + b_k + (bT)_{ijk} + e_{ijkl}$$

Table 2. Analysis variance.

Source	Df	MS	E(MS)
Replication	r – 1	M_1	M1/M3
Genotype (G)	t -1	M_2	M2/M3
Error (e)	(r – 1) (t – 1)	M_3	

4. Results and discussion

Morpho-physiologicaly characteristic appearances of maize genotypes in acidic soils.

A total of 20 F1 cross-breed maize inbred lines of CIMMYT and Maros research center cereal crops were planted in acidic soils in the experimental field in Jasinga, Bogor.

Table 3. Means Square Genotype of Morpho-physiology Character.

Morpho-physiology Character	MS
Anthesis Silk Interval (ASI)	34.22**
The number of grains per ear	7542.87**
The number of row per ear	6.89**
The seed dry weight	2399.19**
The cob dry weight	3518.32**

* = significantly different P>ά = 0.05, ** = significantly different P>ά = 0.01

The results showed that F1 maize inbred lines of CIMMYT and Maros showed significant differences in ASI, the number of grains per ear, the number of rows per ear, seed dry weight, and cob dry weight (Table 3). The analysis of variance results showed that all characteristics showed genetic diversity.

ASI is the best indicator to determine plants tolerance to abiotic stress during flowering. A small ASI value shows that flowering is synchronized, which means the chances of pollination are high. A high ASI value indicates asynchronous flowering, which inhibits pollination [12]. A value of 1 to 3 days represents the best value for improved varieties maximum production [15].

Higher stress results in longer flowering periods. Collet et al. [4], conveyed that Al stress conditions resulted in larger ASI values. Efforts to accelerate the flowering and obtain the lower ASI suspected to be the way overcoming stress conditions Al. There are similarities between the Aluminum stress and drought stress. Root damage caused by Aluminum toxicity leads to reduced water supply, which results in drought at flowering. Hassan et al. [7], conveyed that maize will reduce ASI by accelerating flowering to improve drought stress tolerance.

A large ASI value in stress condition affects the number of grains per ear, and inhibits pollination and seed formation. Collet et al. [4], showed that ASI negatively correlated with corn seed production. Increasing ASI will reduce the seed corn production. Borras et al. [2], stated that the indirect effect resulting in stresses Al is assimilates decreased. The revenue sharing of assimilates will be hampered to the plants that need. Low assimilates will reduce fotosintat flow to the cob (sink).

The crosses obtained differences in male and female flowering dates ranging from 3.97- 15.63 days; cross combination CLA 46 X CLA 84 had the highest ASI, while cross combination NEI 9008 X CLA 106 had the lowest ASI (Table 4). Cross combination CLA 46 X NEI 9008 is a prospective hybrid combination due to its favorable ASI, number of rows per ear, number of grains per ear, seed dry weight, and dry cob weight; it showed high productivity and a low ASI value in acidic soil (Table 4).

NEI 9008 was the best combiner because it had a high GCA value. CLA 46 was not the best combiner for yield, but a combination of CLA 46 X NEI 9008 had a higher GCS value for yield (GCA and GCS data are not shown). Crosses between strains have high GCA and low GCA; they generally give high GCS [8]. Appearance best combination of tolerance characters and good yield can be developed as a potential candidate for a hybrid [9]. It is suspected that the distant relationship between lines CLA 46 (origin CIMMYT) and NEI 9008 (original Maros) caused the F1 offspring to have a better appearance.

Table 4. Means square of ASI, Seed Dry Weight (SDW), Number of Rows per Ear (NRE) , Number of Grains per Ear (NGE) and Cob Dry Weight (CDW).

Genotype	ASI Mean	SDW Mean	NRE Mean	NGE Mean	CDW Mean
CLA 84 X CLA 106	7.77	30.80	10.00	112.00	42.33
CLA 106 X CLA 84	5.20	19.50	11.87	141.87	23.67
CLA 84 X CLA 46	13.67	20.73	12.40	214.27	33.00
CLA 46 X CLA 84	15.63	20.27	9.00	155.27	20.63
CLA 84 X 1042-71	6.87	51.87	12.13	256.80	75.67
1042-71 X CLA 84	5.90	45.07	11.07	177.47	68.67
CLA 84 X NEI 9008	4.67	75.80	11.87	196.40	90.33
NEI 9008 X CLA 84	6.77	69.37	12.67	195.87	71.67
CLA 106 X CLA 46	11.00	14.87	8.00	63.00	16.50
CLA 46 X CLA 106	8.53	36.20	10.67	161.73	56.67
CLA 106 X 1042-71	11.90	14.10	11.87	148.93	23.33
1042-71 X CLA 106	9.10	25.63	12.67	162.67	35.33

CLA 106 X NEI 008	8.53	32.80	12.67	176.67	38.67
NEI 9008 X CLA 106	3.97	41.03	11.87	159.33	94.00
CLA 46 X 1042-71	11.90	26.30	11.30	125.20	23.33
1042-71 X CLA 46	9.10	74.33	13.33	220.40	83.33
CLA 46 X NEI 9008	5.70	108.47	12.40	228.67	123.33
NEI 9008 X CLA 46	4.57	87.13	13.60	225.07	97.33
1042-71 X NEI 9008	8.20	56.67	11.73	160.27	70.00
NEI 9008 X 1042-71	7.90	76.87	12.40	193.20	96.67

5. Conclusion

This study evaluated ASI, number of grains per ear, and seed dry weight to determine that the cross-breed CLA 46 and NEI 9008 are good candidates for planting in acidic soil.

References

[1] Barchia M F 2009 *Agroekosistem Tanah Mineral Masam* (Yogyakarta: Gadjah Mada University Press) P 228

[2] Borras L, Westgate M E, Astini J P 2007 *Physiological processes to understand genotype x environment interactions in maize silking dynamics in scale and complexity in plant systems research Gene-Plant-Crop Relations* ed Spiertz JHJ et al (Wageningen: Springer) pp 105-13

[3] Chohan M S M, Saleem M, Ahsan M and Asghar M 2012 *Genetic analysis of water stress tolerance and various Morpho-physiological traits in Zea mays L. using graphical approach Pak. J.Nutr* **11** 489-500

[4] Collet L, Leon C, Walter J H 2000 *Screening maize for adaptation to acid aluminium-toxic soils of Colombia.* Deutscher Tropentag 2000 in Hohenheim Session: Overcoming Stresses in Crop Production (DE)

[5] Dencic S R K, Kobiljski B and Duggan B 2000 *Evaluation of grain yield and its components in wheat cultivars and landraces under near optimal and drought conditions Euphytica* **113** 43-52

[6] Gaspersz V 1994 *Metode perancangan percobaan. Untuk ilmu-ilmu pertanian,ilmu-ilmu teknik dan biologi* ed Edisi kedua (Bandung : Armico)

[7] Hassan A F, Tardieu F and Turc O 2008 *Drought-induced changes in anthesis-silking interval are related to silk expansion: a spatio-temporal growth analysis in maize plants subjected to soil water deficit Plant Cell and Environment* **31** 1349-60

[8] Iriany R N, Sujiprihati S, Syukur M, Koswara J and Yunus M 2011 *Evaluasi daya gabung dan heterosis lima galur jagung manis (Zea mays var. Saccharata) hasil persilangan dialel J. Agron* **39** 103-11

[9] Iqbal J, Saleem M, Ahsan M and Ali A 2012 *General and specific combining ability analyses in maize under normal and moisture stress conditions J Anim Plant Sci.* **22** 4-8

[10] Lakitan B and Gofar N 2013 *Kebijakan inovasi teknologi untuk pengelolaan lahan suboptimal berkelanjutan.* [Presented on National Sub optimal conference in Palembang, Indonesia 20th-21th September 2013]

[11] Made J M, Matsum D and Marcia P 2011 *Pola heterosis dalam pembentukan varietas unggul bersari bebas dan hibrida* Pusat Penelitian dan Pengembangan Tanaman Pangan Balitseral Maros - Indonesia

[12] Paliwai R L 2000 Tropical maize morphology dalam tropical maize: improvement and production (Rome : Food and Agriculture Organization of the United Nations) pp13-20

[13] Pellet D M, Grunes D L and Kochian L V 1995 Organic acid exudation as an aluminium

tolerance mechanism in maize (Zea mays L.) *Planta* **196** 788-95
[14] Samuel T D, Kucukakyuz K and Zachary M R 1997 Al partitioning pattern and root growth as related to Al-sensitivity and Al-tolerance in wheat *Plant Physiol* **113** 527-34
[15] Wahyudi M H, Setiamihardja R, Baihaki A and Ruswandi D 2006 Evaluasi daya gabung dan heterosis hibrida hasil persilangan dialel lima genotip jagung pada kondisi cekaman kekeringan *Zuriat* **17** 1-9

Survey of Basal Stem Rot Disease on Oil Palms (*Elaeis guineensis* Jacq.) in Kebun Bukit Kijang,North Sumatera, Indonesia

Lisnawita[1,4], H Hanum[2] and A R Tantawi[3]

[1,2] Department of Agroecotechnology, Faculty of Agriculture, University of Sumatera Utara, Medan, 20155, Indonesia

[3] Department of Agrotechnology, Faculty of Agriculture, University of Medan Area, Medan, 20223, Indonesia

Email: itamuis@yahoo.com

Abstract. Basal stem rot disease caused by *Ganoderma* sp. is a significant disease on oil palm plantations in Indonesia, especially in North Sumatera. Currently, the pathogen does not only attack the plants that have produced (old plants) but also attacks the plants that have not produced in the first generation yet. A survey of the distribution of the basal stem rot disease in the plantation of the community has been completed in order to illustrate the distribution and the incidence of the basal stem rot disease in 5 locations of the oil palm plantation of the community in Desa Bukit Kijang, Region of Asahan, North Sumatera, Indonesia. From the research, it is revealed that the basal stem rot disease has spread to all of the observed locations with the level of disease incidence between 0.71% in Kebun Bukit Kijang 3 to 50% in the 17 years old oil palm in Kebun Bukit Kijang 4 and Bukit Kijang 5. The observable symptoms of the basal stem rot disease are chlorotic leaves, the appearance of fruiting body, collapsed plants, and the existence of holes on the basal stem. The incidence of basal stem rot disease is higher on land due to a high sand content (>50%).

1. Introduction

Palm oil (*Elaeis guineensis* Jacq.) is the most significant plantation commodity in Indonesia. At the current time, Indonesia is the biggest oil palm exporter and producer, as the width of the oil palm plantation area was 8.91 million hectares in 2011 [1], and the average production of crude palm oil is around 31 million ton per year [2]. The width of this oil palm plantation will continue to grow, and in 2020, it will be dominated by the plantation of the community and the large private plantation. One of the provinces that has the second largest oil palm plantation area in Indonesia is North Sumatera, where the total area is almost 1.2 million hectares [3]. From the width, more than 405 thousands hectares comprise the oil palm plantations of the people in the area [4].

Currently, a serious problem that emerges on the oil palm plantation, especially the community's plantation, is the basal stem rot disease that is caused by *Ganoderma* sp. Field research has shown that in the oil palms that are planted on the land that was previously a forest or held rubber plants, the symptoms of basal stem rot appear in the 10th – 12th year of the plants' lives with the level of disease incidence around 1 – 2%, and it increases to 25% when the plants reach 25 years of age [5]. On the replanted oil palms with underplanting as the technique, disease incidence can reach 33% on plants that are 15 years of age [6]. Hence, the basal stem rot disease can reduce the oil palm production to 35%.

Though the information about the incidence and the loss of the result caused by the attack of the basal stem rot disease in Indonesia has been frequently reported, information on the distribution and the attack level of the basal stem rot disease has not been published yet. The lack of information in the farmer's knowledge about this disease means that there is no preventive action taken by the farmer.

The objective of this research was to determine the distribution and the incidence of the basal stem rot disease in 5 locations of the community's oil palm plantation in Desa Bukit Kijang, Region of Asahan, North Sumatera, Indonesia.

2. Materials and methods

2.1. Determination of the sampling location

A survey was done in 5 community's oil palm plantations in Desa Bukit Kijang, Gunung Melayu, in North Sumatera, Indonesia. The determination of the sampling location is based on the consideration toward the existence of the plant's age difference. From each chosen location, the plantation width, the total number of the plants, age of the plants, and GPS coordinates were recorded.

2.2. The Observation of the basal stem rot disease incidence

The observation of the basal stem rot disease incidence done for each location is based on the symptoms and a visual diagnostic that uses the scale 0 – 4. This scale is based on the method of Abdullah et al. [7] (table 1).

Table 1. The diagnostic and the symptom of the basal stem rot on the plants that are checked based on the scale of the disease (0 – 4) [7].

Severity Class	Description
0	Healthy looking plants with green leaves without appearance of fungal mycelium on any part of plants.
1	Appearance of white fungal mass on any part of plants, with or without chlorotic leaves and unopened spear leaves at the centre.
2	Appearance of basidiomata on any part of plants with chlorotic leaves, skirt–like appearance of the leaves resulting in collapse of the lower leaves.
3	Formation of well-developed basidiocarp and bole creation.
4	Death of the plant and creation of bare land

Disease incidence is counted by using the formula below [7]:

$$DI = \frac{\text{The number of the plants with scale 1- 4}}{\text{The total number of whole plants observed}} \times 100\%$$

DI = Disease Incidence (%)

2.3. The Soil Characteristics

Soil analysis is done for each plantation location that is surveyed. A soil sample is taken from a plant that shows the symptom of the basal stem rot disease by using the random purposive sampling method. 300-500 grams of soil is taken from 5 locations from a 10 – 20 cm depth that is composited, which is then put into a plastic bag and labeled. In the next step, the soil sample is put into the cooler box and brought to the laboratory center, Faculty of Agriculture, University of Sumatera Utara, Medan, Indonesia, to be analyzed.

IOP Conf. Series: Earth and Environmental Science **41** (2016) 012007 doi:10.1088/1755-1315/41/1/012007

3. Results and discussions

The results of the survey show that the basal stem rot disease has been found in all survey locations, and the incidence of the disease was between 0.71% in Kebun Bukit Kijang 3 and 50% on the 17 years old oil palms in Kebun Bukit Kijang 4 and Bukit Kijang 5 (table 2). Generally, the incidence of the disease was higher on the young plants compared to the old plant (over 30 years). Based on Treu [8] and Susanto *et al.* [9], the basal stem rot disease does not only attack the old oil palms but also attacks the oil palms that have not produced yet.

Table 2. Disease incidence on 5 oil palm locations that are surveyed in desa Bukit Kijang Gunung Melayu, Sumatera Utara, Indonesia.

Locations	Coordinate	Plant age while it is surveyed	The land width (ha)	Year of planting	Number of plants	Disease incidence (%)
Bukit Kijang 1	N 02°42°48.0° E 099°31°50.9°	4	2.7	2011	378	1.42
Bukit Kijang 2	N 02°42°38.1° E 099°31°32.0°	4	1.0	2011	140	4.28
Bukit Kijang 3	N 02°42°37.2° E 099°31°29.3°	33	2.0	1982	280	0.71
Bukit Kijang 4	N 02°42°35.1° E 099°30°40.2°	17	2.0	1998	280	50.0
Bukit Kijang 5	N 02°42°34.6° E 099°30°38.2°	17	2.5	1998	350	50.0

The disease incidence of basal stem rot in Kebun Bukit Kijang 4 and Bukit Kijang 5 was high, which proves that the distribution of the basal stem rot disease during the experimental period is not similar to the reported incidence in previous decades. Now, this disease has spread to the first generation oil palms. The quick infection is related to how contagious the disease is and how the pathogen is distributed; that is, through the contact between the root of the plant and the *Ganoderma* inoculums, weather and basidiospore, and the secunder inokulum in the shape of an alternative house [10] ; [11]. The question remains: why are there a lot of young plants in the first generation (17 years) infected? This is caused by inoculums that have been obtained and started from the seedling. Based on the results of interviews with farmers, it is known that the soil for seedlings is directly taken by the farmers from the oil palm plantation. So, it is predicted that the soil that is used already contains inoculums.

For the plants that are infected by the basal stem rot disease with light symptoms, it is seen that the chlorotic leaves and several lower leaves are dry (figure 1a). For the more serious symptoms, the leaves are dry and broken (figure 1b), the fruiting body is created (figure 1c), the plant collapses (figure 1d), and there are existing holes on the basal stem (figure 1e). The above symptoms are the specific symptoms of the *Ganoderma* basal stem rot disease [12].

From the results of soil sample analysis for all surveyed locations, it is found that the soil pH of the plant with the basal stem rot disease incidence under 50% was 4.30 – 4.93, and the soil pH of the plant with 50% basal stem rot disease incidence is 4.64 – 5.05 (Table 3). The high disease incidence in the Desa Bukit Kijang plantation is associated with the soil texture in this location. The disease incidence is higher on the plantation with the over-50% soil content (Kebun Bukit Kijang 2, 4, and 5). The disease appearance is influenced by the high soil fraction, particularly the ones that are related to the physical character of the soil, such as soil porosity and water circulation.

Figure 1. Symptoms of *Ganoderma* basal stem rot disease in the field: a. chlorotic leaves, b. all leaves dry and the leaf midribs are broken, c. the plant collapses, d. *Ganoderma* fruiting body, e. holes on the basal stem.

Chang [13] reported that land with the sandy soil texture has a higher tendency to be attacked by the basal stem rot disease because this condition influences the natural soil matrix. The *Ganoderma* infection is quicker on the sandy soil because the soil texture has high porosity causing the roots of the plant to reach the inoculum faster. The high soil porosity on the sandy land also has an effect on the *Ganoderma* population. Susanto *et al.* [11] reported that the population of microorganisms in the soil that tends to be sandy is lower than the clay soil texture, but the *Ganoderma* infection in the sandy soil is higher than the clay.

Additionally, the soil's physical characteristics, such as the soil pH, also influence the incidence level of the basal stem rot disease. From table 3, it is clear that the incidence of the basal stem rot disease occurs on an extensive soil pH range, which is from very acidic to acidic. It shows that *Ganoderma* can adapt well to acidic soil. Abadi and Dharmaputra [14] reported that *G. boninense* grows well with the soil pH between 3.0 – 8.5, based on laboratory research.

Table 3. The soil characteristics of the plants that are infected by the basal stem rot disease in the surveyed locations.

Locations	Soil pH	C-org (%)	N-total (%)	Sand (%)	Dust (%)	Clay (%)	Texture
Bukit Kijang1	4.54	1.22	0.13	50.66	18.56	30.78	sandy clay loam
Bukit Kijang2	4.30	2.23	0.18	75.60	18.20	6.20	loamsandy
Bukit Kijang3	4.93	2.82	0.21	40.88	49.16	9.96	silty loam
Bukit Kijang4	4.64	3.12	0.25	72.41	17.69	9.90	sandy loam
Bukit Kijang5	5.05	1.86	0.18	58.88	22.59	18.53	sandy loam

4. Conclusions

This research is the first report of the basal stem rot disease which is caused by *Ganoderma* on the community's plantations in North Sumatera, Medan, Indonesia. The high disease incidence and the lack of knowledge that farmers have about the *Ganoderma* basal stem rot disease on oil palms cause the disease incidence to be high (0.71-50%). The observable symptoms of the basal stem rot disease are chlorotic leaves, the appearance of fruiting body, collapsed plants, and the existence of holes on the basal stem. The incidence of basal stem rot disease is higher on land due to a high sand content (>50%). This condition should be considered by the Indonesian Government to warrant more attention toward the community's plantations, focusing on the oil palms especially and this disease with its increasing infection.

Acknowledgment

This Research is part of The National Competitive Research on Scheme MP3EI (2014) from The Ministry of Research, Technology, and Higher Education of Republic of Indonesia with project number DIPA-023.04.1.673453/2014.

References

[1] http://www.deptan.go.id/kelapa-sawit-informasi-ringkas-komoditas-perkebunan (retrieved on: 13 December 2013)

[2] http://www.indonesia-investments.com. (retrieved on: 20 August 2015)

[3] http://www.deptan.go.id/luas-areal-kelapa-sawit-menurut-Provinsi-di-Indonesia-2008-2012 (retrieved on: 13 December 2013)

[4] www.sumut.bps.go.id/BPS-Propinsi-Sumatera-Utara (retrieved on: 13 December 2015)

[5] Singh G 1991 *Proc of Ganoderma Workshop* (Palm Oil Research Institute of Malaysia, Kuala Lumpur) pp 7 – 35

[6] Khairuddin H 1990 *Agriculture Science Thesis*: Basal stem rot of oil palm: incidence, etiology and control (Universiti Pertanian Malaysia, Selangor, Malaysia)

[7] Abdullah F, Ilias G N M, Nelson M, Nur Ain Izzati M Z and Umi Kalsom Y 2003 Disease assessment and the efficacy of Trichoderma as a biocontrol agent of basal stem rot of oil palms *Research Bulletin Science Putra*. **11**31–3

[8] Treu R 1998 Macro fungi in oil palm plantations of South East Asia *J.Gen Mycol* **12** **(1)** 10-4

[9] Susanto A, Ginting P A, Surianto and Prasetyo A E 2008 Pola penyebaran *Ganoderma boninense* pada perkebunan kelapa sawit (*Elaeis guineensis*) di lahan gambut : studi kasus di PT. Anak Tasik Labuhan Batu Sumatera Utara *Jurnal Penelitian Kelapa Sawit*. **16** 135-46

[10] Turner P D 1981 *Oil Palm Diseases and Disorders* (Kuala Lumpur: Oxford University Press) p 280

[11] Susanto A, Prasetyo A E and Wening S 2013 Laju infeksi Ganoderma pada empat kelas tekstur tanah *Jurnal Fitopatologi Indonesia* **9** (2) 39-46

[12] Abdul R J H, Ahmad K, Ramdhan A S, Idris S, Abdul R, Aminul R and Fauzi I 2003 Mechanical trunk injection for control of Ganoderma *MPOB Information Series* MPOB TT No 215

[13] Chang T T 2003 Effect of soil moisture content on the survival of *Ganoderma* species and other wood inhabiting fungi *Plant Dis* **87** (10) 1201-4

[14] Abadi A L and Dharmaputra O S 1998 Pengaruh pH medium dan suhu terhadap pertumbuhan miselium *Ganoderma boninense Laporan Tahunan Kerjasama Penelitian* Pusat Penelitian Marihat-Biotrop tahun 1988 (Bogor, Biotrop)

IOP Conf. Series: Earth and Environmental Science **41** (2016) 012008 doi:10.1088/1755-1315/41/1/012008

Nitrogen, Phosphor, and Potassium Level in Soil and Oil Palm Tree at various Composition of plant species mixtures grown

C Hanum[1, 4], A Rauf[1], Nasution I[2], D A Fazrin[3] and A R Habibi[3]

[1]Agrotechnology Department, Faculty of Agriculture, Universitas Sumatera Utara. Indonesia

[2] Public Administration, Social and Political Sciencis. Medan Area University.Jalan Kolam No.1. Medan Estate. Sumatera Utara

[3]Student of Agrotechnology Bachelor Program

E-mail: chairani_as@yahoo.com

Abstract. In productive oil palm plantation areas, poor vegetation is generally caused by low light intensity. This condition causes excessive erosion and decreases soil fertility. One of the efforts for soil and water conservation at oil palm plantations is through increased vegetation diversity. The changes of soil and plant nitrogen, phosporus, and potassium content, observed by planting two types of herbs under oil palm tree, with different compositions. Vegetation composition was set as: *Arachis glabrata* 100%; *Stenotaprum secundatum* 100%; *Arachis glabrata* 50% + *Stenotaprum secundatum* 50%; *Arachis glabrata* 75% + *Stenotaprum secundatum* 25%; *Arachis glabrata* 25% + *Stenotaprum secundatum* 75%. The shoot and root fresh/dry weight, nutrient content (nitrogen, phosphor, and potassium) of each cutting were measured at the end of the experiment. Ten of treatment plant were harvested and divided shoots and roots after washing out of soil. Biomass samples were dried at 70 °C for 48 h and weighed. The total N and its proportional concentration (N%) were analyzed with the micro-Kjeldahl method. Potasium analyzing with flamephotometry, and phosphor and from samples was determined by analyzing with spectrophotometry method. The results showed the highest shoot growth of *A.glabarata* if planting was mixed with *S. secundatum*, but the result was different with *S.secundatum* being superior if planted with monoculture system. Combination of interrow cultivation is more recommended for soil conservation and nutrient maintenance in palm oil trees were *A. Glabarata* 75% + *S.secundatum* 25%.

1. Introduction

The success of oil palm cultivation is generally determined by soil conditions and water availability which includes physical, biological, and chemical properties of soil. Productive oil palm (POP) plantation areas are different than young oil palm (YOP), one difference being the light intensity level on the ground. The young oil palm gets enough light intensity, but the POP is not. Light is one of the most important factors mediating plant growth [27] and is a vital regulator of numerous processes [5] Light is the energy source for photosynthesis, producing the ATP and NADPH to assemble carbon atoms into organic molecules. The limited incidence of light in POP interrow causes lowered soil fertility and productivity. This decreased number of plants under the crop will cause fertilizer effectiveness to be reduced. Increasing plant population in POP interrows would prevent the loss of fertilizer applied. In addition, the rate of erosion (*run off*) will be reduced because the smaller root plants are capable of holding and binding the soil aggregates.Therefore, soil conservation efforts are needed to prevent erosion, improve damaged soil, and maintain soil productivity so the soil can be used sustainably. Water supply is a major limiting factor for the production of palm oil. Drought affects all aspects of palm oil, including the quality of the crops such as the oil [6] and sugar contents [17]. The decline in sugar content will result in the plant lacking the energy for growth and development. Inhibited seed filling rate is also one of the reasons for low crop production. Better soil conditions will affect the processes of water and nutrient uptake, root respiration as well as ease of maintenance and harvest.

Not all plants growing on the oil palm plantation area are weeds that should be exterminated. Clean weeding might invite pests, especially caterpillars, and reduce the soil moisture.

Some types of ferns such as Neprolepsis, Casiacubanensis, and Turnera subulata, however, are worth keeping. Neprolepis ferns are useful for retaining soil moisture, and Turnera and Casia could serve as hosts for Cantheconidea (predator of caterpillar). Therefore, other plants that have the ability to survive under low light intensity are also necessary. These plants can be expected to have a dual function: top revent run-off rate and increase soil organic matter. Selection of plants is based on their ability to adapt at low light intensity, and its capability to produce relatively high and stable foliage year-round. Types of legume or shade-tolerant grasses are chosen alternative compared to other ground plants that are considered best for cover crops. One of these legumes is A.glabarata and S. secundatum for the grasses. Both of these are shade resistant and grow easily ([20]. Increasing the plant diversity will affect on soil fertility. Many researchers clarified numerous effects on nutrient concentration and status [12, 13, 31]. Palm oil trees need nitrogen, phosphorus and potassium in large quantities. Nitrogen deficiency is usually associated with topsoil erosion. It will result in general and stiffening of the pinnae which lose their glossy lustre, reduce the number of effective fruit bunches, as well as the bunch size. Phosphor deficient leaves are shorter than well-nourished ones. Bunch size and trunk diameters are reduced. Potassium deficiency symptoms were intensifies these spots turn orange, or reddish-orange, and dessication sets in, starting from the tips and outer margins of the pinnae. In soils with low water holding capacity potassium deficiency can lead to a rapid premature dessication of fronds (http://ipipotash.org/udocs/Nutrient Management of the oil palm.pdf). The availability of soil nutrients will affect to oil palm productivity.

2. Material and methods

The research was conducted in field experiments on 8-years old productive plants. Observation on light intensity under palm oil tree using a lux meter showed that light intensity level was about 40%. Before planting *Stenotaprum secundatum* and *Arachis glabarata,* soil fertility and plants that live in these area were analyzed. Vegetation analysis was done by creating seven plots, each measuring 1x1m. Vegetation analysis studies the composition (component type) of plant communities. Along the POP interrows were planted both *A.glabrata* grass and legume with the following composition: *A. glabrata* 100% ; *S. secundatum* 100%; *A. glabrata* 50% + *S. secundatum* 50%; *A. glabrata* 75% + *S. secundatum* 25%; and *A. glabrata* 25% + *S. secundatum* 75%. The shoot and root fresh/dry weight, nutrient content (nitrogen, phosphor, and potassium) of each cutting were measured at the end of the experiment. Ten of treatment plant were harvested and divided shoots and roots after washing out of soil. Biomass samples were dried at 70 °C for 48 h and weighed. The total N and its proportional concentration (N%) were analyzed with the micro-Kjeldahl method. Potasium analyzing with flamephotometry, and phosphor and from samples was determined by analyzing with spectrophotometry method.

3. Results and discussion

3.1. Analysis vegetation

IOP Conf. Series: Earth and Environmental Science **41** (2016) 012008 doi:10.1088/1755-1315/41/1/012008

Analysis vegetation is research approach for study of spatial patterns of vegetation. It seeks to understand the structure and variation of the vegetation of lanscape. The term gradient in space of variables on environmental factors, species population and characteristic of communities (Cajo and Braak, 1987). Vegetation analysis on POP crop before planting *Stenotaprum secundatum* and *A.glabarata* showed that Kilingia species has the highest growth, followed by *Asistasia* (Table 1).

Table 1. Vegetation analysis on POP area.

VegetationType	ObservationPlot						
	I	II	III	IV	V	VI	VII
Cyperus Kilingia	26	38	37	19	4	28	0
Ageratum conyzoides	12	6	13	16	0	18	20
Asistasia	28	23	6	10	4	12	7
(a) Phyllanthus Niruri L	26	15	31	23	8	29	3
(b) Cyclosorus aridus	0	1	0	2	0	2	0
Cyperaceae	0	42	0	0	0	0	6
Axonopus compresus	0	0	0	24	17	0	0
Broadleaf weeds	0	0	0	0	43	0	0
Total	92	125	87	94	76	89	36

Kilingia (*Cyperus kylingia*) is a type of cyperaceae hat developed with its root and rhizome. It grows and multiplies rapidly, and can survive in low light intensity, making it the most dominant weed. Basic or under growth vegetation are also important components in an ecosystem, being able to support the life of other plant species. Basic vegetation is basic cover crop consisting of herbaceous ground cover, bushes or shrubs, lianas and ferns. The results showed that low diversity of plants on the POP area. Decline in biodiversity with land use change identified as one of the most important drivers. The lost of diversity has a significat on ecosystem fungtioning [33].

Shoot fresh and dry weight of Interrow Plants

A.glabarata that were planted individually or simultaneously showed that each treatment affected herb growth significantly (Table 2). The highest shoot fresh weight on treatment (*S.secundatum* 100%) was 53.56g. Higher fresh weight in *S. secundatum* is possibly caused by the vegetation's better ability to compete in absorbing resources.

Table 2. Shoot fresh and dry weight of interrow plants at various planting composition.

Population Percentage	Shoot fresh weight	Shoot dry weight
	.... g............................	
A. glabrata 100%	19,46e	11.45g
S. secundatum 100%	53.56a	43.61a
A. glabrata 50%	18.46e	10.48g
S. secundatum 50%	44.66c	30.82d
A. glabrata 75%	28.49d	21.85e
S. secundatum 25%	47.74b	38.56b
A. glabrata 25%	19.01e	13.61f
S. secundatum 75%	43.54c	33.41c

Note: The number followed by repeated letter in the same column is insignificant according to Duncan average test (DMRT) at 5% level.

This herb is also shade-tolerant and able to grow under limited light intensity, also found an increase in the growth of *S. secundatum* planted under shade [28]. This grass also has ability to adapt to saline soil [16]. The results of this study indicated that *S. secundatum* have adaptability under limited light intensity. This ability is thought to be caused by its high growth rate, making it excel in competition. *S.secundatum* is a plant belonging to the group of C4 metabolism [24]. This group has a higher CO_2 utilization, water use, and higher radiation use efficiency [26]. Therefore C4 producing more assimilates,, and higher nitrogen use efficiencies because it will require less Rubisco and hence less nitrogen.

Planting one plot with one type of plant (*S. secundatum* 100%) also made it grow better (Table 2). This is because it didn't have to share its nutrients with other plants. The number of plant population per hectare is also an important factor to obtain maximum yield. Higher plant density will create higher competition between plants for nutrient absorption. *Arachis glabarata* has a higher wet and dry weight average in mixed-population plots (*A. glabrata*75%+*S. secundatum* 25%) compared to single-type plots.

In a monoculture, the obtained dry weight was 11.45 but in mixed-population plots it increased to 21.85 (Table 2). This result indicated an increase in *A. glabarata* by mixing the diversity of the vegetation in a plot. Planting more than one type of plant in an area had proven to increase nutrient enrichment condition [7, 18, 23, 30],water and reduce the effects of plow pan. Thus output per unit area increases with mani fold returns to the growers. Increased diversity in one area will affect the micro climate [29], reduce soil hardening effect [32], improvement of soil structure and organic matter content. The effects of such practices on the range of interactions within the plant-soil system are manifest via plant interspecific competition, pest and disease attenuation, soil community composition and structure, nutrient cycling, and soil structural dynamics. Soil biotic communities are affected by plant diversity, which can increase abundance, diversity and activity of functional groups. Attendant rhizosphere located processes can facilitate nutrient uptake between component crops [11]. Differences to monocultures derive from: (i) interplant competitive interactions, which can lead to complementary resource use in time, space and form [10,15]; (ii) mobilisation and transfer of nutrients to component crops within the system [2]; (iii), increase in organic matter inputs [1, 22]. Rhizodeposition transfers energy and recycles organic matter and nutrients between crops, microorganisms and the soil and thus is important for crop productivity [34].

Root fresh and dry weight of Interrow Plants

Various compositions of interrow planting showed significant effect on its root fresh and dry weight. The results in Table 3 showed the highest fresh weight in treatment (*A.glabrata* 75% and *S.secundatum* 25%). The results of this study indicated a complementary interaction between *A.glabarata* with *S.secundatum* if planted in a mixed system (mixcroping). The most commonly accepted reason explaining why it is possible to obtain better yields with crop mixtures is that the component crops differ in their growth requirements.

Table 3. Root fresh and dry weight of interrow plants in various compositions.

Population Percentage	Root Fresh Weight	Root Dry Weight
	.... g...................................	
A. glabrata 100%	13.11c	9.82c
S. secundatum 100%	4.39e	1.09g
A. glabrata 50%	29.42b	17.32b
S. secundatum 50%	4.22f	1.68g
A. glabrata 75%	37.19a	21.49a
S. secundatum 25%	12.56c	4.03e
A. glabrata 25%	11.61d	7.11d
S. secundatum 75%	3.09f	2.88f

Note: The number followed by repeated letter in the same column is insignificant according to Duncan average test (DMRT) at 5% level.

A.glabarata is a legume plant with deep root and many branches, shade tolerance [3], adapt to soil bad drainage, and has the symbiotic capability with the microbial. The adequacy of N in plants will influence the extraction of P which, in turn, will contribute to increased growth of *A.glabarata* roots. *A.glabarata* has high adaptability and shade tolerance which could produce high dry matter [3]. These support by the ability to adapt soil with either good or bad drainage.

3.2. Nitrogen (N), Phosphorus (P) , and Potassium (K) contents in intercrops

The highest N content is showed in monoculture treatment of *A. glabarata*, while P and K levels can be seen in mixed planting (*A. glabarata.* 75% + *S. Secundatum* 25%) (Table 4). High potassium level in intercrops would affect the absorption and translocation of phosphorus nutrient. Potassium has an important role in both plant cells and membrane transport system (Table 4).

Table 4. Nitrogen (N), Phosphorus (P), and Potassium (K) contents in intercrops.

Treatment	Plant nutrient content		
	N (%)	P (ppm)	K (me/100g)
Arachis g. 100%)	2.46	0.250	1.832
Stenotaphrum s. 100%)	2.19	0.294	2.043
Arachis g. 50% + *S. secundatum*50%)	2.20	0.322	1.649
Arachis g. 75% + *S. secundatum*25%)	2.15	0.332	2.660
Arachis g. 25% + *S. secundatum*75%)	2.40	0.319	1.574

Potassium is known to possess a variety of functions such as managing to cell expansion process [9]; nutrient movement and translocation [9]; and stomata behavior [8]. Phosphor and potassium of plant were highest in plantations with poly culture system, nitrogen was highest in *A. glabarata* monoculture (Table 4). The effectiveness of N microbial activity to fixation in extracting nitrogen is the cause of its level to be highest in *A. glabarata.*

3.3. Soil *nitrogen, phosphorus and potassium analysis*

Soil analysis performed before and after treatment of the composition of plant diversity is shown in table 5.

Table 5. Soil Nitrogen (N), phosphorus (P), and potassium (K) level with mixed vegetation treatment.

Treatment	Nutrient content					
	N	Increase %	P	Increase %	K	Increase %
Without vegetation	0.18	0	4.00	0	0.38	0
A. glabarata. 100%	0.27	50	8.47	111.75	0.257	- 47.85
*S. secundatum*100%	0.18	0	8.51	112.75	0.643	69.21
A. glabarata. 50% + *S.secundatum*50%	0.80	344.4	9.26	131.5	0.516	35.78
*A. glabarata*75% + *S.secundatum*25%	0,62	244.4	8.60	115	0.622	63.68
A. Glabarata 25% + *S. secundatum*75%	0.54	200	7.63	90.75	0.548	44.21

Generally, increasing the diversity of the vegetation would make persentage change in soil nutrients with different percentage (Table 5). Highest nitrogen level reached 344.4% and phosphorus was 131.5% in mixed planting (*A.glabarata* 50% + *S.secundatum*50%). Potassium level percentage (69.21%), however, increased in a single-type plot. (*S.secundatum* 100%). Composition and diversity may affect soil fertility through differential species effect on nutrient inputs, may also affect fertility through differential species effect on nutrien retention. Soil nutrient heterogeneity interacted with nutrient availability and plant functional diversity to determine productivity and nutrient cycling responses. Above and below ground productivity increased under heterogeneous nutrient supply.

4. Conclusion

1. *Growth of A.glabarata* was highest if planted in a mixed/polyculture system, while *S.secundatum* was highest in a monoculture;
2. The best root and shoot growth were obtained in mixed vegetation with treatment in *A. glabarata*75% + *S. secundatum* 25%;
3. The highest N level in treatment by *A.glabarata*, while phosphor and potassium in treatment by mixed vegetation *A. glabarata* 75% + *S.secundatum.* 25%;
4. Combination of interrow cultivation is more recommended for soil conservation and nutrient maintenance in palm oil trees were *A. Glabarata* 75% + *S.secundatum* 25%.

Acknowledgement
Appreciation towards Directorate General of Higher Education, Ministry of Education and Culture.In accordance with the Implementation Agreement Assignment Penelitian Unggulan Perguruan Tinggi (PUPT) No. 1076 / UN.5.1.R / KEU / 2014.

References

[1] Bambrick A D, Whalen J K, Bradley R L, Cogliastro A, Gordon A M, Olivier A and Thevathasan N V 2010 Spatial heterogenity of soil organic carbon in tree-based intercropping systems in Quebec and Ontario, Canada *AgroforSyst* 79 343–53
[2] Betencourt E, Duputel M, Colomb B, Desclaux D and Hinsinger P 2012 Intercropping promotes the ability of durum wheat and chickpea to increase rhizosphere phosphorus availability in a low P soil *Soil BiolBiochem* 46 181–90
[3] Bowman A M and Wilson G P M 1996 Persistence and yield of forage peanuts (Arachisspp.) on the New South Wales north coast *Tropical Grassland* **30** 402–6
[4] Cajo J F and Ter B 1987 The analysis of vegetation environment relationship by canocical correspondence analysis vegetation *Vegetation* **69** 69-77
[5] Carrigan O Andrew. Elizabeth Hinde, Nicole LU, Xin-Qin Xu, Honglang Duan, Guamin Huang, Michelle Mak, Bill Bellotti, and Zhong-Huan Chen. 2014 Effects of light irradiance on stomatal regulation and growth of tomato *Environ. Exp. Bot.* **98** 65–73
[6] Cornaire B, Daniel C, Lamade E and Fodil Z 1994 Comportamiento de la palma de aceite bajo estrés hídrico *Palmas* **15** 61-70
[7] Dala R C 1974 Effects of Intercropping Maize With Pigeon Peas on Grain Yield and Nutrient Uptake *Exp. Agr.* **10** 219
[8] Dietrich P, Sanders D and Hedrich R 2001 The role of ion channels in light-dependent stomatal opening *J. Exp. Bot.* **52** 159–96
[9] Dolan L and Davies J 2004 Cell expansion in roots *CurrOpin. Plant Biol.* **7** 33–9
[10] El Dessougi H, ZuDreele A, Claassen N 2003 Growth and phosphorus uptake of maize cultivated alone, in mixed culture with other crops or after incorporation of their residues *J. Plant Nutr. Soil Sci.* **166** 254–61
[11] Ehrmann J and R Karl 2013 Plant: soil interactions in temperate multi-cropping production systems *Plant Soil* **3 76**1-29
[12] Jose S, Merritt S and Ramsey C L 2003 Growth, nutrition, photosynthesis and transpiration

responses of longleaf pine seedlings to light, water and nitrogen *Forest Ecol. Manag.* **180** 335-44

[13] Keutgen A J and Pawelzik E 2009 Impacts of NaCl stress on plant growth and mineral nutrient assimilation in two cultivars of strawberry *Environ. Exp. Bot.* 65 170-6

[14] Koenraad V M, Helsen K and Hermy M 2014 Impact of land-use intensity on the conservation of functional and phylogenetic diversity in temperate semi-natural plant communities *Biodiversity and Conservation* 23 2259-27

[15] Liebman M 1995 *Polyculture cropping systems* ed Altieri M A Agroecology, the science of sustainable agriculture, 2nd edn. (Colorado: Westview) pp 205–18

[16] Marcum K B and Murdoch C L 1990 Growth responses, ion relations, osmotic adaptations of eleven C4 turfgrasses to salinity *Agron. J.* 82 892-6

[17] Menddez Yurany Dayanna Rivera, Leonardo Moreno Chacón, Cristihian Jarry Bayona, Hernán Mauricio Romero. 2012. Physiological response of oil palm interspecific hybrids (Elaeis oleifera H.B.K. Cortes versus Elaeis guineensis Jacq.) to water deficit. Braz. J. Plant Physiol. vol.24 no.4.

[18] Mishra B P 2010 A study on the micro-environment, litter accumulation on forest floor and available nutrients in the soils of broad-leaved, mixed pine and pine forests at two distinct altitude in Meghalaya, northeast India *Curr. Sci.* **90**(12) 1829-33

[19] Moran N 2007 Osmoregulation of leaf motor cells *FEBS Lett.* 581 2337–47

[20] Mullen B F and Shelton H M 1996 Stenotaprum secundatum: a valuable forage species for shaded environments *Tropical Grassland* 30 289-97

[21] Ng S K *Nutrition and Nutrient Management of the Oil Palm New Thrust for the Future Perspective* http://ipipotash.org/udocs/Nutrient Management of the Oil Palm.pdf

[22] Oelbermann M and Echarte L 2011 Evaluating soil carbon and nitrogen dynamics in recently established maize-soyabean inter-cropping systems *Eur J. Soil Sci.* 62 35–41

[23] Perala D A and Alban D H 1982 Biomass, nutrient distribution and litterfall in Populus, Pinus and Picea stands on different soils in Minnesola *Plant Soil* 64 177-92

[24] Rien A 1998 Interspecific competition in natural plant communities: mechanisms, trade-offs and plant-soil feedbacks *Journal of Experimental Botany* **50**(330) 29-37

[25] Rivera M Y D, Chacón L M, Bayona C J and Romero H M 2012 Physiological response of oil palm interspecific hybrids (Elaeis oleifera H.B.K. Cortes versus Elaeis guineensis Jacq.) to water deficit *Braz. J. Plant Physiol.* 2 4 4

[26] Rizal G, Karki S, Thakur V, Chatterjee J, Coe RA, Wanchana S and Quick WP 2012 Towards a C4 rice Asian *J. Cell Biol.* **7** 13–31

[27] Rodríguez-García E and Bravo F 2013 Plasticity in Pinus pinaster populations of diverse origins: Comparative seedling responses to light and Nitrogen availability *Forest Ecol. Manag.* 307 196–205

[28] Samarakoon S P, Wilson J R and Shelton H M 1990 Growth, morphology and nutritive quality of shaded Stenotaphrumsecundatum, Axonopuscompressus and Pennisetumclandestinum *The Journal of Agricultural Science* 114 161-9

[29] Sharma S D, Tripathy S and Gurung P 1998 *Case study Jeypore tract, Orissa In: Gender Dimensions in Biodiversity management* ed M S Swaminathan (New Delhi: Konark Publisher) pp 123-38

[30] Singh J N and Mudgal V 2000 Assessment of mineral content of tree leaf litter on Nokrek Biosphere Reserve and its impact on soil properties *Trop. Ecol.* 41 225-32

[31] Slamaa I, Ghnaya T, Savouré A and Abdelly C 2008 Combined effects of long-term salinity and soil drying on growth, water relations, nutrient status and proline accumulation of Sesuvium portulacastrum *C. R. Biol.* 331 442-5

[32] Sullivan P 2003 *Intercropping principles and production practices* Retrieved October 5 2010 from http://attra.ncat.org/attra-pub/intercrop.html

[33] Van Meerbeek Koenraad, Kenny Helsen, Martin Hermy 2014 Impact of land-use intensity on

the conservation of functional and phylogenetic diversity in temperate semi-natural plant communities *Biodiversity and Conservation* **23** 2259-227
[34] Zhang E, Li L, Huang G, Huang P and Chai Q 2002 Regulation of fertilizer application on yield and root growth of spring wheat-faba bean intercropping system *Chin J Appl. Ecol.* **13** 939–4

Assessment of ecological security based on soil and water conservation: a case study from Gansu Province, China

Z J Li[1, 2], Q Tian[1] and L L Song[1]

[1] College of Forestry, Gansu Agricultural University, Lanzhou 730070, China

Email:lzjie314@163.com

Abstract. In this study, the analytic hierarchy process (ahp), ecological security index(S) and PSR model were used to evaluate Gansu's ecological security based on the large number of survey data. The results indicated that Gansu's ecological security index increased from 0.31 in 1986 to 0.66 in 2013, which reflected ecological security was in sensitive state ($0.7 > S \geqslant 0.5$). The main reason was that national policy on protecting the ecological environment has played a crucial role, especially the national project of returning farmland to forest and grass carried out in recent years. Moreover, the environmental issues such as the higher PM2.5, sand storms and climate extremes, had significantly improved people's environmental awareness in the study area. The regional difference of ecological security index was significant in Gansu Provinces, and the part of Yangtze river basin was higher than the part of Yellow River basin, whiles the in-land river basin was the lowest value. In a world, Gansu's ecological security had improved in recent years, but there was strong need for paying more attention to policy for ecological environment protection and increasing the propaganda to ensure the Gansu's ecological security in the future. This study will provide a scientific basis for the sustainable development of regional social economy and ecological environment.

1. Introduction

In 1989, the concept of "ecological security" and the global ecological security status monitoring were firstly put forward by the international institute for applied systems analysis (IIASA) [1-6]. National security strategy report of the United States also proposed the ecological environment security will bring into the category of national security in 1991[7-9]. The concept of ecological security has caught extensive debate from the international society and since then the world has begun to study ecological security [10-13]. Ecological security is not only confined to the ecosystem security, but the national security constituted by political security, military security and economic security [14-15]. However, political security, military security as well as economic security are the important guarantee for ecological security [16-18]. Worldwide ecological environment situation is not optimistic and many countries and regions have seen the deterioration of the environment [4]. And this eventually led to the destruction of the normal order of production and life and caused the serious economic losses. Thus to ensure the ecological security has become a pressing social needs, and the extensive study of ecological security has also become common in the world [19].

To some extent, the ecological safety is human survival safety. Jiang [20] et al (2011) thought that ecological safety was a state of healthy and sustainable development for human survival environment, whiles Yao [21] believed that the ecological security was a stable and sustainable ecosystem system, but which played a supporting role in human society and economy system. Excessive interference or disaster will damage the structure and function of ecosystem, so the ecological security situation will be threatened. Norton [22] founded that ecological security could maintain the survival and development of its normal state and it is the human survival environment. Considering from the perspective of national ecological safety, ecological security problem is a wider range of disorders in ecological structure, and it has affected the normal operation[23]. In addition, Rapport[24] argued ecological security and sustainable development of human society might have common sense, and the goal was to provide continuity and good ecological service function for sustainable development.

Gansu Province is located in northwestern China with complex terrain and climate, where located in loess plateau, mainly including the part of Yangzi river basin, Yellow river basin and Chinese in-land river basin. The loess geomorphy leads to cultivated soil resources become less and less. Finally, the problem of ecological security is becoming increasingly serious. However, the research on regional ecological security is rare. In recent decades, there has been many achievements about ecological security research at home and abroad focused on big cities of ecological security and ecosystem security, and the evaluation index system involves social, economic, natural, etc., and these researches have various forms and characteristics[25-29] . Selecting Gansu Province as the research area, this research employed soil and water conservation measures as index for evaluating the ecological security situation. This study firstly calculated the ecological security index in three sub-basins, respectively, and then analyzed the Gansu's ecological safety from the angle of time and space. Furthermore, the research put forwards the soil and water conservation measures in order to improve the regional ecological environment of Gansu Province. Finally, this research discussed the relationship between ecological safety and sustainable development, and argued that the sustainable development provides standardized guidelines for achieving human security. The quantitative evaluation for Gansu's ecological security will provide the scientific evidence for improving ecological environment in the future. At the same time, the result can provide a direction for government and lay a solid foundation for further research of ecological security.

2. Study area

Gansu Province(92°-108°E, 32°-42°N) located in the northwestern China, which lied in the interaction belt among loess plateau, the Mongolia plateau and the Tibet Plateau. Stretching 1,665 km from west to east and extending 530 km from south to north, with an area of 454,400km^2, the territory has high terrain and topography is complex. Mountain, plateau, valley, desert and Gobi desert staggered, hilly and mountains area accounts for more than two-thirds of the total area. Altitude ranges from 1,000 m to 5000 m. Terrain is tilt from the northwest to southeast, and the natural landscape exhibited an obviously vertical change with the significantly horizontal zonally. The regional differences of climate is significant characterized by the arid climate, little rain and abundant sunshine. The variational rate of precipitation and temperature is very large. The multiyear average temperature of the whole Province ranges from 0 to 14 °C, whiles annual precipitation is 280.6 mm. Vegetation from south to north showed the obviously latitude zonal distribution influenced by natural factors such as latitude, climate and landform. Area of forest vegetation is narrow and mainly distributed in certain high level for the Qilian mountains, Longnan mountains and Gannan grassland, whiles it is grassland and desert steppe below the forest belt and is alpine meadow and sub ice sparse vegetation and alpine snow belt above the forest belt. The study region can be divided into the parts of Yellow river basin, the Yangtze River basin, and in-land river basin. The area of Yellow river basin part is 145,523 km^2, and water erosion area is 112,534 km^2. The area for Yangtze River basin is 38,364 km^2, and water erosion area is 23,278 km^2. The in-land river basin area is 269,807 km^2, and the water erosion area is 35,986 km2 and the wind erosion area is 223,642 km^2. Soil and water loss has seriously affected and restricted the regionally economic and social sustainable development and the smooth realization of the comprehensive construction well-off society goal.

Gansu is the most serious soil and water loss in China, soil erosion, wide area, multi type, heavy damage and hard treatment. Over the years, Gansu has organized a small watershed comprehensive management and Terrace construction engineering based on the situation in the Province. It has played a positive role in prevention and control of soil and water loss, improving the ecological environment, natural disaster reduction, and improving the land bearing capacity. But soil erosion governance progress slowly, without restraint use has not been completely reversed, redevelopment, light protection, heavy eyes, light long-term traditional concept still exists. The fundamental reasons lie in the lack of management system, lower cost of resource usage as well as breaking the law. ambiguous property rights of the soil and water maintain facilities and uncertainty soil and water conservation ecological function region division. The system of paid use of land and water resources has not yet been established, soil and water conservation ecological compensation mechanism has not been formed. On the basis of analyzing the concept innovation, mechanism innovation and measure innovation of ecological civilization construction, the concrete measures to strengthen soil and water conservation and promote ecological civilization are put forward.

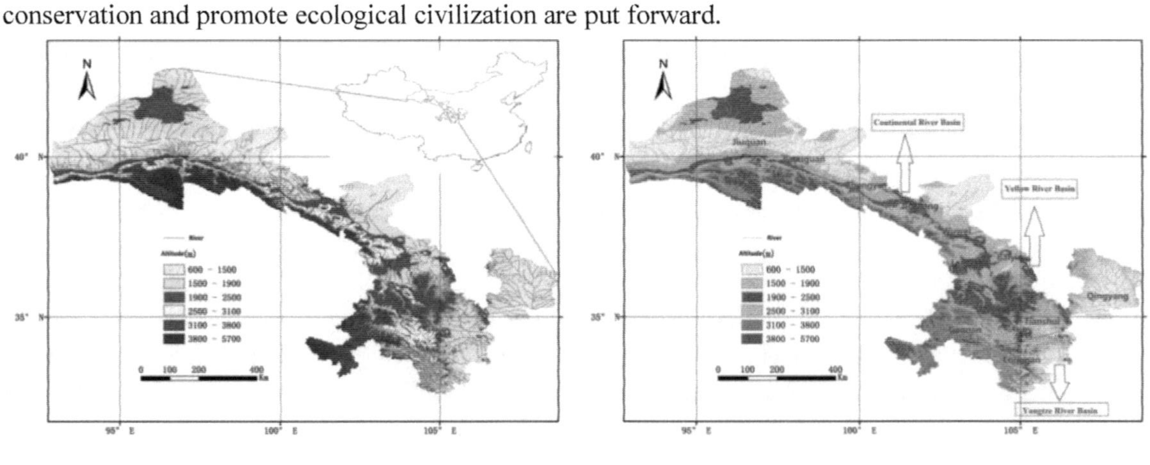

Figure 1. Sketch map of the Gansu Province.

IOP Conf. Series: Earth and Environmental Science **41** (2016) 012009 doi:10.1088/1755-1315/41/1/012009

3. Materials and methods

3.1 Data Collection

Soil and water conservation data came from the database "National water resources census and National soil erosion remote sensing investigation" and "Bulletin of Soil and Water Conservation data in Gansu Province in China". Other data were from "Statistical Yearbook of Gansu" and "Rural Economy Yearbook of China"(Table 1). All the index values were calculated based on the data collected from these databases.

Table 1. The data sources of this study.

Index layer	Data source	Time
The area of cultivated land per capita(hm^2) P1	Statistical Yearbook of Gansu and Rural Economy Yearbook of Gansu	1987-2014
The amount of fertilizer per area (kg/hm^2) P2	Statistical Yearbook of Gansu and Rural Economy Yearbook of Gansu	1987-2014
The amount of pesticide per area(kg/hm^2) P3	Statistical Yearbook of Gansu and Rural Economy Yearbook of Gansu	1987-2014
The ratio of natural disasters inundated area P4	Statistical Yearbook of Gansu and Rural Economy Yearbook of Gansu	1987-2014
The control area ratio of water loss and soil erosion P5	Statistical Yearbook of Gansu and Rural Economy Yearbook of Gansu	1987-2014
The forest area of soil and water conservation(hm^2) P6	Bulletin of Soil and Water Conservation Data in Gansu	1987-2014
The area of artificial grass (hm^2) P7	Bulletin of Soil and Water Conservation Data in Gansu	1987-2014

Index layer	Data source	Time
The area of closing hillsides to facilitate afforestation (hm^2) P8	Bulletin of Soil and Water Conservation Data in Gansu	1987-2014
The area of closing hillsides to facilitate grass(hm^2) P9	Bulletin of Soil and Water Conservation Data in Gansu	1987-2014
The investment amount of soil and water conservation (million) P10	Statistical Yearbook of Gansu and Rural Economy Yearbook of Gansu	1987-2014
The grain output per capita(t) P11	Statistical Yearbook of Gansu and Rural Economy Yearbook of Gansu	1987-2014
The agricultural output value per unit area (million/km^2) P12	Statistical Yearbook of Gansu and Rural Economy Yearbook of Gansu	1987-2014

3.2 Construct evaluation index system

Based on the data of soil and water conservation in Gansu Province, the choice of ecological security evaluation index should not only consider the environment condition of agricultural ecosystem, but also reflect the potential impact on the important factors of soil and water conservation, and what's more, it must reflect the characteristic of regional agriculture ecological security evaluation [30]. Therefore, establishing the ecological security evaluation index system must follow the principle of scientific, objective, comprehensive system, relative independence, pertinence, operability and comparability [31].

Thus, this paper consulted the relevant literature about ecosystem safety evaluation, and combined with implementation status of soil and water conservation from 1986 to 2013 in Gansu Province, the present agriculture resources situation and the trend of rural economy development. Through the analysis of the main factors affecting agriculture and agricultural development, this research selected 12 indexes to build ecological security index system based on soil and water conservation of Gansu Province (Table 2).

3.3 Data analysis

IOP Conf. Series: Earth and Environmental Science **41** (2016) 012009 doi:10.1088/1755-1315/41/1/012009

3.3.1 Index standardization. For the original data in this study, it showed some different dimensions, and the order of magnitude also had the obvious difference. Therefore it is necessary to standardize the original data before analysis. At first, original data was calculated by extreme value method, and then confirmed each index between 0 and 1[32-33]. Considering the influence about each index to the system existed differences, the index value, the better development degree is. For others, the smaller the value is, the better is. The former is positive index, and the latter is negative one. So different evaluation index had different quality and its dimension was also different. In order to eliminate the hard evaluation from different dimensions, original data was adopted to improve the standardization using the normalization method before calculation [15].

For the positive effect index, which means this kind of index has much less risk to ecological security. The standardization values can be calculated by:

$$p_{ij} = \frac{x_{\max} - x_{ij}}{x_{\max} - x_{\min}} \tag{1}$$

For the negative effect index, which means this kind of index has much more risk to ecological security. The standardization values can be calculated by:

$$p_{ij} = \frac{x_{ij} - x_{\min}}{x_{\max} - x_{\min}} \tag{2}$$

Where, P_{ij} is normalized value; X_{ij} as the actual value; X_{max} is biggest value of this index, and X_{min} is the minimum value of this index.

3.3.2 Weight determination. As a kind of objective value assignment method, the method of mean square error has been adopted to evaluate objectively Gansu's ecological security. Its weight is formed by actual data from unit of each index in the evaluation [34-35], whereas it does not depend on man's subjective judgment, and it is strong objectivity. The formula is as following:

$$E(Ai)= \sum_{j=1}^{28} P_{ij} / 28 \tag{3}$$

$$V(Ai)= \sqrt{\sum_{j=1}^{28} \left[(P_{ij} - E_{A_i})^2 \right] / 27} \tag{4}$$

$$\xi = V_{(A_i)} / P \tag{5}$$

$$W_i = \xi \Big/ \sum_{i=1}^{12} \xi \tag{6}$$

$$S=\sum_{i=1}^{12} Wi \times Pij \quad (7)$$

Where S is the value of ecological safety degree, the Ai (i = 1, 2, 3,..., 12) is index series, Gj (j = 1, 2, 3,..., 28) is solution series from 1986 to 2013, respectively, P_{ij} is the attribute values for solution series to index series. Firstly, P_{ij} by equation (1), (2) was calculated and then the average value for the standardized Ai (P) and its variable coefficient (ξ), and the mean square error for Gj (V(Ai)) were calculated; Secondly was to confirm the weight of each index (Wi) using equation (3), (4), (5), and (6); Finally the ecological safety degree was calculated by the equation (7).

Table 2. Weight of ecological security evaluation index system for soil and water conservation of Gansu Province.

Target layer	Rule layer	Index layer	Index weight
The composite index of ecological security(A)	The pressure indicators of ecological environment(G1)	The area of cultivated land per capita(hm^2) P_1	0.0792
		The amount of fertilizer per area (kg/hm^2) P_2	0.0866
		The amount of pesticide per area(kg/hm^2) P_3	0.0976
		The ratio of natural disasters inundated area P_4	0.0577
	The status indicators of ecological environment (G2)	The control area ratio of water loss and soil erosion P_5	0.0807
		The forest area of soil and water conservation(hm^2) P_6	0.0883
		The area of artificial grass (hm^2) P_7	0.0771
		The area of closing hillsides to facilitate afforestation (hm^2) P_8	0.0883
		The area of closing hillsides to facilitate grass(hm^2) P_9	0.0914
	The response indicators of ecological environment(G3)	The investment amount of soil and water conservation (million) P_{10}	0.0915
		The grain output per capita(t) P_{11}	0.0707
		The agricultural output value per unit area (million/km^2) P_{12}	0.0874

3.4 Grading standard of ecological security

It has been reported that it is very hard to directly judge the ecological security situation through calculating the ecological security degree [36]. Therefore, this study divided the ecological security degree (0 to 1) into five ranks, on one hand, this research is based on the existing research results [14,25,36] and consulting experts. On the other hand, this study divided the ecological security degree (0 to 1) into five ranks according to the relative index method referenced to other classification proofs, and in viewing of the present situation of soil and water conservation and ecological environment of the study area. The details for the ecological security ranks were showed in Table 3.

Table 3. Classification benchmarks of ecological safety for soil and water conservation of Gansu Province.

Rank	Safe state	Ecological security degree(S)	Characteristics
I	Safe state	S≥0.8	The function of land ecosystem service is perfect , the agriculture ecosystem environmental without interference, system structure and complete function is strong, the land is fertile, vegetation coverage rate is high,and there is no significant ecological problems
II	Safer state	0.8＞S≥0.7	The function of land ecosystem service is relatively perfect, less the damage to agricultural ecological environment, system structure and complete function is less, the fertility of soil is stronger, the vegetation coverage rate is higher, the value agricultural output is high, the use degree of land is high, and ecological problems does not significant
III	Sensitive state	0.7＞S≥0.5	The function of land ecosystem service begin to degradation, agricultural ecological environment is interference, system structure has been deteriorating, but still can maintain the basic functions, the soil fertility is decrease, vegetation coverage is reduced, ecological disasters have occurred
IV	Risk state	0.5＞S≥0.3	The function of land ecosystem service and agricultural ecological environment have been destroyed. The damage of system structure is bigger, and difficult to recover the interference. The soil fertility is lower, governance difficulties and ecological disasters become more and more significant
V	Deterioratin g state	S＜0.3	The function of land ecosystem service and agriculture ecological environment are destroyed completely, the ecological environment system structure is incomplete and function is loss, ecological restoration and reconstruction is very difficult, the ecological environment problem is very big and ecological disasters often happened

4. Results

4.1 Spatial variation of the ecological security degree

Through the systematic analysis on the three river basins and Gansu Province, the ecological security in most of the area had reached to the safety state in 2011. Figure 2 showed the spatial distribution of the ecological security degree of Gansu Province in 1991, 2001 and 2011, respectively. The degree of ecological security increased from 0.31 in 1986 to 0.66 in 2013 based on the analysis of three major river basins. Ecological safety had improved with the ecological very difficult security job, whiles the average of ecological security changed from 0.36 to 0.59 in 14 cities (state) of Gansu Province. The rank of ecological safety was in the III, IV grade in study basin, which means the degree of ecological security increased, but land ecosystem service function degraded. The system structure had deteriorated after the disturbance for agricultural ecological environment, but can still maintain the basic function with the decreased soil fertility and vegetation coverage. The ecological problems and ecological disasters occurred occasionally. As Figure 2 showing, the Baiyin city, Gannan city and Jiayu-guan city were deteriorating state in 1991, whereas some cities were at risk state, such as Qingyang city, Pingliang city, Dingxi city, Lanzhou city, LinXia city and Jiuquan city, but it showed the sensitive state for Jinchang city, Tianshui city, Longnan city, Wuwei city, Zhangye city. In 2001, Baiyin city, Gannan city and Jiayu-guan city were deterioration, whiles risk cities were Qingyang city, Lanzhou city, LinXia city and Jiuquan city, and the sensitive cities was Dingxi city, Jinchang city, Pingliang, Zhangye and Wuwei, but Tianshui and Longnan were safer. In 2011, Baiyin, Gannan, Jiuquan city and Jiayu-guan city were in risk, whereas it displayed the sensitive state for Qingyang city, Dingxi city, Lanzhou city, LinXia city, Longnan city, Jinchang city and Wuwei city, and it was safety for Tianshui city, Pingliang city and Zhangye city.

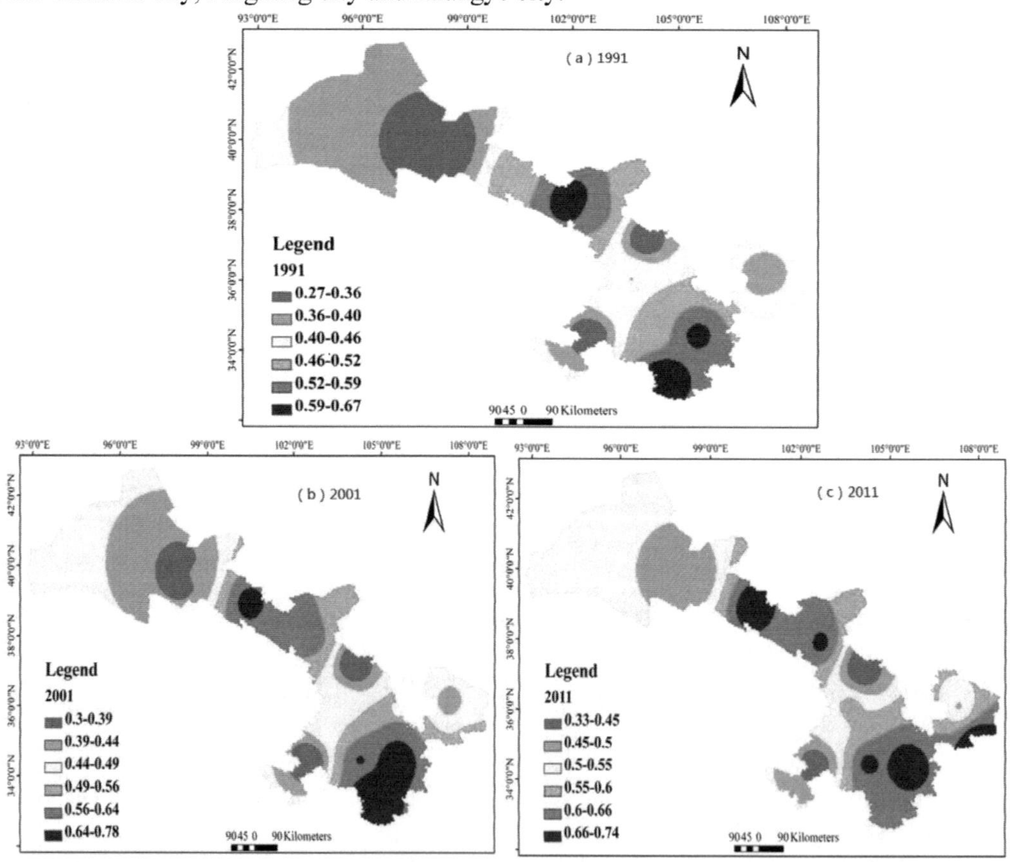

Figure 2. Spatial pattern of ecological security degree in 1991, 1986 and 2011.

4.2 Gansu's Ecological security evaluation

Ecological safety in the whole Province had improved significantly from 1986 to 2013, and there wasn't a city in risk state until 2013. This situation suggested that the ecological deterioration was significantly lower. Ecological security level was better mainly due to the increase of vegetation coverage in Gansu Province and the population control, and it also because of the local intensive production mode. The fluctuation of different level may be related to the importance of ecological environment protection and the way of the regional ecological environment protection. Qingyang city's ecological security increased year by year, changed from the deterioration state in 1991 to the sensitive state in 2011. This conclusion is similar to previous research [37]. Consistent with the study by Yang and Liu [38], Jinchang city' ecological security improved gradually, from the sensitive to the safer state, and this result mainly due to the development of soil and water conservation and the protection of the east river water source conservation forests. Owing to Jiayu-guan city' natural conditions of being limited and located in Gobi desert, its ecological deficit was larger. Therefore the ecological security state had been at risk state, but there was no denying that the ecological environment had changed. From deterioration state in 1991 to risk state in 2013, this was similar to the research by Zhang [39]. In addition, Tianshui city' ecological security was maintained in safer state from 1986 to 2013, its ecological safety was the most stable, and the main reason was that the destruction of the vegetation was less. Vegetation coverage was higher with the well-developing local's soil and water conservation, and its benefit was also remarkable. Longnan city's ecological security reduced from safety in 2001 to sensitive state in 2011, and its degree has declined. Zhangye city and Pingliang city have improved in 2011 and were in safety state. This may mainly due to the differences from each local government policy because it must revise the related policy immediately and ensured ecological security with good sustainable development.

5. Discussions

5.1 Ecological security zone differences in Gansu Province

5.1.1Yellow River basin. Ecological security degree, pressure index, status index and response index from 1986 to 2013 in Yellow River basin were presented in Figure 3.The ecological security status for Yellow River basin can be concluded as following: it was in the risk state from 1986 to 2003, whiles it was in sensitive state during from 2004 to 2010, and it was much better than the former years and stayed in safety state until 2011 to 2013. The overall state was in risk state with the average of index value of 0.48 during the past 28 years.

It exhibited the lowest values (0.31) for ecological security degree in 1990, and the other three index values were kept comparatively low. The main reasons were the rapid growth of population in China leaded to the nervous land at that time. More people wanted to acquire more land, and Gansu Province located in Loess Plateau with the complex and varied topography, and the plains was preciously few. So people transfer their target to the mountains land and begin deforest and reclamation, and a large number of grassland and forest had been destroyed in the study area and caused the serious soil erosion. Until 1999, the destroying forest had obviously protect when the government put forward the policy of returning farmland to forest and grass, and it effectively maintain the water and soil through large-scale project of returning farmland to forest. And the ecological security degree was 0.47 in 1999, lower than in 2000 (0.38) because it was the first year for returning farmland to forest, and the project could result in serious destroy of the surface soil. The situation was gradually improving from 2000 to 2013 for the trend of soil and water conservation and ecological security in the whole basin. As the research by Li [1], Huining County, located in Gansu Province, where the remarkable achievements of returning farmland to forest, and the vegetation coverage has increased significantly under the implement of returning farmland to forest project, whiles the NDVI index increased area of vegetation is 4404.75 km^2 in 2011 comparing with 2000, accounting for 77.74% of the total area of Huining County.

As showing in Fig.3, pressure index showed the decreasing trend year by year, whiles state index and response index showed increasing trend. This is mainly due to the adjustment of national policy. People have become deeply awareness of the importance of protecting ecological environment, and it also closely related to the current control of population. This demonstrated the land ecosystem service function is relatively successful, and the damage to agricultural ecological environment has become less, whiles system structure is complete and the function is stronger with the stronger soil fertilizer, the higher vegetation coverage rate, the high agricultural output value and the high land use degree, but the significant ecological problems had not happened.

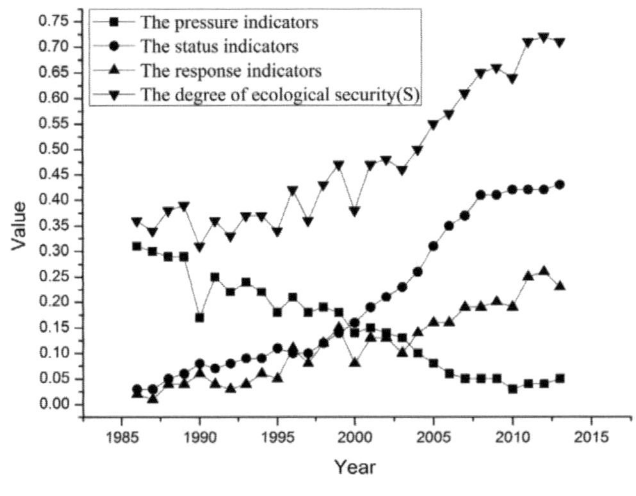

Figure 3. Variation of ecological security degree for Yellow River basin from 1986 to 2013.

5.1.2 Yangtze River basin. It had the lowest value (0.33) in 1986 for ecological security degree in Yangtze River basin, and status index and response index were also kept lowest, but pressure index was the highest (Figure 4). From 1986 to 1992, it was in risk state, while it was in sensitive state during from 1993 to 2010. However, it was much better than the former years and stayed in safer state until 2011 to 2013. For the past 28 years, the overall state was in sensitive state with the average value of 0.57. Ecological security had changed obviously in the Yangtze River basin from 1986 to 2013, and ecological security basically showed an increasing trend over the past 28 years (Figure 4).

Ecological security changed from risk state to safer state in 28 years, and ecological security degree increased by 127.27%, from 0.33 in 1986 to 0.75 in 2013. Land ecosystem service function was relatively perfect with less damage to agricultural ecological environment, and the soil fertility was stronger. It was high for the vegetation coverage rate and the agricultural output value with the high land use degree, and the ecological disasters decreased and agricultural output value increased year by year. And in August 2010, Zhouqu county catastrophic landslides occurred in the Southern Gansu where ecological security was at risk. This result was similar to the former view [40] because the region was in overload condition. Of course, the occurrence of debris flow was not an accident. Main reason was the geographic and geomorphic conditions, but human activity had also increased the likelihood of occurrence. Especially, the forest had been over-exploitation for the development of mineral resources, whereas the original vegetation was being destroyed. As figure.4 showing, pressure index showed the decreasing trend year by year, whereas state index and response index showed the increasing trend. The trend of the Yangtze River basin was consistent with Yellow River basin; the main reasons were people deeply awarded the importance of protecting ecological environment also closely related to the current control of population.

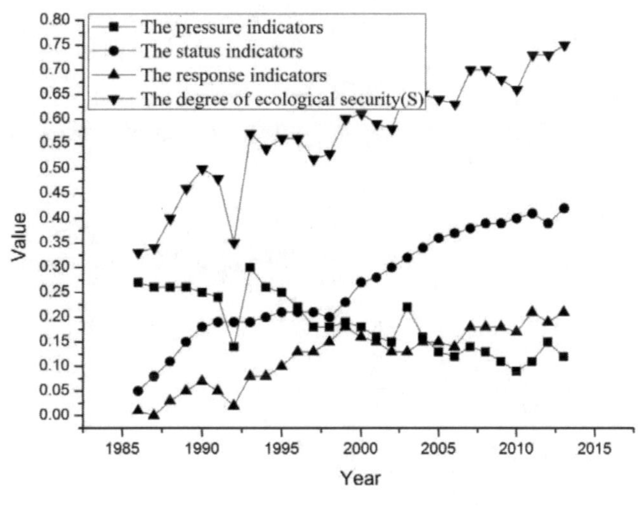

Figure 4. Variation of ecological security degree for Yangtze River basin from 1986 to 2013.

5.1.3 In-land River basin. For in-land river basin, ecological security degree presented a tendency of continuous improvement. And the degree increased by 37.50% from 0.40 in 1986 to 0.55 in 2013 (Figure 5). In generally, ecological security was from the risk stage to the sensitive stage. To some extent, this reflected the retention rate of the soil and water conservation measures was high with remarkable efficiency in in-land river basin. The bearing capacity of the ecological economic system was growing. However, three indexes presented obvious difference: pressure index showed a trend of decline dropped from 0.31 in 1986 to 0.15 in 2013. It showed the reduced trend for the pressure index contribution rate year by year. The main reason was that the regional population growth has caused water and soil resources pressure. Farmland area of per person is 0.43 hm^2 in 1986 but reduced to 0.21

hm2 in 2013 and it had dropped 51.16%. Status indicators also was risen generally with an increase from 0.05 in 1988 to 0.21 in 2013.This is mainly contributed by the result for the project of returning farmland to forest implements and the comprehensive treatment of soil erosion since 2000.. Response index showed an increasing trend, from 0.03(1986) to 0.19(2013). The investment of soil and water conservation has increased from 21.13 Yuan per hm^2 in 1986 to 60.6 Yuan per hm^2 in 2013. The reverse trend for response index caused by two facts: on the one hand, it was due to the increased government investment for soil and water conservation year by year, on the other hand, owing to the output of grain per capita and agricultural output value per unit area caused by agricultural technology progress.

In in-land river basin, ecological security can be concluded as following: it was in risk state from 1986 to 1998, while it was in sensitive state from 1999 to 2013. For the purpose of measuring state during the past 28 years, the average value was 0.47, which means it was in risk state. So ecological security of in-land river basin still had a long way to reach the safety state, and it needs to do a lot of management work in the future.

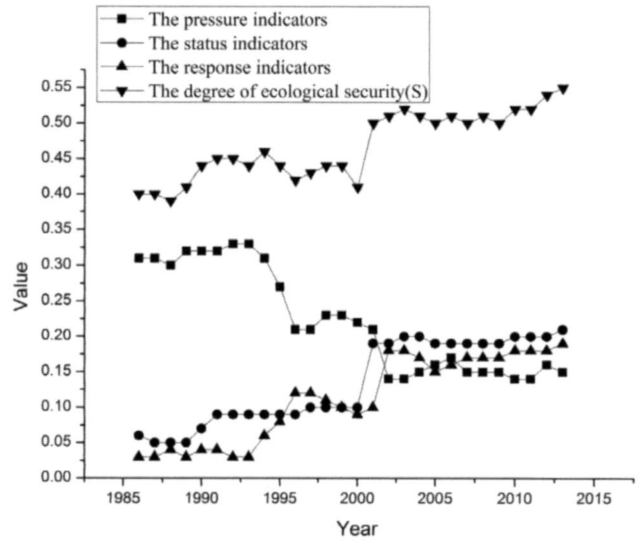

Figure 5. Variation of ecological security degree for In-land river basin from 1986 to 2013.

5.2.1 Controlling the population growth and relieving the population pressure. In recent years, the Gansu provincial government had seriously implemented the national population control policy, and it has exhibited a preliminary results. However, there was a big gap to the international standards of ecological safety, so the government of Gansu Province should continue to strengthen the work of population control, strengthen population policy advocacy and guidance, especially for the cities with the great population pressure, such as Linxia, Tianshui, and Lanzhou city. And what's more, the government should attach great importance to population issues, which combined with regional characteristics and land carrying capacity requirements to develop a reasonable population development planning and control population growth. For example, Linxia Prefecture as a Hui Autonomous Prefecture, population fertility concept is relatively outdated with a large number of

IOP Conf. Series: Earth and Environmental Science **41** (2016) 012009 doi:10.1088/1755-1315/41/1/012009

ethnic minorities, so the important significance of the people to focus on the promotion of fertility policy. For the capital city of Lanzhou, it is to carry out effective guidance with a large population base and a certain population agglomeration effects, which can ease population pressure by controlling the foreign population and other policies.

5.2.2 Pay attention to agricultural technology innovation. At present, the process of agricultural production in Gansu Province is more traditional because agricultural technology is not enough. There are many problems, such as the low production efficiency and the excessive use of chemical fertilizer. These problems not only damage the ecological environment, but also do not adapt to the requirements of modern society for agricultural products. Therefore, it is necessary to introduce new agricultural technology, cultivate agricultural production talents, promote green agriculture, and to reduce the amount of pesticide and chemical fertilizer. It is to arrange for the agricultural production from the region in different climate region in order to give full play to regional natural and resource advantages, such as in Jiuquan city and Gannan city. Owing to strong desertification land and the loess plateau in Gansu Province, soil is relatively poor, less rainfall, drought and water shortage but has plenty of light, heat and wind energy resources advantage. so it can take advantage of abundant energy to develop the plastic greenhouse and drip irrigation technology in the development of regional agriculture, with reasonable for the choose of crop species and planting measures, agriculture introduced the latest research results, in order to improve the plant the fixed conversion efficiency of solar energy.

5.2.3 Increase capital investment, strict supervision mechanism. In recent years, Gansu Province was mainly focused on economic development, which has caused many environmental problems in the process of industrial production with more waste water, waste gas, waste emissions, but with the lower governance rate and comprehensive utilization rate. It showed a continuous growth for GDP in Gansu Province during recent years, but it was in the lack of investment in environmental management. Some companies were focused on economic efficiency, but it ignored the long-term interests with excessive sewage phenomenon. And what's more, environmental governance was not enough to catch up with the speed of pollution, whiles land pollution was becoming increasingly serious. It must reduce the industrial pollution in order to improve the utilization rate and treatment rate of industrial waste recycling.

Government departments should attach great importance to increased capital investment in environmental governance. The financial subsidies should support the enterprises to implement cleaner production engineering and development of raw materials and products. At the same time, government implement emissions of pollutants and sewage permit system for industrial pollution enterprises. It is combining with the characteristics of industrial enterprises, such as Qingyang city and Pingliang city with the coal enterprises, Yumen city with oil production enterprises, Jinchang city with copper and nickel ore enterprises, which are the serious pollution of heavy industry. Government must carry out strict control and supervision. The strict law must be enforcemented when found illegal polluters with the increasing punishment, investigating the corporate responsibility and raising the cost of corporate wrongdoing. These varieties of measures and methods have been executed to prevent and control environmental pollution and to improve "treatment after pollution" phenomenon under economic development.

6. Conclusions

Based on the observation data from 1986 to 2013, the results indicated that Gansu's ecological security degree increased from 0.31 in 1986 to 0.66 in 2013, and the ecological security was in the sensitive state ($S<0.7$). The main reason was that national policy on improving the ecological environment had played a decisive role in recent years, especially the project of returning farmland to forest and grass. These made people's environmental awareness has significantly improved in the study area. Ecological security also had the obvious regional differences. Yellow river's ecological security changed from risk state to safer state in 28 years, and ecological security degree increased from 0.36 in 1986 to 0.71 in 2013. Yangtze River's ecological security changed from risk state to safer state, and ecological security degree increased by 1.3 times. The ecological security degree for in-land river basin increased from 0.40 in 1986 to 0.55 in 2013 with an average value of 0.47. This may be largely due to the difference of geographical environment of the three major rivers of Gansu Provinces. In conclusion, although Gansu's ecological security had improved in recent years, the environment problem is still serious. There is a strong need for paying more attention to policy for ecological environment protection and increasing the propaganda in the future to ensure the Gansu's ecological security. The studies on ecological security are conductive to protect fragile ecological environment and local sustainable development. Ecological Security Assessment is a complex system engineering that involves many factors. The appraisal methods nowadays have their limits on selecting evaluating indicators, and the determination of the target weight may have some subjectivities. All the problems need to be further discussion and study, and the evaluation method is yet to be improved.

Acknowledgments

This study was jointly supported by the National Natural Science Foundation of China (31260122), the open foundation from state key Laboratory(SKLCS-OP-2014-07,SKLFSE201403), the China Postdoctoral Science Foundation (2012M510219, 2013T60899), the postdoctoral fellow at the international exchange plans from China Postdoctoral Association (20140043), and the Youth Innovation Promotion Association, Chinese Academy of Sciences (2012hhs03).

Reference

[1] Li Z J, Yang C H, Ma R, Qi G P, Wang L, Zhang C, Song L L and Wang J G 2014 Analysis on vegetation condition after the implement of returning farmland to forest project in Huining county *Journal of soil and water conservation bulletin* **34** (1) 214-9

[2] Song L L, Tian Q, Li Z J, Kang F L and Wei Y Q 2015 Analysis of ecological security in Yellow River basin in gansu Province based on PSR model *Journal of West China Forestry Science* **44**(2) 126-30

[3] Tian Q, Li Z J, Song L L and Zhang F 2014 Ecological security pattern of soil and water conservation in the HeXi area of Gansu, China in 1986-2011 *Journal of Desert Research* **234**(6) 1692-8

[4] Tian Q, Li Z J, Song L L 2014 Soil and water conservation and ecological security evaluation in Yangtze river basin in gansu Province *Yangtze river* **45**(15) 100-4

[5] Guo M, Xiao D N and Li X 2006 Changes of landscape pattern between 1986 and 2000 in

Jiuquan oasis, Heihe River basin *Acta Ecologica Sinica* **26**(2) 457-66

[6] Herrmann S, Dabbert S and Schwarz-von Raumer H G 2003 Threshold values for nature protection areas as indicators for bio-diversity—a regional evaluation of economic and ecological consequences *Agriculture, ecosystems & environment* **98**(1) 493-506

[7] Iwasa Y, Hakoyama H, Nakamaru M and Nakanishi J 2000 Estimate of population extinction risk and its application to ecological risk management Population ecology **42**(1) 73-80

[8] Jia L Q, Ou Y, Z Y, Zhao T Q and Wang X K 2004 Ecological security assessment of an urban ecosystem *Ecology and Environmen* **13**(4) 592-6

[9] Kang P and Xu L 2010 The urban ecological regulation based on ecological carrying capacity *Procedia Environmental Sciences* **2** 1692-700

[10] Liang P, Liming D and Guijie Y 2010 Ecological security assessment of Beijing based on PSR model *Procedia Environmental Sciences* **2** 832-41

[11] Wang G X, Cheng G D and Qian J 2003 Several problems in ecological security assessment *Chinese Journal of Applied Ecology* **14**(9) 1551- 6

[12] Han B, Liu H and Wang R 2015 Urban ecological security assessment for cities in the Beijing–Tianjin–Hebei metropolitan region based on fuzzy and entropy methods *Ecological Modelling* **318** 217-25

[13] Liu D & Chang Q 2015 Ecological security research progress in China *Acta Ecologica Sinica* **35**(5) 111-21

[14] Grekousis G 2015 An overview of 21 global and 43 regional land cover mapping products *International Journal of Remote Sensing* **36**(21) 5309-53

[15] Zhou K H, Liu Y L, Tan R H and Song Y 2014 Urban dynamics, landscape ecological security, and policy implications: A case study from the Wuhan area of central China *Cities* **41** 141–53

[16] Hope B K and Peterson J A 2000 A procedure for performing population-level ecological risk assessments *Environmental management* **25**(3) 281-9

[17] Liu H, Wang H and Zhang X W 2006 Ecological security assessment review *Journal of ecology* **25**(1)74-8

[18] Gao C B, Chen X G, Wei C H and Xiao C P 2006 Ecological security status and trends of quantitative evaluation of guangdong Province *Acta ecologica sinica* **26**(7) 2191-7

[19] Wan Y, Xu L L, Geng Q F，Leng X，An S Q and Tang J W 2012 Ecological hot topics in global change on the 2nd international young ecologist forum *Acta Ecological Sinica* **32**(17) 5601-8

[20] Jiang Z M, Jiang X B 2011 Chongqing ecological security based on GIS space-time dynamic analysis *The people of the Yangtze river* **15** 81-5

[21] Yao X S and Tian J Y 2007 Ecological security research progress and application *Chinese journal of environmental management cadre institute* **2** 47-50

[22] Norton S B, Rodier D J, van der Schalie W H, Wood W P, Slimak M W and Gentile J H 1992 A framework for ecological risk assessment at the EPA *Environmental toxicology and chemistry* **11**(12) 1663-72

[23] Rogers K S 1997 Ecological security and multinational corporations *Environmental Change and*

Security Project Report **3** 29-36

[24] Rapport D J, Gaudet C, Karr J R, Baron J S, Bohlen C, Jackson W and Pollock M M 1998 Evaluating landscape health: integrating societal goals and biophysical process *Journal of Environmental Management* **53**(1) 1-15

[25] Bai X and Tang J 2010 Ecological security assessment of Tianjin by PSR model *Procedia Environmental Sciences* **2** 881-7

[26] Gong J Z, Liu Y S, Xia B C and Zhao G W 2009 Urban ecological security assessment and forecasting, based on a cellular automata model: A case study of Guangzhou, China *Ecological Modelling* **220**(24) 3612-20

[27] Hu S H, Wu K Y and Wang J Q 2008 Study pre-warning of ecological security on basis of fuzzy optimize in Anhui Province *Journal of Biotechnology* **136** S32

[28] Hua Y E, Yan M A and Limin D O N G 2011 Land ecological security assessment for Bai autonomous prefecture of Dali based using PSR model--with data in 2009 as case *Energy Procedia* **5** 2172-7

[29] Li X, Tian M, Wang H, Wang H and Yu J 2014 Development of an ecological security evaluation method based on the ecological footprint and application to a typical steppe region in China *Ecological Indicators* **39** 153-9

[30] Wang G, Wang L and Wu W 2007 Recognition on regional ecological security definition and assessment system *Acta Ecologica Sinica* **27**(4) 1627-37

[31] Yu R R, Xie W X, Zhao Q S, Xu Zhen and Liu W L 2012 Ecological safety assessment of Dagu estuary wetland in Jiaozhou Bay of Shandong Province, East China based on landscape pattern *Chinese Journal of Ecology* **31**(11) 2891-9

[32] Cui S H, Hong H S, Huang Y F and Xue X Z 2005 Progress of the ecological security research *Acta Ecologica Sinica* **25**(4) 861- 8

[33] Duan J, Kang M Y, Dai C, Jiang Y 2012 Integrated spatial assessment of ecological security in Dongjiang River Basin based on gridding *Chinese Journal of Ecology* **31** (8) 2075-81

[34] Meng Z X, Li C Y and Deng Y L 2009 Ecological security early-warning and it ecological regulatory countermeasures in the Tuojiang River Basin *Journal of Ecology and Rural Environment* **25**(2) 1-8

[35] Mageau M T, Costanza R and Ulanowicz R E 1995 The development and initial testing of a quantitative assessment of ecosystem health *Ecosystem Health* **1**(4) 201-13

[36] Ma K M, Fu B J, Li X Y and Guan W B 2004 The regional pattern for ecological security (RPES): the concept and theoretical basis *Acta Ecologica Sinica* **24**(4) 761-8

[37] Zeng C P, Qiu H Z, Zhang W M and Zhang J 2010 Assessment on ecological security of Qing yang based on PSR model *Journal of Arid Land Resources and Environment* **24**(12) 67-72

[38] Yang K E and Liu J Q 2012 East river water source conservation forests on the influence of the singularity of the ecological environment and its protection countermeasures *Gansu Science and Technology* **28**(6) 146-8

[39] Zhang W F, et al 2010 Ecological security evaluation of urban ecosystem in arid area *Journal of An hui Agri.Sci.* **38**(29) 16419-20, 16479

[40] Zhu X L, Li W L , Xue Z Z, Xu D D and Wang Hao 2012 Evaluation of ecological carrying

capacities in Gannan based on PSR model *Prataculturral Science* **29**(2) 198-203

Agronomic threshold of soil available phosphorus in grey desert soils in Xinjiang, China

B Wang[1,4], H Liu[1], X Y Hao[2], X H Wang[1], J S Sun[1,4], J M Li[3] and Y B Ma[3,4]

[1] Institute of Soil Fertilizer and Agricultural Water Saving, Xinjiang Academy of Agricultural Sciences, Urumqi 830091, China

[2] Agriculture and Agri–Food Canada, Lethbridge Research Centre, 5403 1st Ave. S., Lethbridge, AB T1J 4B1, Canada

[3] National Soil Fertility and Fertilizer Effects Long–term Monitoring Network, Institute of Agricultural Resources and Regional Planning, Chinese Academy of Agricultural Sciences, Beijing 100081, China

E–mail: wbx_wm@126.com (B. Wang); 2436405336@qq.com (J. Sun); ybma@caas.ac.cn (Y. Ma)

Abstract. Based on 23 years of data, yields of maize, wheat and cotton were modelled under different fertilizer management practices and at different levels of available phosphorus (Olsen-P) in soil. Three types of threshold models were used, namely linear-linear (LL), linear-plateau (LP), and Mitscherlich type exponential (Exp). The agronomic thresholds of available phosphorus were 25.4 mg·kg^{-1} for cotton, 14.8 mg·kg^{-1} for wheat, 13.1 mg·kg^{-1} for maize and 25.4 mg·kg^{-1} for the grey desert soil regions of Xinjiang in China as a whole.

1. Introduction

Phosphorus (P) is one of the essential nutrients for plant growth as well as the key nutrient for high yield. To boost yields, farmers in China tend to apply P in excess of the recommended doses[8], which results in accumulation of P in farmland topsoil to form a huge pool of P[17]. The minimum amount of soil available P for maximum crop yield is referred to as its agronomically critical value or the threshold value of soil available phosphorus[16], which is the value used by researchers as the criterion for deciding whether to continue applying P[1,2].

Because of its unique location and climate, Xinjiang farming system is only one crop a year. Cotton, wheat, and maize are the main three produced crops in Xinjiang, accounting for 73% of the total region' s cropped area (cotton, 35%; wheat, 21%; and maize, 17%)[20]. The agronomic thresholds of soil P for these three crops have not been reported for the grey desert soil region of Xinjiang for the single-crop system mentioned above—a lacuna the present study sought to fill. We used the data available from a long-term location trial to study the effect of different doses of P and soil available P on the yields of cotton, wheat, and maize and eventually determined the agronomic thresholds of soil P.

2. Materials and methods

2.1. Overview of the experimental site

The experimental site was in Xinjiang Grey Desert Soil Fertility and Fertilizer Efficiency Long-term Monitoring Station of which was one site of the National Long-term Monitoring Network of Soil Fertility and Fertilizer Effects, about 25 km north of Urumqi City, Xinjiang (43°58′23″ N, 87°25′58″ E) in China. The station is under the National Modern Agricultural Science and Technology Demonstration Zone of the Xinjiang Academy of Agricultural Sciences. The soil of the experimental site is the typical grey desert soil which is a major soil type in the hinterland of Eurasia, in the temperate inland region, and also in the arid desert areas of north-western China, where they are spread over 1.8×10^6 ha. About 80% of that area is in Xinjiang. The terrain of the experimental site is uneven, with higher elevations in the east and the south and lower elevations in the west and the north (the gradient is 1/100—1/70), the average being 600 m. The water table is at a depth of 30 m. The relevant weather parameters (annual averages) are as follows: precipitation, 310 mm; evaporation, 2570 mm; temperature, 7.7 °C; sunshine hours, 2594, and frost-free days, 156. The basic physical and chemical properties of topsoil (0–20 cm) were listed as Table 1.

Table 1. The basic physical and chemical properties of topsoil (0–20 cm) of Xinjiang grey desert soil.

Items	Value
Organic matter ($g \cdot kg^{-1}$)	15.2
Total N ($g \cdot N \cdot kg^{-1}$)	0.87
Total P($g \cdot P \cdot kg^{-1}$)	0.67
Total K ($g \cdot K \cdot kg^{-1}$)	19.8
Alkali-hydrolysable N ($mg \cdot N \cdot kg^{-1}$)	55.2
Available P ($mg \cdot P \cdot kg^{-1}$)	3.9
Soil pH	8.1
Soil bulk density ($g \cdot cm^{-3}$)	1.25

Note: Table 1 is the basic physical and chemical properties of topsoil (0-20 cm) of Xinjiang grey desert soil in 1989. Uniformity trials of the grey desert soil were conducted during 1988–1989, and soil fertility and fertilizer efficiency have been monitored since 1990. Each test plot in the monitoring station is 468 m^2. The plots are separated by cement boards up to a depth of 1.2 m.

2.2. Experimental design

Seven treatments from the long-term fertilizer experiments were selected for analysis: (1) CK (control; no fertilizer), (2) NK (N and K fertilizers), (3) NP (N and P fertilizers), (4) NPK (N, P, and K fertilizers), (5) NPKM2 (NPK fertilizers with a regular dose of manure), (6) NPKM1 (NPK fertilizers with overdose of manure), and (7) NPKS (NPK fertilizers with straw returned to the soil). Sheep dung was used as manure, applied once each fall at the time of ploughing, and all the straw from the current season was crushed before being ploughed back. Crop rotation and one crop a year followed the established requirements for each treatment, with maize grown in 1990, 1993, 1996, 2000, 2003, 2005, 2008 and 2010; spring wheat was in 1991, 1994, 2002 and 2006; winter wheat was in 1992, 1995, 1997, 1998, 2001, 2004, 2007 and 2011; and cotton was in 1999, 2009 and 2012. The topsoil was tested annually after harvest. The average annual nutrient inputs for each treatment are shown in Table 2.

Table 2. Average annual nutrient inputs from different sources during 1990–2015.

Treatment	Chemical fertilizer			Manure	Straw
	N ($kg \cdot ha^{-1}$)	P ($kg \cdot ha^{-1}$)	K ($kg \cdot ha^{-1}$)	M ($t \cdot ha^{-1}$)	S ($t \cdot ha^{-1}$)
CK	0	0	0	0	0
NK	99.4–241.5	0	18.8–50.4	0	0
NP	99.4–241.5	29.2–60.2	0	0	0
NPK	99.4–241.5	29.2–60.2	18.8–50.4	0	0
NPKM2	29.8–84.9	8.7–22.4	6.7–10.1	30	0
NPKM1	59.6–151.9	17.4–39.4	13.4–15.5	60	0
NPKS	89.4–216.7	24.5–50.8	16.9–42.3	0	4.5–9

Note: The application rates of N, P and K ($kg\ ha^{-1}$) for chemical fertilizer were 99.4 $kg \cdot N \cdot ha^{-1}$, 29.2 $kg \cdot P \cdot ha^{-1}$ and 18.8 $kg \cdot K \cdot ha^{-1}$ in 1990–1994 and 241.5 $kg \cdot N \cdot ha^{-1}$, 60.2 $kg \cdot P \cdot ha^{-1}$, and 50.4 $kg \cdot K \cdot ha^{-1}$ in 1995–2015 in the NPK treatment. In NPKM1 and NPKM2, the organic manure was sheep dung, which was assumed to contribute 15% of the total nutrients in the current year; the straw was assumed to contribute 10% of the total nutrients in the current year. In these treatments, the fertilizer dose was reduced proportionately, such that the total inputs of N, P, and K remained the same in all the treatments except CK. All the P and K and 60% of N were given as the basal dose, and the remaining N was given as a top dressing.

2.3. Methods of measurements

The soil available P (Olsen-P) was extracted by 0.5 $mol \cdot L^{-1}$ sodium bicarbonate ($NaHCO_3$) at pH 8.5, and crop yields were determined from the whole plot area.

2.4. Modeling procedures

The agronomic threshold of soil Olsen-P was determined by using the functional relationship between the extent of increase in yield following long-term application of fertilizers and soil Olsen-P content. In the present study, the functional relationship between the increase in the yield of different crops (cotton, wheat, and maize) and Olsen-P content of the grey desert soil was fitted as both linear (linear-linear (LL) or linear-plateau (LP)) and nonlinear equations (Mitscherlich (Exp)). The cut-off point of the LL and LP models divided the reaction curves of the crops to soil Olsen-P content into two sections, and this cut-off point was the agronomic threshold for soil Olsen-P. In the Exp model, the point of 90% of the maximum yield on the curve served as the threshold value

The effect of externally applied P on the yield of cotton, wheat, and maize was assessed in terms of absolute yield in the present study. The absolute yield was calculated using the following equation:

$$Y_A = Y_{NPK} - Y_{CK} \tag{1}$$

where Y_A is the absolute yield, Y_{NPK} and Y_{CK} represent the yield ($kg \cdot ha^{-1}$) in the NPK and the CK treatments, respectively; thus the absolute yield is the difference in yield between the two treatments and eliminates any variations in the annual yield due to such factors as crop varieties.

The LL model can be represented as equations (2) and (3):

$$Y_1 = b_1X + a_1 \text{ when } X < C \tag{2}$$

$$Y_2 = b_2X + a_2 \text{ when } X \geq C \tag{3}$$

where Y_2 and Y_1 are the predicted absolute yields, a_2 and a_1 are the intercept parameters, b_2 and b_1 are the slope parameters, X is the soil available P (Olsen-P) content ($mg \cdot P \cdot kg^{-1}$), and C is the soil available P (Olsen-P) threshold ($mg \cdot P \cdot kg^{-1}$).

The LP model can be represented as equations (4) and (5):

$$Z = bX + a \quad \text{when } X < C \tag{4}$$

$$Z = Q \qquad \text{when } X \geq C \tag{5}$$

where b is the slope parameter, Z is the predicted absolute yield, X is the soil available P (Olsen-P) content ($mg \cdot P \cdot kg^{-1}$), a is the intercept parameter, C is the soil available P (Olsen-P) threshold ($mg \cdot P \cdot kg^{-1}$) and Q is the predicted plateau of the increase in absolute yield.

The Exp model can be represented as equation (6):

$$W = \left[1 - e^{-c(X+b)}\right]A \tag{6}$$

where W is the predicted absolute yield, A is the maximum absolute yield, c is the effective factor, b is the soil's capacity to supply P, and X is the soil available P (Olsen-P) content ($mg \cdot P \cdot kg^{-1}$).

In fitting with either of the linear models (LL or LP), the threshold for Olsen-P is the intersection of two lines (b_1 and b_2 in equations (2) and (3) or b and Q in equations (4) and (5) in part of modeling procedures)[15,19]. Whereas, the threshold of non-linear model (Exp) is the significant response point of Olsen-P content for 90% of the total yield. Colomb et al. (2007) discuss thoroughly the simulation process and the standard deviation calculation method of the Exp model.

2.5. Data analysis

Statistical analysis was performed using the nonlinear (NLIN) procedure in SAS software[13]. Marquardt procedure was used for the LL and LP models[15]; and the Gauss–Newton procedure was used for Mitscherlich model[19].

3. Results and analyses: threshold for soil Olsen-P

The response of the yields of cotton, wheat and maize to soil Olsen-P content following long-term (23 years) application of fertilizers to the grey desert soil was fitted using three separate models (Table 3). Of the three fitted curves, the Exp model presented the highest threshold value, followed, in that order, by LP and LL (Figure 1). The soil Olsen-P threshold values for the three crops as determined by the models are given in Table 4. The soil Olsen-P threshold values of wheat were 10.3 $mg \cdot kg^{-1}$, 14.4 $mg \cdot kg^{-1}$ and 19.8 $mg \cdot kg^{-1}$ for LL, LP and Exp, respectively; maize were 9.48 $mg \cdot kg^{-1}$, 10.4 $mg \cdot kg^{-1}$ and 19.5 $mg \cdot kg^{-1}$ for LL, LP and Exp, respectively; and cotton were 18.0 $mg \cdot kg^{-1}$, 20.1 $mg \cdot kg^{-1}$ and 38.2 $mg \cdot kg^{-1}$ for LL, LP and Exp, respectively. The average threshold values of soil Olsen-P for wheat (14.8 $mg \cdot kg^{-1}$) and maize (13.1 $mg \cdot kg^{-1}$) were very similar (a difference of only 1.7 $mg \cdot kg^{-1}$, or less than 12%) whereas the average threshold value of soil Olsen-P for cotton (25.4 $mg \cdot kg^{-1}$) was significantly higher than that for the other two crops (the difference was as much as 12.3 $mg \cdot kg^{-1}$, or more than 48%).

Table 3. The models about the thresholds of soil available P (Olsen-P) and yields for maize, wheat and cotton.

Crop	Model	Equation [a]	R^2 [b]	N [c]
Wheat	LL [d]	$Y_1 = -0.886+0.472X$ $Y_2 = 3.68+0.0306X$	0.763***	75
	LP [d]	$Z = -0.399+0.368X$ $Z = 4.91$	0.726***	
	Exp [d]	$W = 5.26(1-e^{-0.130(X-2.10)})$	0.717***	
Maize	LL	$Y_1 = -0.204+0.403X$ $Y_2 = 3.278+0.0360X$	0.709***	46
	LP	$Z = -0.267+0.421X$ $Z = 4.10$	0.651***	
	Exp	$W = 4.92(1-e^{-0.124(X-1.01)})$	0.720***	
Cotton	LL	$Y_1 = -0.214+0.163X$ $Y_2 = 2.61+0.00621X$	0.897**	12
	LP	$Z = -0.214+0.163X$ $Z = 3.06$	0.895***	
	Exp	$W = 3.16(1-e^{-0.0625(X-1.34)})$	0.838***	

[a] The letters "Y, Z and W" are the predicted absolute yields and X is the soil available P (Olsen-P) content ($mg \cdot P \cdot kg^{-1}$).

[b] R^2 is determination coefficient.

[c] N is the sample size.

[d] LP, LL and Exp are short for linear-plateau, linear-linear and Mitscherlich models, respectively.

Table 4. Mean absolute yield and threshold of soil available P (Olsen-P) (mg·P·kg^{-1}).

Crop	Mean absolute yield [a] (t·ha^{-1})	Threshold of soil Olsen-P			Mean (mg·P·kg^{-1})
		LL [b] (mg·P·kg^{-1})	LP [b] (mg·P·kg^{-1})	Exp [b] (mg·P·kg^{-1})	
Wheat	4.40 ± 0.10	10.3	14.4	19.8	14.8
Maize	3.94 ± 0.09	9.48	10.4	19.5	13.1
Cotton	2.84 ± 0.04	18.0	20.1	38.2	25.4

[a] Mean ± SD.

[b] LP, LL and Exp are short for linear-plateau, linear-linear and Mitscherlich models, respectively.

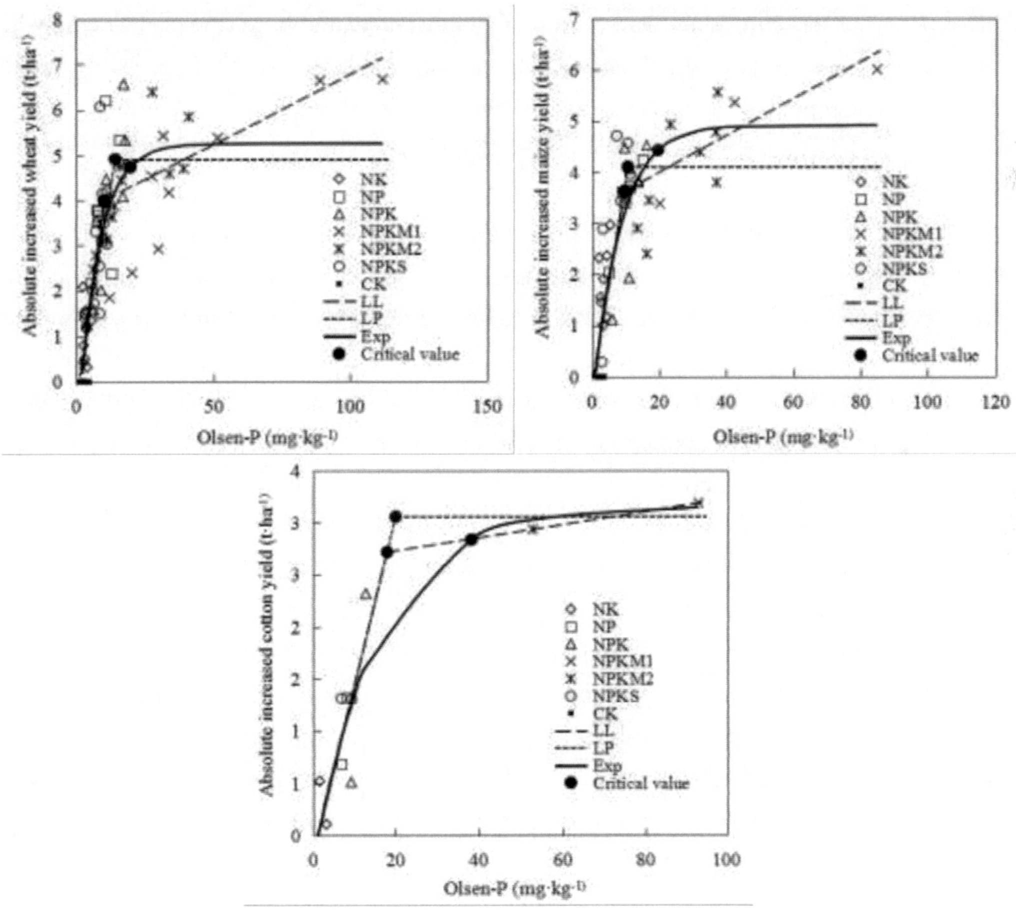

Figure 1. Response of cotton, wheat, and maize yield to soil Olsen-P fitted by LL, LP and Exp models. Note: The solid line and the dashed lines represent the curves fitted using three different models. The dots represent the agronomic threshold of Olsen-P in grey desert soil. The data for the curves are given in Table 3 and the threshold values are shown in Table 4. Abbreviations: LP, LL and Exp are Linear-Plateau, Linear-Linear and Mitscherlich model, respectively.

4. Discussion

Soil Olsen-P threshold values from the two linear models (LL and LP) were smaller than those from the non-linear model (Exp), and the average value of soil available P (Olsen-P) threshold for maize was smaller than that for wheat. All these results (the soil Olsen-P threshold values of LL and LP were smaller than Exp, and that of maize was smaller than wheat) are consistent with those of an earlier report[16]. Because the crops and the soil types in the present were different from those in the earlier report[16], the soil Olsen-P threshold value as determined from the Exp model was 1.9- to 2.1-fold of that determined by the LL model and 1.4- to 1.9-fold of that determined by the LP model. However, the soil Olsen-P threshold value was 1.4- to 2.1-fold and 1.3- to 1.9-fold for LL and LP, respectively, reported by Tang[16]. In the present study, the threshold value for wheat was 42% less than that of cotton, and the value of maize was the lowest among three crops, being 11% lower than that of wheat.

The average soil Olsen-P threshold values in the one-crop-a-year system (the crops being wheat or maize) in the grey desert soils of Xinjiang were lower those[16] proposed for the two-crops-a-year system (a rotation comprising wheat and maize) common in mainland China, indicating that the threshold values are influenced by the annual inputs of fertilizers and crop yields for different areas of crops system.

If the soil Olsen-P concentration is lower than the threshold value, yields will increase as the soil Olsen-P content increases, P being the limiting factor for high yields in this situation. Therefore, in order to increase crop yields, more P fertilizer should be applied to make the soil more fertile quickly.

If the soil Olsen-P concentration is equal to the threshold value, yields will not increase despite any increase in the soil Olsen-P content, P no longer being the limiting factor for high yields. Under these conditions, P fertilizers should be applied only to maintain the soil Olsen-P content (P input should be the same as P uptake) to ensure high yields.

Lastly, if the soil Olsen-P concentration is greater than the threshold value, yields will not respond to the addition of P, because P is not the limiting factor for high yields. Continuing application of P will result in rapid accumulation of P in soil, which will have a negative effect on production, and may even become a serious threat to groundwater and to rivers, lakes, and other freshwater bodies[5,6,9,12,14,16,17,18]. External application of P should be discontinued forthwith. The average threshold values for wheat and maize obtained in the present study fall within the ranges of 3.9–15.0 $mg \cdot P \cdot kg^{-1}$ for maize[4,10,11] and 4.9–20.0 $mg \cdot P \cdot kg^{-1}$ for wheat[3,4,7]. However, the threshold for cotton (25.4 $mg \cdot kg^{-1}$) was established in this study for the first time and has not been reported earlier.

5. Conclusion

The agronomic threshold of soil Olsen-P in the grey desert soil regions of Xinjiang was 25.4 $mg \cdot kg^{-1}$ for cotton, 14.8 $mg \cdot kg^{-1}$ for wheat, and 13.1 $mg \cdot kg^{-1}$ for maize. By relating these thresholds values to yields and measured concentrations of soil available P (Olsen-P) in Xinjiang, the agronomic threshold of soil P for the region as a whole was determined as 25.4 $mg \cdot kg^{-1}$.

Acknowledgments
We thank the National Natural Science Foundation of China (41161054) and Science and Technology Project Be Supportted Xinjiang(2016E02020) for providing fundings for this work. We also thank all the staff associated with long-term soil fertility and fertilizer effects location monitoring network of Xinjiang Grey Desert Soil in China for supplying the historic dataset.

References

[1] Aulakh M S, Garg A K and Kabba B S 2007 Phosphorus accumulation, leaching and residual effects on crop yields from long–term applications in the subtropics *Soil Use Manage* **23** 417-27

[2] Bai Z H, et al 2013 The critical soil P levels for crop yield, soil fertility and environment safety in different soil types *Plant Soil* **372** 27-37

[3] Bollons H M and Barraclough P B 1999 Assessing the phosphorus status of winter wheat crops: inorganic orthophosphate in whole shoots *J. Agr. Sci-Cambridge* **133** 285-95

[4] Colomb B, Debaeke P, Jouany C and Nolot J M 2007 Phosphorus management in low input stockless cropping systems: crop and soil responses to contrasting P regimes in a 36-year experiment in southern France *Eur. J. Agron.* **26** 154-65

[5] Heckrath G, Brookes P C, Poulton P R and Goulding K W T 1995 Phosphorus leaching from soils containing different phosphorus concentrations in the broadbalk experiment *J. Environ. Qual.* **24** 904-10

[6] Huang S, Ma Y, Bao D, Guo D and Zhang S 2011 Manures behave similar to superphosphate in phosphorus accumulation in long-term field soils *International Journal of Plant Production* **5(2)** 135-46

[7] Johnston A E, Lane P W, Mattingly G E G, Poulton P R and Hewitt M V 1986 Effects of soil and fertilizer P on yields of potatoes, sugar beer, barley and winter wheat on a sandy clay loam soil at Sazmundham, Suffolk *J. Agr. Sci-Cambridge* **106** 155-67

[8] Li J M, Gao J S, Liu J, Xu M G and Ma Y B 2010 A predictive model for phosphorus accumulation in paddy soils with long-term inorganic fertilization. *Commun Soil Sci. Plan* 1-19

[9] Li Q, Li J M, Cui X L, Wei D P and Ma Y B 2012 On-farm assessment of biosolids effects on nitrogen and phosphorus accumulation in soil. *J. Integr Agr.* **11** 1545-54

[10] Mallarino A P and Blackmer A M 1992 Comparison of methods for determining critical concentrations of soil test phosphorus for corn. *Agron. J.* **84** 850-6

[11] Mallarino A P and Atia A M 2005 Correlation of a resin membrane soil phosphorus test with corn yield and routine soil tests *Soil Sci. Soc. Am. J.* **69** 266-72

[12] Pote D H, Daniel T C, Sharpley A N, Moore P A, Edwards D R and Nichols D J 1996 Relating extractable soil phosphorus to phosphorus losses in runoff *Soil Sci. Soc. Am. J.* **60** 855-9

[13] SAS Institute 2003 *The SAS system for Windows, release 9.1* SAS Institute Inc, Cary, NC

[14] Shepherd M A and Withers P J 1999 Applications of poultry litter and triple superphosphate

fertilizer to a sandy soil: effects on soil phosphorus status and profile distribution *Nutr. Cycl. Agroecosyst* **54** 233-42

[15] Shuai X F, Zhou Z J and Yost R S 2003 Using segmented regression models to fit soil nutrient and soybean grain yield changes due to liming *J. Agric. Biol .Environ Stat.* **8(2)** 240-52

[16] Tang X, Ma Y B, Hao X Y, Li X Y, Li J M, Huang S M and Yang X Y 2009 Determining critical values of soil Olsen-P for maize and winter wheat from long-term experiments in China *Plant Soil* **323** 143-51

[17] Wang B, Liu H, Wang X H, Li J M, Ma Y B and Ma X W 2015 Soil phosphorus accumulation model for an arid area of Northwest China with 3-year rotation of wheat, maize and cotton *J. Agr. Sci-Cambridge* **153(7)** 1247-56

[18] Wang B, Li J M, Ren Y, Xin J S, Hao X Y, Ma Y B and Ma X W 2015 Validation of a soil phosphorus accumulation model in the wheat-maize rotation production areas of China *Field Crop Res.* **178** 42-8

[19] Waugh D L, Cate R B and Nelson L A 1973 Discontinuous models for rapid correlation, interpretation, and utilization of soil analysis and fertilizer response data. Techn. Bull. 7 *International Soil Fertility Evaluation and Improvement Program* North Carolina State University, Raleigh, NC

[20] National bureau of statistics of the People's Republic of China edit 2015 *China Statistical Yearbook* (Beijing: China Statistics Press)

On-line Vis-Nir sensor determination of soil variations of sodium, potassium and magnesium

Y Tekin[1, 4], Z Tümsavaş[2], Y Ulusoy[1] and A M Mouazen[3]

[1] Vocational School of Technical Sciences. Uludag University, Bursa, Turkey

[2] Agricultural Faculty. Uludag University, Bursa, Turkey

[3] Cranfield Soil and AgriFood Institute. Cranfield University, United Kingdom

E-mail: ytekin@uludag.edu.tr

Abstract. Among proximal measurement methods, visible and near infrared (Vis-Nir) spectroscopy probably has the greatest potential for determining the physico-chemical properties of different natural resources, including soils. This study was conducted to determine the sodium, potassium and magnesium variations in a 10. Ha field located in Karacabey district (Bursa Province, Turkey) using an on-line Vis-Nir sensor. A total of 92 soil samples were collected from the field. The performance and accuracy of the Na, K and Mg calibration models was evaluated in cross-validation and independent validation. Three categories of maps were developed: 1) reference laboratory analyses maps based on 92 points 2) Full-data point maps based on all 6486 on-line points Vis-Nir predicted in 2013 and 3) full-data point maps based on all 2496 on-line points Vis-Nir predicted in 2015. Results showed that the prediction performance in the validation set was successful, with average R2 values of 0.82 for Na, 0.70 for K, and 0.79 for Mg, average root mean square error of prediction (RMSEP) values of 0.02% (Na), 0.20% (K), and 1.32% (Mg) and average residual prediction deviation (RPD) values of 2.13 (Na), 0.97 (K), and 2.20 (Mg). On-line field measurement was also proven to be successful with validation results showing average R2 values of 0.78 (Na), 0.64 (K), and 0.60 (Mg), average RMSEP values of 0.04% (Na), 0.13% (K), and 2.19% (Mg) and average RPD values of 1.57 (Na) 1.68 (K) and 1.56 (Mg). Based on 3297 points, maps of Na, K and Mg were produced after N, P, K and organic fertilizer applications, and these maps were then compared to the corresponding maps from the previous year. The comparison showed a variation in soil properties that was attributed to the variable rate of fertilization implemented in the preceding year.

1. Introduction

In the last two decades, the number of studies evaluating other Vis-Nir spectroscopy applications in soil science and agronomy has increased rapidly, with a primary focus on measuring various basic properties of soils, such as the organic matter content, clay content and, more recently, chemical properties [1]. Current studies have shown that Vis-Nir spectroscopy is capable of providing accurate quantification of the main physical and chemical soil properties, and that it is a useful tool for digital soil mapping and for precision agriculture applications [2-5].

There is a natural accumulation of sodium (Na) in soil as a result of fertilizers, runoff from shallow, salt-laden waters, irrigation water and the breakdown of minerals which release salt. The excess sodium in the soil is taken up by plant roots and can cause serious vitality problems an addition to disrupting the soil structure. Potassium (K), as one of the macronutrients, has many vital physiological, metabolic and biochemical functions in crop plants which increase their crop yield and quality. Potassium increases root growth, improves drought and cold resistance, affects the time of harvest, improves the availability of nitrogen and helps to increase resistance to disease. On the other hand, a potassium deficiency affects shoot and root growth [6]. Magnesium (Mg) is a secondary nutrient which maize and other crops require to capture solar energy for growth and production through photosynthesis. Magnesium is an essential component of chlorophyll molecules, each of which contains 6.7% Mg. It has been suggested that the interaction of Mg with common bean varieties has a significant effect on the crop yield, foliar content, and yield components [7]. The conventional analytical methods used for the determination of Na, K and Mg are expensive, complex and time-consuming. Consequently, researchers have been attempting to find alternative solutions that are fast, simple and cost-effective. Vis-Nir spectroscopy is one of the main methods that have been explored. This can be attributed to the fact that, by using suitable chemometric methods, large sets of spectral information can be extracted from the Vis-Nir spectra of soils. The complex relationship between spectral signatures and soil properties can be better modeled via multivariate regression methods, which have an advantage over the simple bivariate relationships, e.g., those based on peak intensity measurements [8]. Partial least squares (PLS) regression is the most common technique adopted today to model the relationships between the infrared spectral intensity characteristics of the soil components and the soil properties through derived PLS loadings, scores, and regression coefficients [9]. The PLS regression establishes a series of components or latent vectors that provide a simultaneous reduction or decomposition of X and Y such that these components explain, as much as possible, the covariance between X and Y [10]. One of the advantages of PLS regression compared to other chemometric methods, such as principal component regression analysis, is the possibility of interpreting the first few latent variables, because these show the correlations between the property values and the spectral features [11]. The calibration samples should cover the variability expected in the full sample set, and the future unknown data and the validation (test) set must be independent of the calibration set in order to avoid an optimistic assessment of predictive performance [8, 12, 13].

The aim of this study was to explore the potential of a Vis-Nir on-line sensor to measure Na, K and Mg. Laboratory-measured, laboratory Vis-Nir-predicted and on-line Vis-Nir-predicted maps were produced and used with independent validation sample sets. Based on 3297 points, maps of Na, K and Mg were produced after N, P, K and organic fertilizer applications, and these maps were then compared to corresponding maps from the previous year. This comparison showed variations in soil properties that were attributed to the variable rate of fertilization implemented in the preceding year.

2. Material and methods

2.1. On-line sensor

A simple metal frame for the on-line sensor was manufactured at Uludag University using the patented design [14] of A.M. Mouazen (Figure 1). The optical unit was attached to the backside of the subsoiler chisel in order to acquire soil spectra from the smooth bottom of the trench in the diffuse reflectance mode. The subsoiler (acting as a soil-cutting tool) and the optical probe were set on the metal frame. The on-line soil sensor was then mounted on the three-point linkage of a tractor for collecting soil spectra under mobile conditions. The sensor was equipped with an AgroSpec mobile, fiber-type vis–NIR spectrophotometer (Tec5 Technology for Spectroscopy, Germany) to measure the soil spectra. A differential global positioning system (DGPS) (EZ-Guide 250, Trimble, USA) was used to record the position of the on-line-measured spectra with sub-meter accuracy. The frame and the on-line sensor were tested at the Uludag University farm before performing the actual field measurements in order to avoid unexpected malfunctions of either software or cable connections during the field measurements.

Figure 1. The on-line Vis-Nir soil sensor, as designed by Mouazen [14].

2.2. Experimental site

Karacabey and its environs lie in one of the most active tectonic areas of Turkey due to their location on the Northern Anatolia Fault Zone. This fault was a primary cause of important geomorphic environmental changes in this area during the Quaternary period. The Karacabey district has many creeks and a depression basin. The parallel and/or subparallel drainage patterns of the creeks that reach the Karacabey basin from the north have evolved as a result of the lithological character of the area.

This study was carried out in an irrigated field of 10 ha, cultivated for sweet corn (maize), having soil classified as clay-type (sand 26%, silt 30.4%, clay 43%) and located in a semiarid environment (Figure 2). This field was chosen because of the soil variability of different zones according to yield. This field was commissioned after the crop was harvested. In past years, generally, fertilizers composed of N, P and K were employed on the field. For laboratory analysis, a total of 92 soil samples were collected from the field, respectively, from the bottom of the trench opened by the subsoiler (Figure 2a). In the year 2013, raw spectra were collected along parallel transects at a speed of approximately 3 km h^{-1} (Figure 2b). Raw spectra on soil were collected along with parallel transects at a speed of approximately 3 km h^{-1}. The sampling positions were recorded with the DGPS. Sampling lines and sampling positions were applied again in 2015 (Figure 2c).

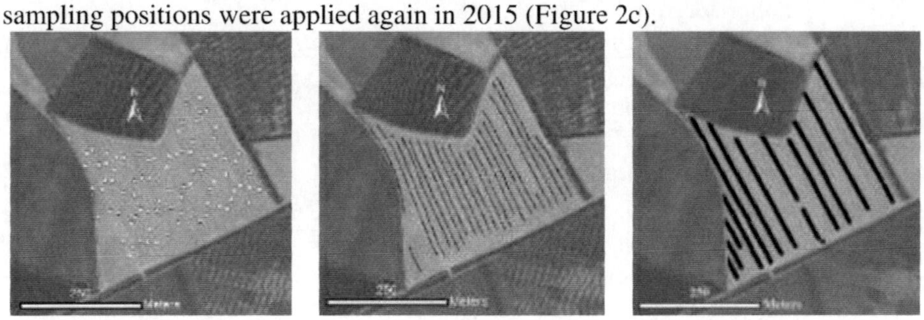

(a) (b) (c)

Figure 2. Field location in Karacabey, Turkey: (a) soil sampling positions, (b) on-line soil measurement points (2013), and (c) on-line soil measurement points (2015).

2.3. Laboratory analyses

The 92 soil samples (Figure 2a) were equally divided into two parts. The first half was used for laboratory reference measurements of Na, K, Mg and particle size distribution (PSD), and the second half was used for optical scanning in the laboratory. Exchangeable Na, K and Mg ions were determined after extraction with 1 N ammonium acetate (pH 7.0). The PSD of the soil was measured using a hydrometer [15]. Soil texture classification was determined according to the United State Department of Agriculture system. Soil samples were scanned in the laboratory using the same Vis-Nir spectrophotometer employed during the on-line field measurements. Each sample was put into three plastic cups (1.2 cm deep and 1.2 cm in diameter) and carefully levelled to form a smooth scanning surface [2]. A white reference was scanned before the soil scanning, which was repeated every 30 min. Each cup was scanned 10 times, and the readings were averaged. The final spectrum for each sample, to be used for further analysis, was the average of the three spectra obtained for the three cups.

2.4. Pre-processing of spectra

First, soil spectra of 305-370 and 2151-2200 nm were deleted to eliminate noise at the edges of each spectrum. Then, the spectra between 371 and 1000 nm were reduced by 3, and between 1001 and 2150 nm reduced by 6. Maximum normalization was followed, which is typically used to get all data to approximately the same scale, or to get a more even distribution of the variances and the average values. The spectra were then subjected to the Savitzky-Golay [16] first derivation to enable the computation of the first or higher order derivatives, including a smoothing factor, which determined how many adjacent variables would be used to estimate the polynomial approximation for derivatives. The second order polynomial approximation was selected. A 2:2 smoothing was carried out after the first derivative [17]. The pre-processing of the spectra and PLS regression with one-leave-out cross validation were carried out using the calibration set in order to develop a calibration model via Unscrambler 7.8 software (Camo Inc., Oslo, Norway).

2.5. Modeling

Detailed statistical information about the laboratory-measured chemical analyses is provided in Table 1.

Table 1. Sample statistics of laboratory and on-line-measured chemical properties of the field calibration and prediction sets.

Nutrients	Sample sets	Sample number	Min.	Max.	Mean	SD
Na, ppm	All samples	92	0.26	0.57	0.45	0.07
	Cross-validation set	74	0.28	0.56	0.45	0.06
	Laboratory prediction set	18	0.3	0.58	0.44	0.07
	On-line prediction set	18	0.27	0.50	0.41	0.06
K, cmol kg^{-1}	All samples	92	0.39	1.79	1	0.25
	Cross-validation set	207	0.07	1.60	0.77	0.32
	Laboratory prediction set	18	0.39	1.52	0.88	0.28
	On-line prediction set	18	0.6	1.46	0.97	0.22
Mg, cmol kg^{-1}	All samples	92	5.53	18.45	11.37	2.80
	Cross-validation set	73	5.53	18.45	11.29	2.95
	Laboratory prediction set	18	5.54	16.94	11.10	3.33
	On-line prediction set	18	5.54	17.11	10.70	3.43

SD: Standard deviation

The performance and accuracy of the chemical property calibration model was evaluated in cross-validation and prediction. The model performance was evaluated by means of the coefficient of determination (R^2), root mean square error of prediction (RMSEP) and ratio of prediction deviation (RPD), which is the standard deviation divided by the RMSEP.

2.6. Fertilizer applications

Amount of N, P_2O_5, K_2O (kg) applied for the entire field are shown in Table 3.

Table 2. Total amount of N, P_2O_5, K_2O application for the entire field.

Years	N(kg)	P_2O_5 (kg)	K_2O (kg)
2013	3264	712	950
2014	9740	2710	0
2015	8700	6600	2100

2.7. Map development

All maps were developed using ArcGis 10 (ESRI, USA) software. Three maps were used for the comparison of chemical properties. The first one was for the laboratory measurement points based on 92 soil samples measured in 2013. The second one was for the on-line measurement in 2013 based on 6486 points. The inverse distance weighing (IDW) interpolation method was used to develop the laboratory-measured maps. IDW method is based on the extent of similarity of cells, while methods, such as trend fitting of a smooth surface, are defined by mathematical function. Robinson and Metternicht [18] concluded that interpolation methods give similar RMSEP values, using the cross validation technique for evaluation. The full-point maps were developed via the Kriging interpolation. Kriging is a statistical method used in diverse application modeling. Kriging is most appropriate when you know there is a spatially correlated distance or directional bias in the data.

3. Results and discussion

3.1. Model performance

The model performances in cross-validation, laboratory and on-line predictions of chemical properties for the field are shown in Table 4. Viscarra Rossel et al., [19] classified RPD values as follows: RPD < 1.0 indicates very poor model predictions and their use is not recommended; RPD between 1.0 and 1.4 indicates poor model predictions, where only high and low values are distinguishable; RPD between 1.4 and 1.8 indicates fair model predictions, which may be used for assessment and correlation; RPD values between 1.8 and 2.0 indicates good model predictions, where quantitative predictions are possible; RPD between 2.0 and 2.5 indicates very good, quantitative model predictions; and RPD > 2.5 indicates excellent model predictions. This classification system was adopted in this study.

Table 3. Summary of model performances in cross-validation, laboratory and on-line predictions of chemical properties.

Nutrients	Sample sets	R^2	RMSEP	RPD	Slope
Na	Cross-validation	0.82	0.02	2.13	0.99
	Laboratory prediction	0.82	0.03	2.33	0.92
	On-line prediction	0.78	0.04	1.57	0.85
K	Cross-validation	0.70	0.20	0.97	0.96
	Laboratory prediction	0.80	0.12	2.27	0.74
	On-line prediction	0.64	0.13	1.68	0.70
Mg	Cross-validation	0.79	1.32	2.20	0.69
	Laboratory prediction	0.88	1.14	2.90	0.98
	On-line prediction	0.60	2.19	1.56	0.60

RMSEP: Root mean square error of prediction
RPD: Residual prediction deviation

Literature [20-21] proves that the worst properties to be measured with NIR are K and Na. Measurement of pH, Ca, and Mg were reported to be more successful as compared to K and Na, but underperformed those properties with direct spectral response in NIR [22]. Chang et al., [23] concluded that exchangeable Na could not be predicted using the NIR spectroscopy technique since $R^2<0.50$. On the contrary, in this research R^2 was found 0.82 while RPD is 2.13 for Na in cross validation.

The calibration set of the field consisted of 74 randomly selected soil samples, whereas the remaining 18 samples were used as the Na prediction set. Figure 3 shows the scatter plots of measured versus predicted Na in cross- validation, laboratory validation and on-line validation.

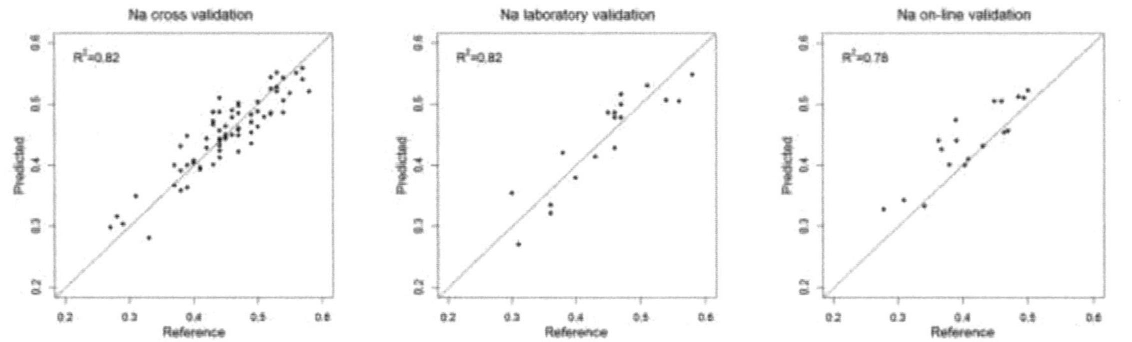

(a) (b) (c)

Figure 3. The scatter plots of measured versus predicted Na for laboratory and on-line predictions.

The Vis–NIR predicted K was chosen as the exhaustive variable, because it was the most correlated with the soil variables [24]. 72 samples from the field were pooled together in one matrix with other 135 samples, collected previously from UK (62 samples), Denmark (28 samples), from Czech Republic (22 samples), Netherland (23 samples). The remaining 18 samples were used for validation of the on-line measurement. Figure 4 shows the scatter plots of measured versus predicted K in cross-validation, laboratory validation and on-line validation.

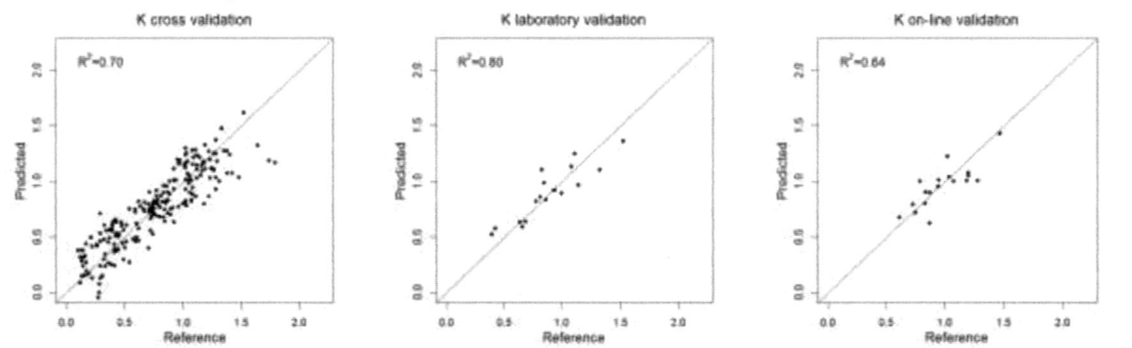

(a) (b) (c)

Figure 4. The scatter plots of measured versus predicted Na for laboratory and on-line predictions.

Udelhoven et al. [25] evaluated the ability of NIR spectroscopy to estimate soil Fe, Mn, Ca, Mg, and K, and they found that only Ca and Mg (R^2 =0.67 and 0.69, respectively) were predictable under in situ conditions. A total of 92 samples were collected from F1. 74 samples of 92 were designated for Cross-validation. Figure 5 shows the scatter plots of measured versus predicted Mg in cross-validation, laboratory validation and on-line validation.

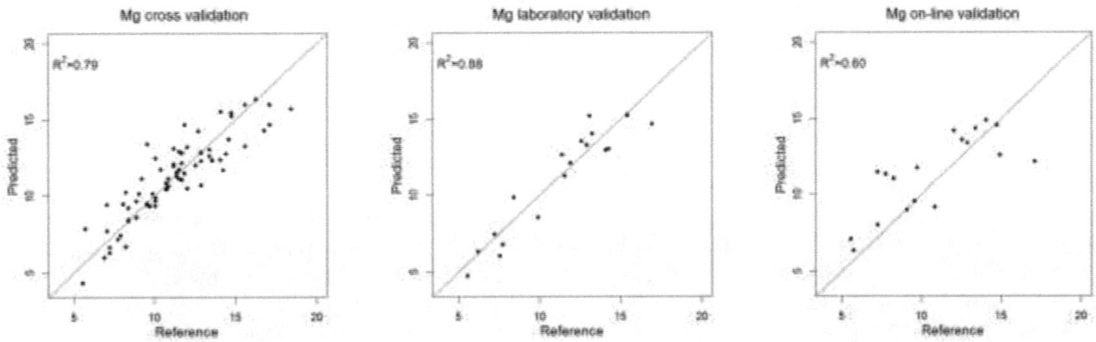

(a) (b) (c)

Figure 5. The scatter plots of measured versus predicted Na for laboratory and on-line predictions.

3.2. Mapping

As can be seen in Figure 6, the Na content of the soil was increased in 2015. The reason for this was that the ratio Na in the irrigation water had been high for the two previous years.

(a) (b) (c)

Figure 6. Na Comparison maps: (a) laboratory-measured, based on 92 samples, (b) on-line Vis-Nir predicted (6486 points) measured in 2013, and (c) on-line Vis-Nir predicted (2496 points) measured in 2015.

Figure 7 shows K variation maps. K content of the soil has been decreased in 2015. The reason can be attributed that two years corn grain were grown. Corn takes up high amount of K from the soil.

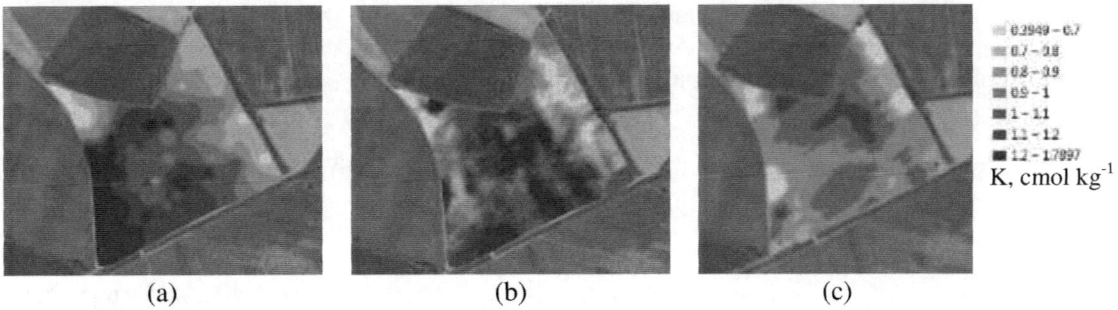

(a) (b) (c)

Figure 7. K Comparison maps: (a) laboratory-measured based on 92 samples, (b) on-line Vis-Nir predicted (6486 points) measured in 2013, and (c) on-line Vis-Nir predicted (2496 points) measured in 2015.

Magnesium is one of thirteen mineral nutrients coming from soil that, when dissolved in water, is absorbed through plant roots. Magnesium is the main component of the chlorophyll that gives green color to the leaves; therefore, it is a very important element for chlorophyll formation. Heavy rains can cause a Mg deficiency by leaching the Mg out of sandy or acidic soil. In addition, if soil contains high amounts of K, plants may take this up instead of Mg, again, leading to a deficiency. Figure 5 shows the Mg variation maps. In the spring of 2015, before the measurements were taken, there were heavy rains in Karacabey region. According to the Turkish Meteorological Station, the Marmara Region received 25% more rainfall than average at that time.

(a) (b) (c)

Figure 8. Mg Comparison maps: (a) laboratory-measured based on 92 samples, (b) on-line Vis-Nir predicted (6486 points) measured in 2013, and (c) on-line Vis-Nir predicted (2496 points) measured in 2015.

4. Conclusion

This study described the potential of visible and near infrared (Vis-Nir) spectroscopy for the measurement of soil Na, K and Mg in fields with clay soils in a semiarid environment in Turkey. The results were evaluated under laboratory and on-line field measurement conditions. It was concluded that:

1- Vis-Nir spectroscopy can be successfully used for the measurement of Na, K and Mg in clay soils under semiarid conditions.

2- The Vis-Nir prediction maps of soil Na, K and Mg were similar to the corresponding measured maps. However, with the full-point maps, more details showing different spatial distribution were observed compared to the maps developed with a limited number of points.

Acknowledgment

The research leading to these results received funding from various sources. The authors are grateful for the funding received for the FarmFUSE project from the ICT-AGRI (ERA-NET scheme of the European Commission under the 7th Framework Programme), the Scientific and Technological Research Council of Turkey (TUBITAK, contract no: 112O471) and the UK Department of Environment, Food and Rural Affairs (contract no: IF0208).

References

[1] Stenberg B, Viscarra R R A, Mouazen A M and Wetterlind J 2010 Visible and Near Infrared Spectroscopy in Soil Science 2. In Advances in Agronomy chap. 5 ed D L Sparks (London: Academic Press)

[2] Mouazen A M, Maleki M R, De Baerdemaeker J and Ramon H 2007 On-line measurement of some selected soil properties using a VIS–NIR sensor *Soil and Tillage Research* **93**(1) 13-27

[3] Rossel R V and Chen C 2011 Digitally mapping the information content of visible–near infrared spectra of surficial Australian soils *Remote Sensing of Environment* **115**(6) 1443-55

[4] Tekin Y, Kuang B and Mouazen A M 2013 Potential of on-line visible and near infrared spectroscopy for measurement of pH for deriving variable rate lime recommendations *Sensors* **13**(8) 10177-90

[5] Kodaira M and Shibusawa S 2013 Using a mobile real-time soil visible-near infrared sensor for

high resolution soil property mapping *Geoderma* **199** 64-79

[6] Cakmak I, Hengeler C and Marschner H 1994 Partitioning of shoot and root dry matter and carbohydrates in bean plants suffering from phosphorus, potassium and magnesium deficiency *J. of Experimental Botany* **45**(9) 1245-50

[7] Canizella B T, Moreira A, Moraes L A C and Fageria N K 2015 Efficiency of Magnesium Use by Common Bean Varieties Regarding Yield, Physiological Components, and Nutritional Status of Plants *Communications in Soil Science and Plant Analysis* **46**(11) 1376-90

[8] Soriano-Disla J M, Janik L J, Viscarra R R A, Macdonald L M and McLaughlin M J 2014 The performance of visible, near-, and mid-infrared reflectance spectroscopy for prediction of soil physical, chemical, and biological properties *Applied Spectroscopy Reviews* **49**(2) 139-86

[9] Janik L J, Forrester S T and Rawson A 2009 The prediction of soil chemical and physical properties from mid-infrared spectroscopy and combined partial least-squares regression and neural networks (PLS-NN) analysis *Chemometrics and Intelligent Laboratory Systems* **97**(2) 179-88

[10] Summers D, Lewis M, Ostendorf B and Chittleborough D 2011 Visible near-infrared reflectance spectroscopy as a predictive indicator of soil properties *Ecological Indicators* **11**(1) 123-31

[11] Yang H and Mouazen A M 2012 Vis/near and mid-infrared spectroscopy for predicting soil N and C at a farm scale *Infrared Spectroscopy—Life and Biomedical Sciences* 185-210

[12] Dardenne P, Sinnaeve G and Baeten V 2000 Multivariate calibration and chemometrics for near infrared spectroscopy: which method? *Journal of Near Infrared Spectroscopy* **8**(4) 229-38

[13] Brown D J, Bricklemyer R S and Miller P R 2005 Validation requirements for diffuse reflectance soil characterization models with a case study of VNIR soil C prediction in Montana *Geoderma* **129**(3) 251-67

[14] Mouazen A M 2006 Soil Survey Device. International publication published under the patent cooperation treaty (PCT). World Intellectual Property Organization, International Bureau. International Publication Number: WO2006/015463, PCT/BE2005/000129, IPC: G01N21/00, G01N21/00

[15] Gee G W and Bauder J W 1986 Particle size analysis ed A Klute *Methods of Soil Analysis* Part 1 2nd edn. Madison WI:Argon Monogr 9 ASA pp 383–411

[16] Savitzky A and Golay M J 1964 Smoothing and differentiation of data by simplified least squares procedures *Analytical chemistry* **36**(8) 1627-39

[17] Mouazen A M, Kuang B, De Baerdemaeker J and Ramon H 2010 Comparison among principal component, partial least squares and back propagation neural network analyses for accuracy of measurement of selected soil properties with visible and near infrared spectroscopy *Geoderma* **158**(1) 23-31

[18] Robinson T P and Metternicht G 2006 Testing the performance of spatial interpolation techniques for mapping soil properties *Computers and Electronics in Agriculture* **50**(2) 97-108

[19] Rossel R V, McGlynn R N and McBratney A B 2006 Determining the composition of mineral-organic mixes using UV–vis–NIR diffuse reflectance spectroscopy *Geoderma* **137**(1) 70-82

[20] Mouazen A M, De Baerdemaeker J and Ramon H 2006 Effect of wavelength range on the measurement accuracy of some selected soil constituents using visual-near infrared spectroscopy *Journal of Near Infrared Spectroscopy* **14**(3) 189

[21] Cozzolino D and Moron A 2003 The potential of near-infrared reflectance spectroscopy to analyse soil chemical and physical characteristics *The Journal of Agricultural Science* **140**(01) 65-71

[22] Kuang B, Mahmood H S, Quraishi M Z, Hoogmoed W B, Mouazen A M and van Henten E J 2012 4 Sensing Soil Properties in the Laboratory, In Situ, and On-Line: A Review *Advances in Agronomy* **114**(1) 155-223

[23] Chang C W, Laird D A, Mausbach M J and Hurburgh C R 2001 Near-infrared reflectance spectroscopy–principal components regression analyses of soil properties *Soil Science Society of America Journal* **65**(2) 480-90

[24] Shaddad S M, Madrau S, Castrignanò A and Mouazen A M 2016 Data fusion techniques for delineation of site-specific management zones in a field in UK *Precision Agriculture* **17** 200-17

[25] Udelhoven T, Emmerling C and Jarmer T 2003 Quantitative analysis of soil chemical properties with diffuse reflectance spectrometry and partial least-square regression: A feasibility study *Plant and soil* **251**(2) 319-29

Relationship in between Chemical Oxidation and Browning of Flavanols

X Dong, Y L Zhang, F Wang, M X Pang and J H Qi*

School of food science and Engineering, Beijing University of Agriculture, Beijing102206, China

Detection and Control of Spoilage Organisms and Pesticide Residue in Agricultural Products, Beijing Laboratory of Food Quality and Safety, Beijing University of Agriculture, Beijing 102206, China

Pig processing team of modern agricultural industry technology system Beijing innovation team

Email: abc960718@sina.com

Abstract. Catechin, epicatechin and chlorogenic acid are widely distributed in the plant kingdom. At present, influencing factors of phenol chemical oxidation is little research. In order to study non-enzymatic browning factors, this research utilized catechin, epicatechin and chlorogenic acid to establish simulation systems. The browning degree and products of flavanols were investigated by transmittance and high-performance liquid chromatography (HPLC). The main results and conclusions were follows. The brown generation is increased after phenols of chemical oxidation at 50℃ and at pH 3.7 phosphate buffered saline, the sequence of influencing factor of browning is pH ＞ kind of phenol ＞ temperature. Oxidation of compounds of catechin and epicatechin results in formation of their Methylene quinone or *o*-Quinones. In addition, oxidation products of catechin, epicatechin and chlorogenic acid were mixture of different molecular sizes. The research has showed that brown generation correlated well with chemical oxidation of phenols and chemical oxidation reaction generated larger molecular weight polymers.

Keywords: catechin, epicatechin, chlorogenic acid, browning, chemical oxidation

1. Introduction

Polyphenols are broadly distributed in the plant kingdom, especially fruits and vegetables. Phenolic oxidation mainly includes chemical oxidation and enzymatic oxidation. In fact, sterilization and heat treatment of fruit and vegetable processing have made phenoloxidase passivation, so for processed fruits and vegetables, phenolic reaction mainly is chemical oxidation. The research only is applicable to oxidation of phenols that have a greater influence on the browning of fruits and vegetables.

Catechin, epicatechin and chlorogenic acid are widely distributed in the plant kingdom. Recently, lots of researches have been showed that phenolic compounds largely contribute to oxidative browning of fruits and vegetables. The phenol oxidation polymerization is the cause of producing browning of lotus root [1], and the results showed that pH had the biggest biggest effect on browning of polyphenols, followed by kind of polyphenols, the temperature had less effect on browning [2]. Phenolic compounds are directly related to the browning phenomenon of white wines [3]. The study found that catechin and epicatechin play crucial role at the early stages of cider making [4]. The degree of browning chlorogenic acid is associated with the browning of fruit juice [5]. Catechin, epicatechin and chlorogenic acid were more potent than others in fruit wine in enhancing browning [6-8]. However, at present, influencing factors of phenol chemical oxidation is little research. This research utilized catechin, epicatechin and chlorogenic acid to develop model systems in order to study the influence factors of phenols oxidation on browning degree. In this way, it can help to provide theoretical support for better control of browning of fruit and vegetable products in the future.

2 .Materials and methods

2.1 Chemical oxidation

Catechin (Sigma), epicatechin (Sigma) and chlorogenic acid (Sigma) were dissolved in phosphate buffer solution at different pH 3.7, 7, 10 (0.02M) by heating on a steam bath. After the mixture was filtered with 0.45 μm membrane filters, 20 mL of phenols solution at different pH's Phosphate buffer solution was added to test tube with plug. The solution was then spared with sufficient oxygen with oxygen cylinder. The test tube were kept at the a water bath with different temperature.

In the experiment, oxidation of catechin, epicatechin, chlorogenic acid and mixture of epicatechin and chlorogenic acid were carried out at pH 3.7, 7, 10 and at temperature 37, 50 and 90℃. Phenols oxidation reaction were investigated according to $L_9\ (3)^4$ orthogonal design (Table 1).

Table 1. $L_9\,(3)^4$ orthogonal design of chemical oxidation about phenolics.

	pH	temperature(℃)	Phenols
1	3.7	37	epicatechin
2	7	50	chlorogenic acid
3	10	90	catechin

2.2 Sampling

The reaction was terminated by adding 0.5 μL of concentrated sulfuric acid to 0.75 mL sample. To prevent precipitation and any loss of the oxidation products, 0.75 mL of 100% acetonitrile was added and the vial sealed.

2.3 HPLC

Shimadzu SCL-10AVP HPLC (SHIMADZU, Japan) was used for analytical separation. Column temperature was set at 40℃. The volume injected was 100 μL. The wavelengths 280 nm was monitored simultaneously, and scanning was done from 230-650 nm by spectrophotometer (721B, China). The mobile phase was filtered with 0.45 μm membrane filters and degassed. Mobile phase A was 40% acetonitrile in double distilled water. A linear gradient was run from 5%-100% mobile phase A. Mobile phase B was 5% acetic acid. A linear gradient was run from 95%-0% mobile phase B in 30 min.

2.4 Browning

Transmittance was measured at 420 nm with double beam UV-visible light spectrophotometer. The experiment was repeated 3 times.

3. Results and discussion

3.1 The influence factors of phenol chemical oxidation

Table 2. Results and analysis of $L_9(3)^4$ orthogonal design of chemical oxidation about phenolics.

	pH	temperature(℃)	phenols	T_{420nm}(%)
1	3.7	37	epicatechin	96.30
2	3.7	50	chlorogenic acid	98.20
3	3.7	90	catechin	96.00
4	7	37	chlorogenic acid	97.90
5	7	50	catechin	96.00
6	7	90	epicatechin	66.70
7	10	37	catechin	16.40
8	10	50	epicatechin	65.10
9	10	90	chlorogenic acid	80.80
Range(browning degree)	42.1	16.23	22.83	

Xidizing reaction continued for 5 h. Phenolate anion concentrations was 200 mg/L

Oxidation of catechin, epicatechin, chlorogenic acid were carried out separately at pH 3.7, 7, 10 and at temperature 37, 50 and 90℃ (Table 2). Table 2 shows that the sequence of influencing factor of browning is pH>kind of phenols>temperature. This phenomenon is in conformity with previous works [2]. The intensity of brown produced was highest at pH 10. The intensity of brown produced was lowest at 50℃ with the largest T_{420nm}. The effect of oxidation of different phenols on browning was different. Intensity of brown produced was higher after catechin and epicatechin oxidation than chlorogenic acid. The means were compared by analysis of variance at a significance level of 0.05.The analysis of the effects of pH, temperature and kind of phenol on browning. The results showed that only pH has a significant influence (F=3.135>0.117) on the browning. In this experiment the browning degree was investigated by measuring T_{420nm}, the greater the transmittance, the lighter the browning.

3.2 Effect of pH on browning degree of chemical oxidative products

In order to study effect of pH on browning degree of chemical oxidative products, oxidation of catechin, epicatechin, chlorogenic acid were carried out separately in various values of pH at 50℃ (Figure 1, Figure 2, Figure 3). It can be observed that the amount of pigmentation produced was higher at pH 10, 7 with small T_{420nm} after oxidizing catechin and epicatechin. For chlorogenic acid, browning degree was bigger at pH 10 than pH 7, 3.7 significantly. This suggests that phenolic compounds of chemical oxidation is more likely to occur in alkaline solution. This phenomenon is in conformity with Cilliers and Singleton's works [9]. So we can think that the effect of the phenol chemical oxidation on the browning of fruit juice (pH 3-4) is small.

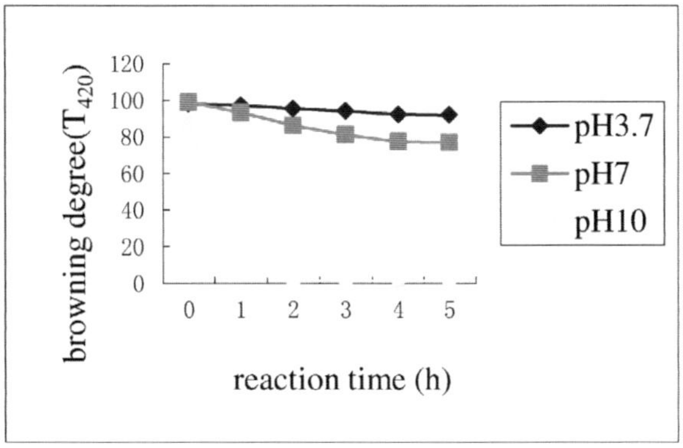

Figure 1. Browning degree of chemical oxidative products by catechin (500 mg/L) in various values of pH（50℃）.

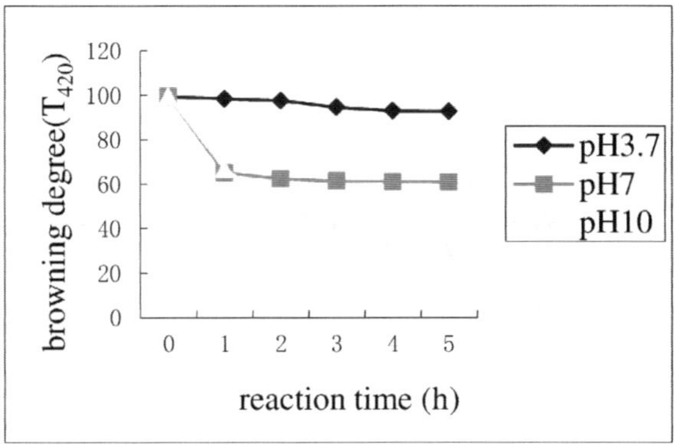

Figure 2. Browning degree of chemical oxidative products by epicatechin (500 mg/L) in various values of pH（50℃）.

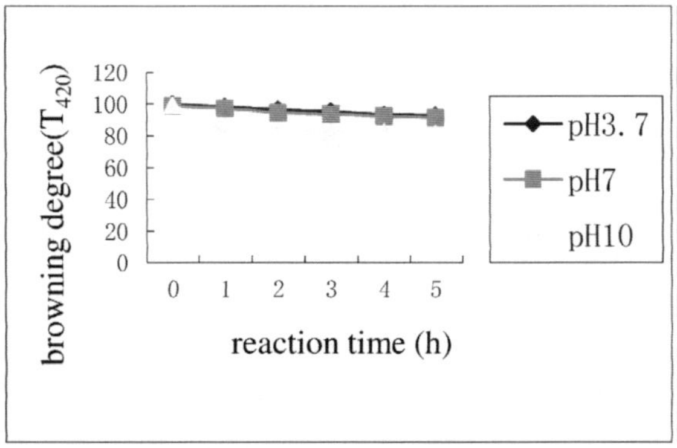

Figure 3. Browning degree of chemical oxidative products by chlorogenic acid (500 mg/L) in different pH（50℃）.

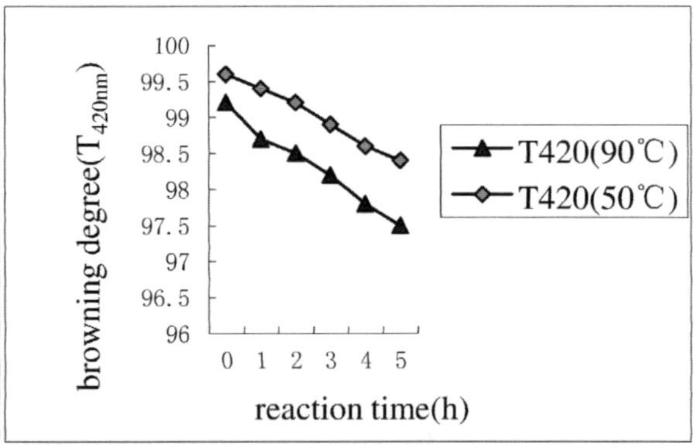

Figure 4. Browning degree of chemical oxidative products by 500mg/L catechin in various temperature (pH3.7).

3.3 Effect of temperature on browning degree of chemical oxidative products

In order to study effect of temperature on browning degree of chemical oxidative products, oxidation of catechin, epicatechin, chlorogenic acid were carried out separately in various temperature at pH 3.7. (Figure 4, Figure 5, Figure 6). It can be seen that with the extension of reaction time, browning degree of three kinds of phenols oxidation products is more serious at 90℃ than 50℃. This suggests that oxidation condensation reaction of phenols are more likely to occur at higher temperatures . Liu Jing studied the the influence of temperature on the jujube juice concentrate browning in the softening process. Liu Jing's research have found that total phenol content gradually decreases with the increase of temperature , the higher the temperature, the faster the decline[10].

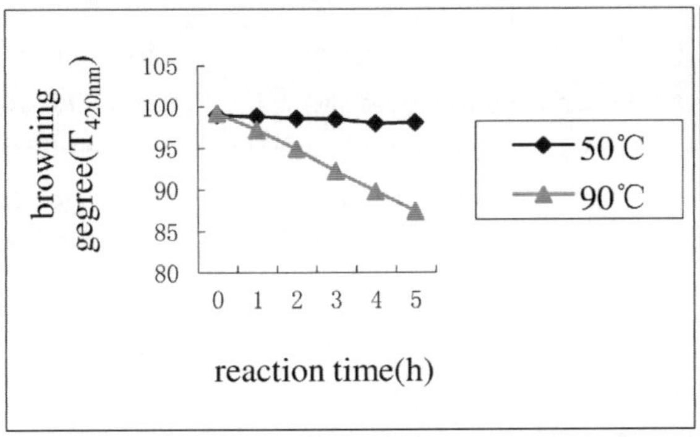

Figure 5. Browning degree of chemical oxidative products by 500mg/L epicatechin in various temperature（pH3.7）.

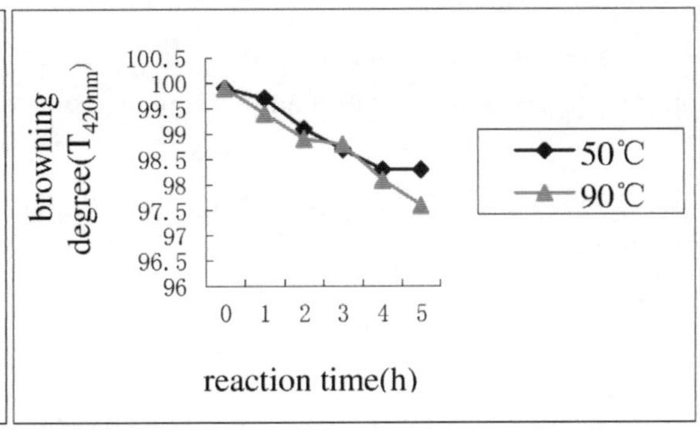

Figure 6. Browning degree of chemical oxidative products by 500 mg/L chlorogenic acid in various temperature（pH3.7）.

3.4 Chemical oxidation of phenol reaction rate

In order to indicate the rate of oxidation reaction of catechin, epicatechin and chlorogenic acid, phenolate anion concentrations was monitored at various reaction time by HPLC.

Table3 shows rate constants（$\times 10^{-5}$ h^{-1}）for the first-order reaction of catechin, epicatechin and chlorogenic acid in various values of pH (200 mg/L). It can be seen that the reaction rate of epicatechin was the fastest at 50℃ and pH 3.7, followed by catechin and then chlorogenic acid. The chemical oxidation rate of chlorogenic acid is relatively small. Oxidation of catechin, epicatechin and chlorogenic acid was faster at pH 10 than pH 3.7. This suggests that phenolic oxidation can occur rapidly under alkaline conditions. Reaction rate was higher at 90℃ than at 50℃, which is consistent with the change trend of browning.

Table 3. Rate Constants（$\times 10^{-5} h^{-1}$）for the First-Order Reaction of phenols in various values of pH .

	catechin		epicatechin		chlorogenic acid	
	50℃（r）	90℃（r）	50℃（r）	90℃（r）	50℃（r）	90℃（r）
pH3.7	33.05(0.99)	35.5(0.98)	35.62(0.94)	51.5(0.99)	28.96(0.92)	44.5(0.81)
pH10	74.15(0.97)		61.2(0.96)		75(0.85)	

number of data points used in calculating the first-order rate constant is 8. r=correlation coefficient of the straight line fitting.

3.5 Effect of oxidation of mixture on browning degree of products

3.5.1 Browning degree of mixed phenols. Browning rate of mixtures of chlorogenic acid and epicatechin is larger than pure 500 mg/L chlorogenic acid or 500 mg/L epicatechin by chemical oxidation at pH 3.7 and at 50℃ (Figure 7). The possible explanation is that oxidative condensation reaction is more likely to happen due to the interaction between different phenolics to produce brown pigment than pure phenol.

3.5.2 Reaction rate of mixed phenols. In chlorogenic acid-epicatechin mixtures, the reaction rate of epicatechin was much faster than pure epicatechin, whereas chlorogenic acid rate was slower than that of single oxidation (Table4). The possible explanation is that the rate of epicatechin is enhanced by chlorogenic acid quinone that chlorogenic acid oxidation generate, with reduction of the chlorogenic acid quinone back to chlorogenic acid. At the same time, it can also explain the phenomenon of the increase of browning in chlorogenic acid-epicatechin mixtures.

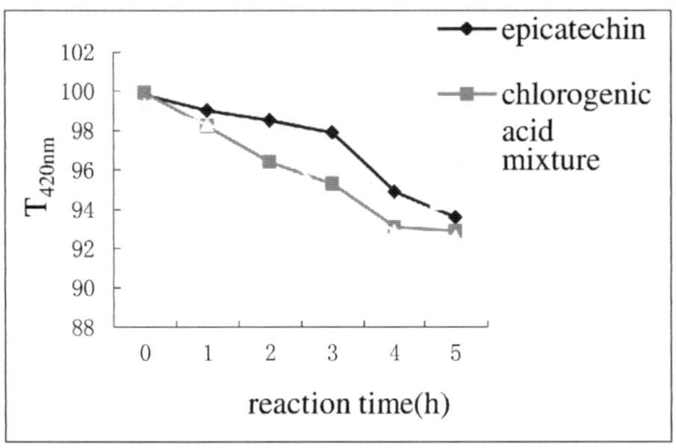

Figure 7. Browning rate of 500 mg/L chlorogenic acid, 500 mg/L epicatechin and mixture by chemical oxidation (pH3.7, 50℃).

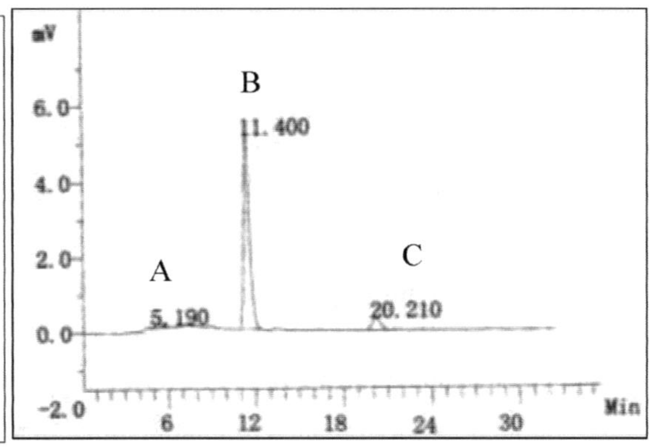

Figure 8. Chemical oxidative products of catechin（500 mg/L）at pH 3.7 at 50℃ after 12h in 100% oxygen.

Table 4. Rate Constants（$\times10^{-5}$ h^{-1}）for the First-Order Reaction of 500 mg/L chlorogenic acid, 500 mg/L epicatechin and the mixture.

pH	epicatechin (r)	chlorogenic acid (r)	mixture	
			epicatechin (r)	chlorogenic acid (r)
3.7	35.62(0.942)	28.96(0.923)	49.64(0.925)	19.75(0.938)

number of data points used in calculating the first-order rate constant is 8. r=correlation coefficient of the straight line fitting.

3.6 Research on phenols oxidation products

3.6.1 Research on chemical oxidative products of catechin. Chemical oxidative products of catechin at pH 3.7 at 50℃ after 12h in 100% oxygen was monitored by HPLC (Figure 8). Peak A has chromatographic elution time shorter than that of catechin (peak B). This would indicate that the oxidation products are more polar than catechin. Meanwhile, Peak C has chromatographic elution time longer than that of catechin. This would indicate that the oxidation products are less polar than catechin. These phenomenon demonstrate that oxidation products are mixture. Chemical oxidative products of catechin were scanned from 200-600 nm (Figure 9). We can think that oxidation products with absorbance at 305 nm have the structure of Methylene quinone. This indicate that catechin with absorbance at 280 nm can react with Methylene quinone to yield a condensation product. Oxidation products with a shoulder peak at 400-500 nm have the structure of *o*-Quinones. These data show that oxidation products have the structure of Methylene quinone and *o*-Quinones after catechin was oxidized, then these products have potential to react further with catechin and other phenols. Customarily, polymer molecular with molecular weight 500-3000 is known as condensed tannins, and polymer of larger molecular weight called phenolic acids [11].

3.6.2 Research on chemical oxidative products of epicatechin. Chemical oxidative products of epicatechin（500 mg/L）at pH 3.7 at 50℃ after 12h in 100% oxygen was monitored by HPLC (Figure 10). Peak A has chromatographic elution time shorter than that of epicatechin (peak B). In all other things being equal, the longer the time of peak flowing out, the bigger the molecular weight. This would indicate that the oxidation products are more polar and smaller molecular weight than epicatechin. Meanwhile, Peak C has chromatographic elution time longer than that of epicatechin. This would indicate that the oxidation products are less polar and bigger molecular weight than epicatechin. Oxidation products of epicatechin were scanned from 200-600 nm (Figure 11). From Figure 11, it can be concluded that oxidation products have the structure of Methylene quinone and o -Quinones after epicatechin was oxidized.

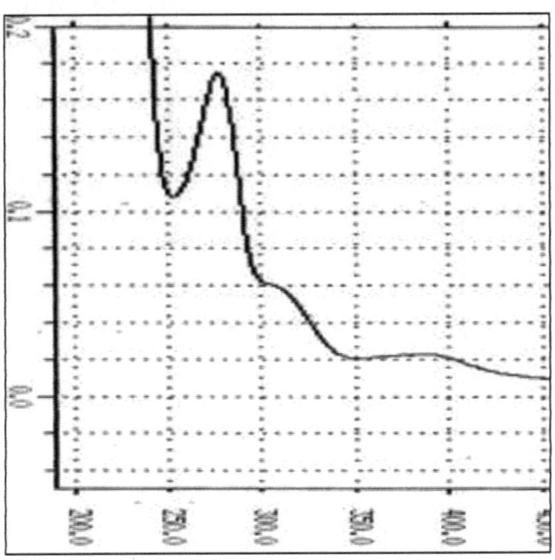

Figure 9. Absorbtion curve of chemical oxidative products of catechin（500 mg/L）at pH 3.7 at 50℃ after 12 h in 100% oxygen.

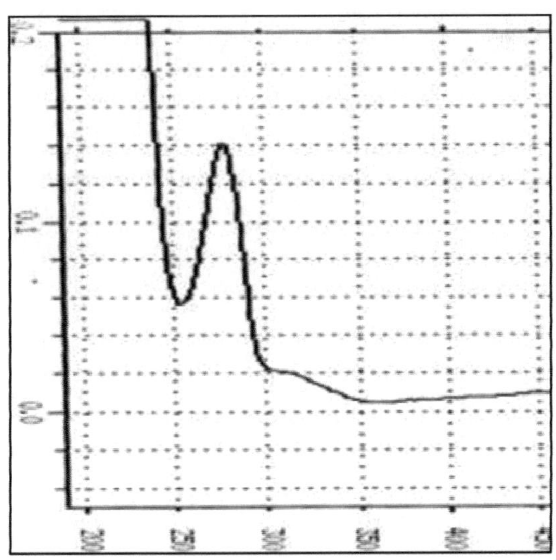

Figure 11. Absorbtion curve of chemical oxidative products of epicatechin（500 mg/L）at pH 3.7 at 50℃ after 12 h in 100% oxygen.

3.6.3 Research on chemical oxidative products of chlorogenic acid. More products were produced after chlorogenic acid was oxidized (Figure12). Peak A, B has chromatographic elution time shorter than that of chlorogenic acid (peak C). This would indicate that the oxidation products are more polar and smaller molecular weight than chlorogenic acid. Meanwhile, Peak D, E has chromatographic elution time longer. This would indicate that the oxidation products are less polar and bigger molecular weight than chlorogenic acid. Typical absorption peak was not found by scanning the each peak.

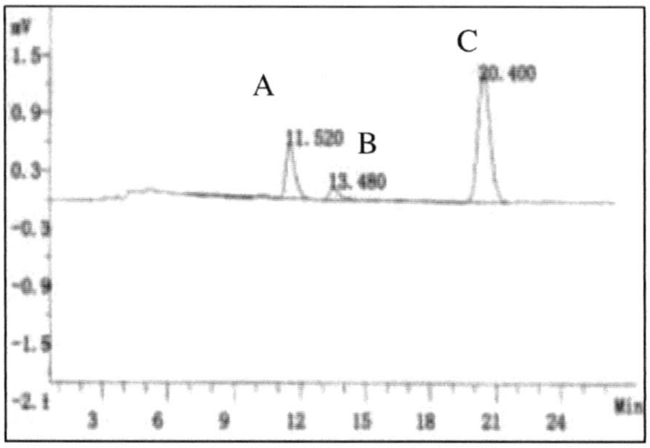

Figure 10. Chemical oxidative products of epicatechin(500 mg/L)at pH 3.7 at 50℃ after 12 h in 100% oxygen.

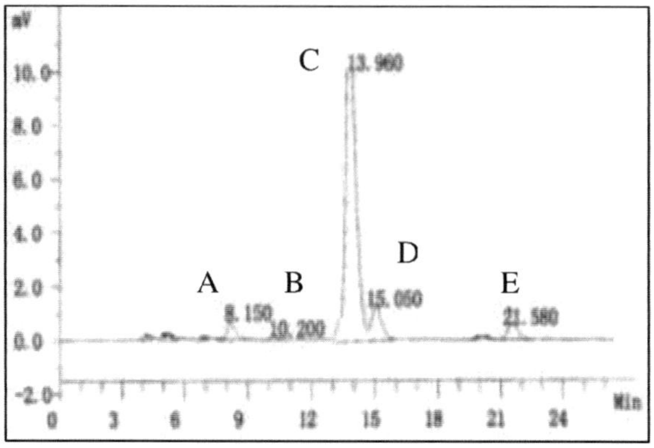

Figure 12. Chemical oxidation of chlorogenic acid(500 mg/L)at pH 3.7 at 50℃ after 12 h in 100% oxygen.

4. Conclusion

The brown generation is increased after phenols of chemical oxidation, the sequence of influencing factor of browning is pH > kind of phenol > temperature.

Phenols are more likely to occur chemical oxidation under alkaline conditions. Epicatechin had the fastest oxidation reaction rate than catechin and chlorogenic acid at pH 3.7 and at 50℃.

Catechin and epicatechin generated oxidative products with structure of methylene quinone and *o*-quinones after catechin and epicatechin were oxidized in phosphate buffer pH 3.7 and at 50 ℃.

The oxidative products of catechin, epicatechin and chlorogenic acid were the mixture with different molecular weight after occurring chemical oxidation in phosphate buffer pH 3.7 and at 50 ℃.

Acknowledgments

The authors thank Beijing pig innovation team for providing technical assistance and financial support.

References

[1] Hu Y 2008 Research on browning mechanism of boiled lotus root (Wuhan: Hua zhong Agricultural University)

[2] Hu Y and Chen Z J 2013 Simulation expeiment of chemical oxidation of polyphenols in Lotus *Hubei Agricultural Sciences* **13** 3128-30

[3] Angeles F R, Ana S M and Lourdes G M 2006 The effect of time and storage conditions on the phenolic composition and colour of white wine *Food Research International* **39** 220-9

[4] Liu W W 2006 Study on the oxidative polymerization of phenolic compounds in apple wine (Wuxi: Jiangnan University)

[5] Tada-aki, Wakayama and Chang Y L 1987 The influence of simple and condensed phenolics on the clarification of apple juice by honey *J. Sci. Food Agric.* **40** 275-81

[6] Chen J S, Yan Y H, Zhan J H and Tan Y L 2009 Research progress on phenolic compounds and non-enzymatic browning in fruit wine *Food science* **7** 281-4

[7] Wan L, Zhai H L, Hu Y and Yang Y H 2014 A Study of Non-enzymatic Browning of (+) – catechin in a Simulated Litchi Wine *Journal of South China Agricultural University* **01** 100-5

[8] Yan Z Z 2013 Research on browning mechanism and brewing technology of cherry wine (Taian: Shandong Agricultural University)

[9] Hocking M B and Intibar D J 1985 Oxidation of phenol by aqueous hydrogen peroxide catalysed by ferric ion-catechol complexes *J. Chem. Tech. Biotechnol.* **35A** 365-81

[10] Liu J 2011 Nonenzymatic browning in jujube juice concentrate during processing and storage (Xian: Northwest University)

[11] Shi B 2000 Plant polyphenols (Beijing: Science Press) p15

Optimization of ultrasound-assisted extraction of flavonoids compounds from *Chenopodium hybridum* L. stem with resp-onse surface methodology

Y Wu[1,3], H B Hu[1], C L Wang[2], S R Ma[2] and L L Zhang[1]

[1]College of Chemistry and Chemical Engineering, Longdong University, Qingyang, China

[2] College of Life Science and Technology, Longdong University, Qingyang, China

E-mail: 405436718@qq.com

Abstract. Ultrasound-assisted extraction (UAE) of flavonoids compounds (FC) from the stem of *Chenopodium hybridum* L.(*C.hybridum* L.) was investigated in this paper. Significant technological parameters were screened and optimized by using Plackett-Burman (PB) design, Steepest ascent method and Box-Behnken (BB) design, respectively. A mathematical model with high correlation coefficient (R^2=0.9896) was developed and showed good consistency between the experimental and predicted values. The optimum conditions for UAE were obtained by res- ponse surface methodology (RSM) as follows: volumn fraction of ethanol 76.62 %, extractive temperature 78.69℃, and liquid to solid ratio 58.43 for 30 min. Under these conditions, total flavo- noid content (TFC) of 9.4701 mg RE/100g were gained and it was closely related with predi- cted value (9.4640 mg RE/100g) and indicated the suitability of the developed model.

1. Introduction

Genus *Chenopodium* belonging to the family Chenppodiaceae includes herbaceous, strongly fragrant annual plants and spreads widely worldwide, mainly in the moderate and subtropical zone [1]. In China, there are 19 species of *Chenopodium*. Many species of *Chenopodium* were reported to possess numerous medicinal properties used as folk medicine. Modern pharmaceutical research has also confirmed potent antipruritic, antibacterial and anticancer activities of these plants [2, 3]. Recently, The plants belonging to *Chenopodium* were known to be a rich source of flavonoids, phenolic acids and terpenoids attracted special attention[4,5], among which flavonoids was one of the most effective and important components, because of their antoxidant, free radical scavenging properties and other biological activities[6,7]. Several studies have been carried out to evaluate *C.hybridum* L. which was rich in flavonoids in our previous work [8,9].

Ultrasound-assisted extraction promised to extract the bioactive components at lower temperature in a shorter time, and its operation more easily than other extraction technologies and methods [10,11]. These features could be due to the cavitation phenomenon generated in the ultrasonic propagation. To our knowledge, none of the published study reported on chemical composition and extraction flavonoids compounds (FC) from *C.hybridum* L.. Therefore, the purpose of present study was to establish optimum ultrasonic-assisted extraction technology for flavonoids from the stem of *C.hybridum* L.. In this paper, UAE of FC from the stem of *C.hybridum* L. was investigated and the significant technological parameters were screened and optimized by using Plackett-Burman (PB) design, Steepest ascent method and Box-Behnken (BB) design.

2. Material and method

2.1. *Plant material*

The stems of *C.hybridum* L. were collected from a local region (Qingyang, Gansu, China) in July 2014, and authenticated by the corresponding author. The stems were dried naturally and ground in a domestic grinder (Beijing ZTE Albert Instrument Co., Ltd., model FZ102) to pass through a 100-mesh screen, and then put in the dryer for experiment.

2.2. *Chemicals and reagents*

Potassium hydroxide (analytical grade) and anhydrous ethanol (analytical grade) were purchased from Xian Chemical Reagent Company (Xian, China).

2.3. *Ultrasound-assisted extraction of FC*

A certain amount of plant material was placed into Erlenmeyer flasks (250 mL), and mixed with different concentrations of ethanol aqueous solution. The extraction process was performed with a domestic ultrasonic device (Kunshan Hechuang Ultrasonic Instrument Co., Ltd., model KH-250E) equipped with timer and temperature controller. After ultrasonic extraction, the mixture was filtered under diminished pressure and the supernatants were collected and put in the refrigerator until further analysis.

2.4. *Determination of total flavonoid content (TFC)*

TFC of the extracting solution was determined by KOH assay method. A known volume of supernatants gained from “2.3” was mixted with 70 % ethanol aqueous solution to make 6 mL solution, and 0.5 mL 10 % KOH was added. After 5 min, the total volume was made up to 10 mL with 70 % ethanol aqueous solution. The solution was mixed well and the absorbance was measured at 403 nm on UV-Vis spectrophotometer (Analytik Jena, model SPECORD-50). TFC was calculated by the calibration curve of rutin mentioned in our previous work [9] and expressed as rutin equivalents (mg RE/ 100g). All samples were analyzed in triplicate and the average values were calculated.

2.5. *Experiment design and data analysis*

2.5.1. Plackett-Burman (PB) design for screening of significant variables. PB design performed in 12 runs was used in this study to screen the important variables that significantly influenced UAE. Each variable was detected at the low and high levels which were denoted -1 and +1, respectively. The variables and their levels were presented in table1, two levels of variables were set by the single factor experimental results. PB design assumes that the main effects of the variables without interaction and first-order polynomial model can be set up.

The Plackett-Burman design matrix and the experimental results were shown in table 2. The significant variables ($p<0.05$) were considered to have an obvious impacts on the UAE of FC and would be selected and optimized. Minitab version 17.1.0 software (Minitaba Inc., PA, USA) was used for the experimental design and data analysis.

Table 1. Variables and levels tested in PB design.

Variables	Coded Variables	Levels	
		-1	+1
volumn fraction of ethanol (%)	x_1	30	50
extraction time (min)	x_2	20	30
extractive temperature (℃)	x_3	50	70
liquid to solid ratio (mL/g)	x_4	30	40

Table 2. PB design matrix of variables with TFC as response.

No.	x_1	x_2	x_3	x_4	y=TFC(mgRE/100 g)
1	+1	-1	+1	-1	6.400
2	+1	+1	-1	+1	6.538
3	-1	+1	+1	-1	4.516
4	+1	-1	+1	+1	6.259
5	+1	+1	-1	+1	6.587
6	+1	+1	+1	-1	6.260
7	-1	+1	+1	+1	5.361
8	-1	-1	+1	+1	5.173
9	-1	-1	-1	+1	4.817
10	+1	-1	-1	-1	5.718
11	-1	+1	-1	-1	4.012
12	-1	-1	-1	-1	3.761

2.5.2. Steepest ascent method. Steepest ascent method is a process moving along the maximum increase of the response value[12]. In this study, the direction of steepest ascent was determined according to that the response value increase most rapidly by increasing or decreasing the values of significant variables, and the step size in the direction was determined by the estimated coefficient from *equation* (1) and practical experience.

2.5.3. Box-Behnken(BB) design for optimization of UAE conditions. For BB design, a total of 17 experiments including five repetitions at the center point were adopted to optimize the variables and their interactions that were screened from PB experiment results. The chosen variables were volumn fraction of ethanol, extractive temperature and liquid to solid ratio coded as z_1, z_2, and z_3 in this design, respectively. Independent variables and corresponding levels were shown in table 3. In this part, Design-Expert version 8.0.6.1 software (Stat-Ease, Inc., Minneapolis, MN, USA) was adopted to design experiment matrix, drawing and data analysis. The experimental matrix and results of BB design were present in table 4. Using response surface regression analysis, BB design experimental results were fitted out a second-order polynomial equation. Meanwhile, contour plots and 3 d surface graphs were used to reflect the effects of coded variables to the response value. The correctness of the model was obtained by compareing the experimental results with predicted values under the optimal conditions.

Table 3. Significant variables and corresponding levels in BB design.

Independent variables	Coded variables	levels of independent variables		
		-1	0	+1
volumn fraction of ethanol(%)	z_1	70	80	90
extractive temperature(℃)	z_2	75	80	85
liquid to solid ratio (mL/g)	z_3	45	55	65

Table 4. The experimental matrix and results of BB design.

Run No.	z_1	z_2	z_3	TFC(mgRE/100g)
1	0	0	0	9.295
2	-1	0	+1	9.014
3	+1	0	+1	7.931
4	0	-1	-1	8.423
5	-1	+1	0	8.784
6	+1	-1	0	8.152
7	-1	-1	0	8.934
8	0	+1	-1	8.623
9	-1	0	-1	8.452
10	+1	0	-1	7.683
11	0	0	0	9.382
12	0	-1	+1	9.134
13	0	0	0	9.457
14	0	0	0	9.189
15	0	+1	+1	8.742
16	0	0	0	9.401
17	+1	+1	0	7.902

3. Results and discussion

3.1. PB design and variables screening

In PB design, a coded linear equation was obtained by data analysis, as follow:

$$y = 5.4502 + 0.8435\, x_1 + 0.0955\, x_2 + 0.2113\, x_3 + 0.3390\, x_4 \quad (1)$$

According to the analysis of variance (ANOVA) presented in table 5, the model F-value of 38.72 indicated the model to be significants ($p<0.05$). Meanwhile, the lack of fit ($p=0.094$, $p>0.05$) and regression coefficient $R^2=0.9568$ also implied that the resulting model could accurately predict the response variable and was adequate for the extraction of FC from C.hybridum L. stem.

In general, variables with a p-value less than 0.05 are considered to be significant factors. Our results implied that the volumn fraction of ethanol (x_1), etraction tempreture(x_3) and liquid to solid ratio (x_4) were significant for TFC measured by spectrophotometric method.

Further analysis using Pareto charts of the standardized effect (Figure 1) comfirmed that the terms beyond the vertical red line were considered to have significant impacts on response value, meanwhile, the similar results were gained by the Normal plot (Figure 2) that factors of A (x_1), C (x_3), D (x_4) were the most effective factors. Main effects plot (Figure 3) demonstrated that specific variables had positive or negative effects on response value [13].

Table 5. ANOVA results of PB design.

Source	DF	Adj SS	Adj MS	F-value	P-value
Model	4	10.5623	2.64059	38.72	0.000
Linear	4	10.5623	2.64059	38.72	0.000
volumn fraction of	1	8.5379	8.53791	125.21	0.000
ethanol	1	0.1094	0.10944	1.60	0.246
extraction time	1	0.5359	0.53594	7.86	0.026
extractive temperature	1	1.3791	1.37905	20.22	0.003
liquid to solid ratio	7	0.4773	0.06815		
Error					
Lack-of-Fit	6		0.07936	66.10	0.094
Pure Error	1	0.4761	0.00120		
Total	11	0.0012			
		11.0397			
S 0.261133			R^2(adjusted)		0.9321
R^2 0.9568			R^2(Predicted)		0.8729

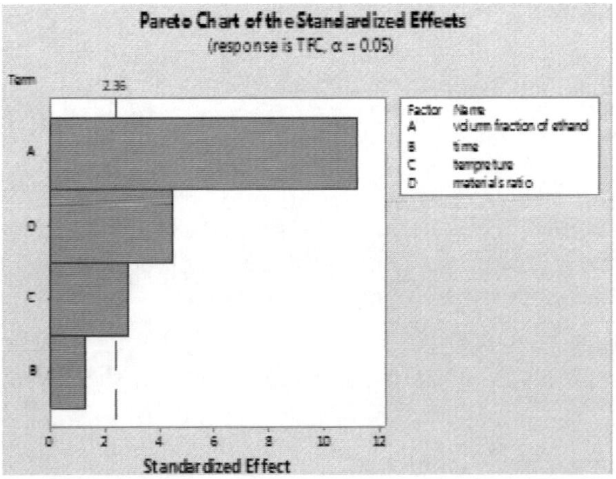

Figure 1. Patet chart in PB design.

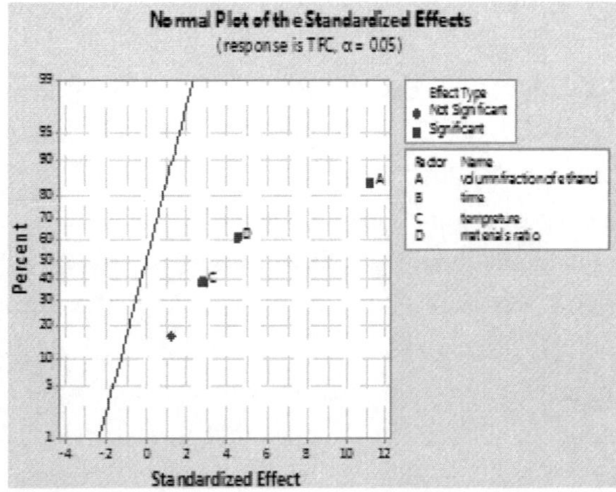

Figure 2. Nomal plot in PB design.

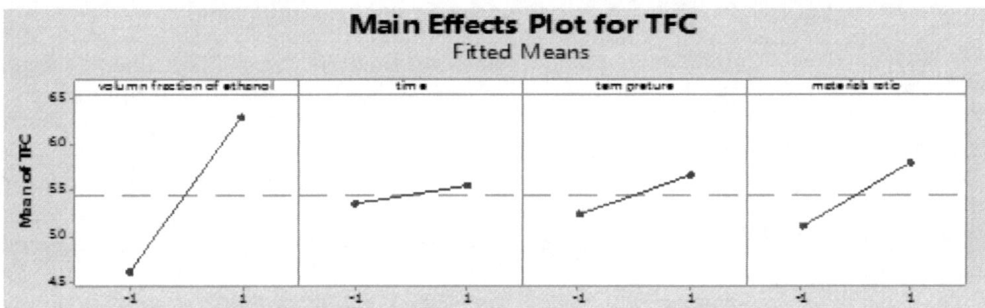

Figure 3. Main effects plot showing factors in PB design.

3.2. Steepest ascent method

According to the results of PB design, experimental points in this section moved along the direction in which volumn fraction of ethanol, extractive temperature and liquid to solid ratio increased. The experimental design and results were presented in table 6, The peak response value was up to 9.189 mg RE /100g with volumn fraction of ethanol 80%, temperature 80℃, liquid to solid ratio was 55 and extraction time kept unchanged for 30 minutes. This point was considered to be near the optimal point and was selected for further optimization.

Table 6. Experimental design and results of the method of steepest ascent.

Run No.	volumn fraction of ethanol(%)	extractive temperature (℃)	Liquid to solid ratio	TFC(mgRE/100g)
1	40	60	35	5.443
2	50	65	40	5.507
3	60	70	45	6.052
4	70	75	50	6.556
5	80	80	55	9.189
6	90	85	60	7.455

3.3. Optimizaton by Response Surface Methodology (RSM)

3.3.1. Second-order polynomial model. In BB design, a quadratic polynomial equation containing only significant terms was derived from multiple regression analysis to the experimental data represented in table 4, as follow:

$$y=9.34-0.44 z_1-0.074 z_2+0.21 z_3-0.15 z_2z_3-0.68 z_1^2-0.22 z_2^2-0.39 z_3^2 \quad (2)$$

The results of the ANOVA shown in table 7 were used to check the adequacy of the quadratic polynomial equation. According to table 7, the quadratic model was highly significant (F= 73.86, P< 0.0001; "Lack of Fit P-value" of 0.8249, P>0.05), and the correlation coefficient (R^2) was 0.9896, indicating good consistency between the experimental and predicted values. Meanwhile, the lower value of CV (1.00) denoted that the experimental results were highly reliable. Thus the quadratic model was adequate for prediction within the range of variables employed.

Table 7. ANOVA for the quadratic polynomial model.

source	Sum of squares	df	Mean square	F-value	P-value [a] Prob F
Model	5.11	9	0.57	73.86	<0.0001
z_1	1.55	1	1.55	201.21	<0.0001
z_2	0.044	1	0.044	5.70	0.0483
z_3	0.34	1	0.34	43.78	0.0003
z_1z_2	2.500×10^{-3}	1	2.500×10^{-3}	0.33	0.5682
z_1z_3	0.025	1	0.025	3.21	0.1163
z_2z_3	0.088	1	0.088	11.41	0.0118
z_1^2	1.95	1	1.95	254.36	<0.0001
z_2^2	0.20	1	0.20	26.69	0.0013
z_3^2	0.65	1	0.65	84.96	<0.0001
Residual	0.054	7	7.680×10^{-3}		
Lack of fit	9.876×10^{-3}	3	3.292×10^{-3}	0.30	0.8249
Pure error	0.044	4	0.011		
Cor total	5.16	16			
R^2	0.9896		Pred R^2	0.9561	
Adj R^2	0.9762		C.V.%	1.00	

[a] P>0.05 not significant, 0.01<P<0.05 significant, P<0.01 highly significant

*3.3.2. RSM analysis.*The P-values gained by Design-expert 8.0.6.1 software were used to distinguish the significance of the variables on UAE. The results from table 7 revealed that z_1, z_2, z_3, z_2z_3, z_1^2, z_2^2, z_3^2 were significant terms (P<0.05) for UAE of FC from the stem of *C.hybridum* L.. Response surface graph and contour plot for TFC as a response value of volumn fraction of ethanol and ectraction temperature with a fixed liquid to solid ratio at zero level were gained and depicted in figure 4.

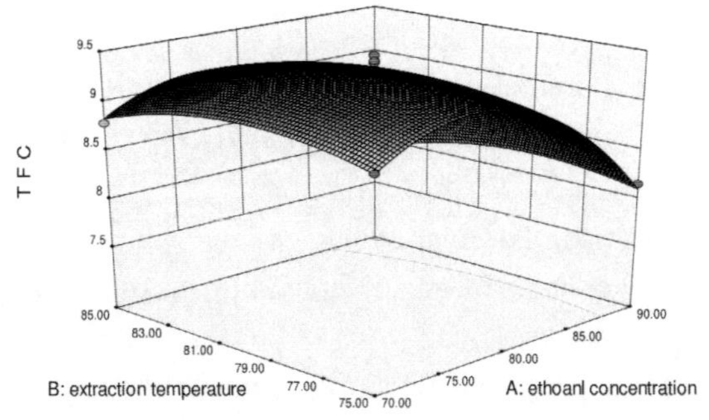

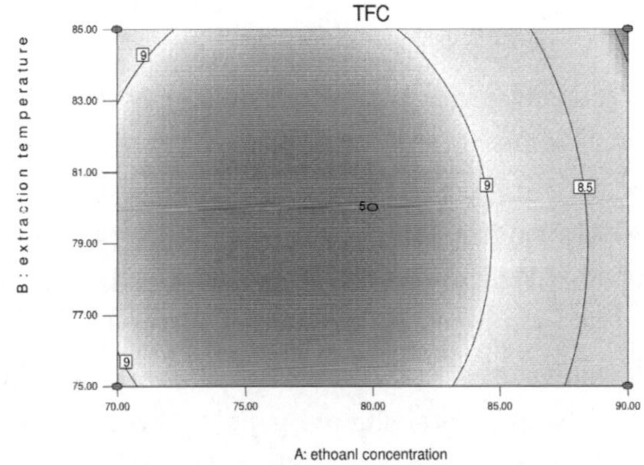

Figure 4. Response surface graph and contour plot for TFC as a response value of volumn fraction of ethanol and ectraction temperature with a fixed liquid to solid ratio at zero level.

The response surface graph of TFC refering to volumn fraction of ethanol and extractive temperature is presented in figure 4, with liquid to solid ratio kept at zero level (55). An increasing in the extractive temperature resulted in the TFC of extraction solution increased firstly and was up to a maximum and then decreased. Liquid to solid ratio and volumn fraction of ethanol had similar effects on TFC were describled by the corresponding figures, and maximal TFC could be obtained at optimal value between every two variables.

The optimal conditions from each variable and the maximum predictive value of TFC were all gained by Design-expert software; the results were consistent with the figures analyzed above. The optimal conditions were listed as follows: volumn fraction of ethanol 76.62 %, extractive temperature 78.69℃ and liquid to solid ratio 58.43, meanwhile, the predictive value of TFC was gained as 9.4640. Futher trials were conducted under the optimal conditions to confirm the model by comparison between the predictive value of TFC and the practical result. An average value of 9.4701(n=3) obtained from real experiments validated the suitability of the developed model.

4. Conclusion

The results of this study suggested that UAE was effective, low-cost and feasible for extracting flavonoids compounds from stem of *C.hybridum* L.. PB design showed that volumn fraction of ethanol, extractive temperature and liquid to solid ratio were the significant variables in the UAE of FC. In addition, the optimal conditions were obtained by response surface methodology: volumn fraction of ethanol 76.62 %, extractive temperature 78.69℃, liquid to solid ratio 58.43 and 30 min of extraction time.

Acknowledgments

This research was supported by the Natural Science Foundation of Qingyang (No.ZJ201304), Applied Chemistry Key Subject of Gansu Province (No. GSACK 2013 0113), and Youth Foundation of Longdong University (No. XYLK1302).

References

[1] Nowak R, Szewczyk K, Gawlik-Dziki U, Rzymowska J and Komsta Ł 2016 Antioxidative and

cytoto- xic potential of some Chenopodium L. species growing in Poland *Saudi J. Biol. Sci*. **23** 15-23

[2] Gawlik-Dziki U, Świeca M, Sułkowski M, Dziki D, Baraniak B and Czyż J 2013 Antioxidant and anticancer activities of *Chenopodium quinoa* leaves extracts-in vitro study *Food Chem. Toxicol* **57** 154-60

[3] Miranda M, Delatorre-Herrera J, Vega-Gálvez A, Jorquera E, Quispe-Fuentes I and Martínez E 2014 Antimicrobial potential and phytochemical content of six diverse sources of quinoa seeds (*Chenopodium quinoa* Willd.) *Agriculture Sciences* **5** 1015-24

[4] Gohar A A and Elmazar M M A 1997 Isolation of hypotensive flavonids from Chenopodium species growing in Egypt *Phytotherapy Research* **11** 564-7

[5] Repo-Carrasco-Valencia R, Hellström J K , Pihlava J M and Mattila P H 2010 Flavonoids and other phenolic compounds in Andean indigenous grains: Quinoa (*Chenopodium quinoa*), Kañiwa (*Chenopodium pallidicaule*) and Kiwicha (*Amaranthus caudatus*) *Food Chem.* **120** 128 -33

[6] Kao E S, Wang C J, Lin W L, Yin Y F, Wang C P and Tseng T H 2005 Anti-inflamatory potential of flavonoid contents from dried fruit of Crataegus pinnatifida in vitro and in vivo *J. Agric. Food Chem.* **53** 430-6

[7] Luo Y P 2010 Advances in pharmacological research of flavonoids *Asia-Pacific Traditional Medicine* **4** 126-8

[8] Wu Y, Han M H and Wang L P 2016 Preliminary study and GC-MS Analysis on the chemical components of the stem of *Chenopodium hybridum* L. *Journal of Longdong University* **27** 48-52

[9] Wu Y, Ma S R, Wang C L, Hu H B and Zhang X W 2015 Determination of total flavone in the leaf of *Chenopodium hybridum* L. *Journal of Anhui Agricultural Sciences* **43** 164-5

[10] Wang J, Sun B G, Cao Y P, Tian Y and Li X H 2008 Optimisation of ultrasound-assisted extra-ction of phenolic compounds from wheat bran *Food Chem.* **106** 804-10

[11] Zhang H F, Yang X H, Zhao L D, and Wang Y 2009 Ultrasonic-assisted extraction of epimedin C from fresh leaves of Epimedium and extraction mechanisum *Innov. Food Sci. Emerg.* **10** 54-60

[12] Chen X C, Bai J X, Cao J M, Li Z J, Xiong J, Zhang L, Hong Y and Ying H J 2009 Medium optimization for the production of cyclic adenosine 3', 5'-monophosphate by microbacterium sp.no.205 using response surface methodology *Bioesource Techology* **100** 919-24

[13] Daneshvand B, Ara K M and Raofie F 2012 Cpmparison of supercritical fluid extraction and ulsound-assisted extraction of fatty acid from quince(Cydonia oblonga Miller) seed using response surface methodology and central composite design *Journal of Chromatography A* **1252** 1-7

Anthraquinones isolated from the browned Chinese chestnut kernels (*Castanea mollissima blume*)

Y L Zhang[1,2], J H Qi[1,2], L Qin[3], F Wang[1,2,4] and M X Pang[1,2]

[1] College of Food Science and Engineering, Beijing Key Laboratory of Detection and Control of Spoilage Organisms and Pesticide Residue in Agricultural Products, Beijing Laboratory of Food Quality and Safety, Beijing University of Agriculture, Beijing 102206, China

[2] Pig processing team, Modern agricultural industry technology system Beijing innovation team, Beijing 100102, China

[3] College of Plant Science, Beijing University of Agriculture, Beijing 102206, China

Email: fangwang1973@sina.com

Abstract. Anthraquinones (AQS) represent a group of secondary metallic products in plants. AQS are often naturally occurring in plants and microorganisms. In a previous study, we found that AQS were produced by enzymatic browning reaction in Chinese chestnut kernels. To find out whether non-enzymatic browning reaction in the kernels could produce AQS too, AQS were extracted from three groups of chestnut kernels: fresh kernels, non-enzymatic browned kernels, and browned kernels, and the contents of AQS were determined. High performance liquid chromatography (HPLC) and nuclear magnetic resonance (NMR) methods were used to identify two compounds of AQS, rehein(1) and emodin(2). AQS were barely exists in the fresh kernels, while both browned kernel groups sample contained a high amount of AQS. Thus, we comfirmed that AQS could be produced during both enzymatic and non-enzymatic browning process. Rhein and emodin were the main components of AQS in the browned kernels.

1. Introduction

The Chinese chestnut, *Castanea Mollissima Blume*, has been utilized as food since ancient times in northern China, and are now widely distributed and consumed around the world [1]. In the processing and storage of chestnuts, browning is a major problem as it brings undesirable sensory properties changes and may even produce some harmful products [2]. It is very important to get farther understanding of the browning reaction. Currently, the main substrates and enzymes in the enzymatically browned kernels were well studied [3-9]. Phenolic compounds in chestnuts are mediated by polyphenol oxidase (PPO) and peroxidase (POD) intially producing quinones, which rapidly condense to produce relatively insoluble brown polymers [10]. But there were few studies referring to the components and the change of components in browned kernels.

Anthraquinones (AQS) represent a group of secondary metallic products in plants. AQS derivatives and AQS-containing plants are widely used as mild laxatives [11]. AQS are also characterized by their antibacterial, antiviral [12-13], antifungal [14-15], antioxidant [16], and anticancer [17] properties. AQS are often naturally occurring in plants and microorganisms.

In our previous study, we investigated the components in the enzymatic browned Chinese chestnut kernels and found that rhein and emodin were the main components of AQS in the browned kernels [18]. It was the first time that AQS were found in the enzymatic browned chestnut kernels.

In the present study, to find out whether AQS could be produced by non-enzymatic browning reaction, AQS were extracted from three groups of Chinese chestnut:

a. fresh kernels group, which was set as control group to confirm that AQS were produced by browning rather than already exited in the fresh chestnut kernels;

b. non-enzymatic browning kernels group, which was set to find out whether this type of browning could produce AQS too;

c. browned kernels, which were browned by two pathways: enzymatic and non-enzymatic browning.

Then the AQS contents of these three groups were determined. HPLC and NMR were used to identify AQS in group c.

The result of this study might be helpful to understand the safety of the browned products in Chinese chestnuts kernels and the formation of AQS in browning reactions.

2. Materials and methods

2.1. Plant materials

Chinese chestnuts (*Castanea mollissima*), called Yanshan Red Chestnuts, were purchased from the Miyun District, Beijing.

2.2. Reagents

All solvents used for extraction of AQS were of analytical grade. Rhein (>99%) and emodin (>99%) used for high-performance liquid chromatography (HPLC) analysis were of HPLC grade and were purchased from the National Institutes for Food and Drug Control (BJ, China). Water used for HPLC analysis was purified by a Milliporeultra-pure water system (0.4-mm filter; Watford, UK).

2.3. Extraction of AQS

Fresh kernels of Chinese chestnuts were powderized (~20 mesh). The powder was separated into 3 groups: fresh kernels, non-enzymatic browned kernels and browned kernels as discribed in the instruction part. The powder for non-enzymatic browned kernels group was processed in a microwave

IOP Conf. Series: Earth and Environmental Science **41** (2016) 012014 doi:10.1088/1755-1315/41/1/012014

(640W) for 60 second to inactivate its enzymes. Then two browning groups were incubated at 45°C for browning reactions. Then all three samples were used for extracting AQS.

25g of each sample was placed into a 250mL flask with a reflux condenser. After adding 50mL 75% ethanol, the flask was heated at 45°C for 60 min. The alcoholic solution was recycled with a rotary vacuum evaporator at 70°C under 31kpa. The sediments were hydrolyzed with 50mL 2.3 M sulfuric acid in a 250mL flask followed by heating at 100°C for 60 min. After cooling down, the solution was extracted three times with trichloromethane. The combined extracts were evaporated at 45°C under 12kpa. The residue in browned paste state was resuspended in methanol (total volume: 50mL) for AQS analyses.

2.4. AQS contents

The AQS content of the extracts were determined by the modified method of Yuan Y [19]. Briefly, 10mL extract solution was mixed with 0.5g/100mL acetic acid-magnesium-methanol (2mL) and diluted with methanol to a total volume of 50mL. The absorbance was measured against methanol at 510nm by spectrophotometry (TU-1810, Beijing, China). AQS contents were calculated from the calibration curve of 1,8-hydroxyl AQS and expressed as 1,8-hydroxyl AQS equivalents.

2.5 Separation and purification of AQS

The browned kernels sample was used to separate and purify AQS according to the method of Yang XY [20]. AQS was sequentially extracted with 5g/100mL aqueous sodium bicarbonate. The extraction was completed in three cycles with a 3-h static period. The extracts were collected and acidified with 2.3 M hydrochloric acid. The yellow sediments were rinsed with distilled water. These extracts were crystallized three times with acetic acid, and crystals of Compound 1 (purity of 92%) were collected. Compound 2 was extracted with 5% aqueous sodium carbonate from the AQS after extracted with 5% aqueous sodium bicarbonate. Acidification and crystallization were the same as Compound 1. The resulting crystals were Compound 2 (purity of 92.5%).

2.6 HPLC Analysis

Compound 1 and 2 were analyzed by HPLC (Shimadzu SCL-10AVP, Japan) equipped with a diode array detector (UV at 254nm); the column was a Keystone (Bellefonte, PA) ODS/3v 250 mm, 4.0 mm i.d., 5 μm; the oven was set to 30°C; the mobile phase was 75% methanol/25% water with an isocratic flow rate of 1.0mL/min; the injection volume was 10μL, and the run length was 6 min. The elution time of rhein was approximately 2.8 min, and the elution time of emodin was approximately 3.5 min.

2.7 NMR spectrometer Analysis

The structures of the two compounds were determined using an ECA-600 spectrometer (JEOL, Tokyo, Japan). Both ^{1}H- and ^{13}C-NMR spectra were determined in DMSO-d6 solvent, using TMS as an internal standard. The chemical shifts are expressed in ppm values, referenced to TMS ($\delta_H = 2.5$ ppm) for the ^{1}H-NMR spectra and the residual solvent signal ($\delta_C = 39.5$ ppm) for the ^{13}C-NMR spectra. All spectra were recorded at ambient temperature. ^{1}H-NMR was measured using a frequency of 600 MHz, a 90° pulse supplied from the decoupler, a 1.455-s acquisition time, and a 1-s relaxation delay.

^{13}C-NMR was measured using a frequency of 150 MHz, a 90° pulse supplied from the decoupler, a 0.692-s acquisition time, and a 2-s relaxation delay.

2.8 Statistical analysis

The experiments for determination of AQS contents were performed in triplicate and the results were expressed as the mean ± standard deviation. Statistical comparisons were made using the Student's t-test. Statistical analyses were carried out by the Statistical Analysis System (SAS, 1985) software package. Analysis of variance was performed by analysis of variance procedures.

3. Results and discussion

The contents of AQS in browned Chinese chestnut kernels were summarized in Table 1. The total amount of AQS was significantly greater in browned kernels (11.18 μg/mL) than in fresh kernels (0.08 μg/mL) and in non-enzymatic browned kernels (0.53 μg/mL). These results illustrated that AQS could be produced during both enzymatic and non-enzymatic browning reaction in the kernels of Chinese chestnuts.

Table 1. AQS contents in three samples (μg/mL).

Samples	AQS
Extracts in fresh kernels	$0.08{\pm}0.02^{A}$
Extracts in non-enzymatic browned kernels	$0.53{\pm}0.12^{B}$
Extracts in browned kernels	$11.18{\pm}3.78^{C}$

$^{A\text{–}C}$ Differences within columns (samples connected by the different capital letter are statistically different at $p <$ 0.05). Values are means ± standard deviation ($n = 3$).

The HPLC/UV analysis showed that the AQS profile of browned kernels was composed of emodin and rhein. The two compounds isolated from the brownd kernels were rhein (Compound 1, Figure 1) and emodin(Compound 2, Figure 2). A robust baseline separation was achieved in 6 min. The retention time of Compound 1 was 2.8 (±0.54) min and the retention time of Compound 2 which was 3.5 (±0.24) min.

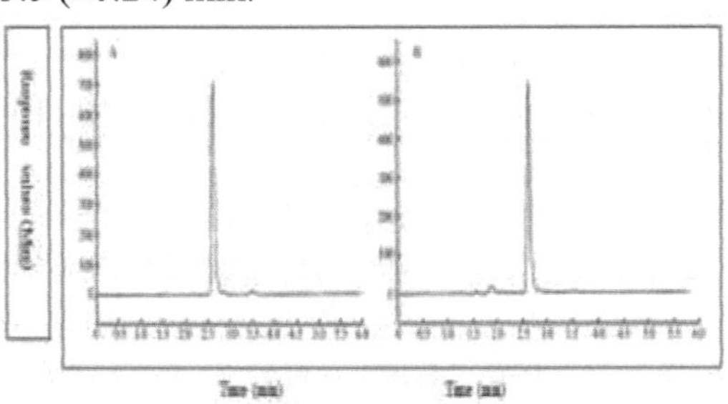

Figure 1. HPLC chromatogram of standard rhein (A) and Compound 1 in 5% sodium bicarbonate solution (B).

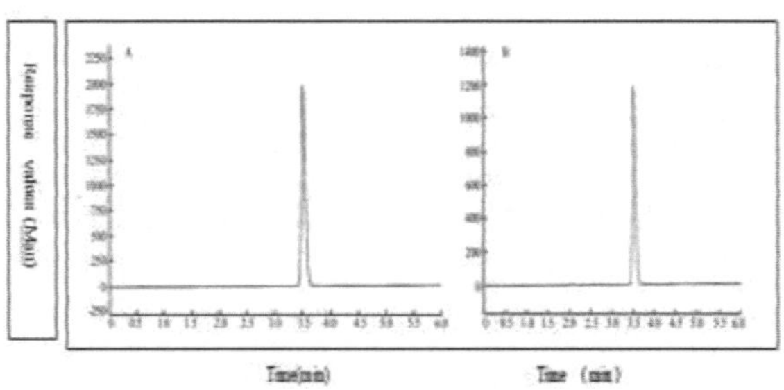

Figure 2. HPLC chromatogram of standard emodin (A) and Compound 2 in 5 g/100 mL sodium carbonate solution (B).

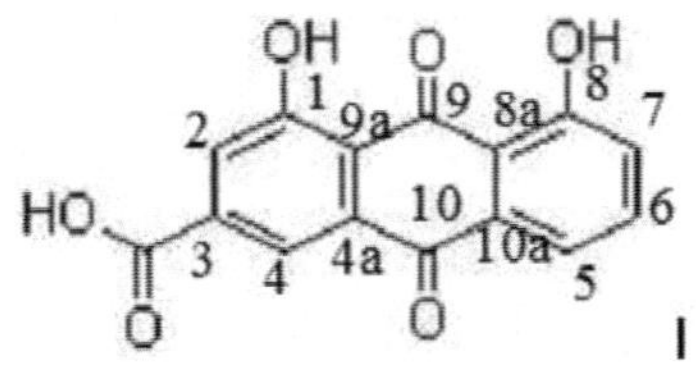

Figure 3. The structure of compound 1.

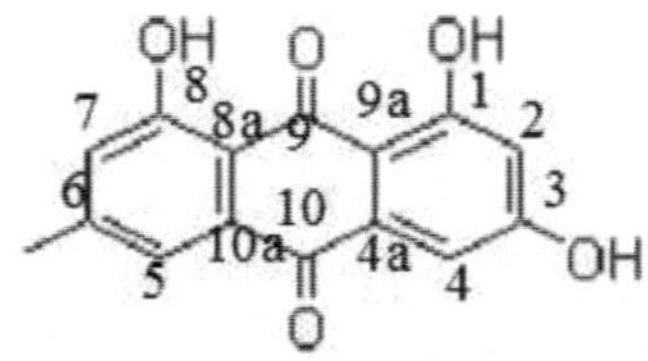

Figure 4. The structure of compound 2.

Table 2. ^{1}H and ^{13}C-NMR data for Compound 1 (600 and 150MHZ, DMSO-d6, Jin HZ and δ in ppm).

No.	δ_H	δ_C
1	/	161.52
2	7.75 (^{1}H, d, *J*2.5)	124.80
3	/	148.20
4	8.16 (^{1}H, d, *J*2.5)	125.70
5	7.72(^{1}H, dd, *J*7.5,11)	118.40
6	7.82(^{1}H, dd, *J*7.5,11)	134.50
7	7.41(^{1}H, dd, *J*8.22,11)	125.70
8	/	161.12
9	/	190.82
10	/	182.31
4a	/	126.60
8a	/	126.30
9a	/	126.10
10a	/	126.40
1-OH	11.89	/
8-OH	11.89	/
3-OOH	13.59	/

Table 3. ^{1}H and ^{13}C-NMR data for Compound 2 (600 and 150MHZ, DMSO-d6, Jin HZ and δ in ppm).

No.	δ_H	δ_C
1	/	162.1
2	6.52 (^{1}H, d, *J*2.1)	126.7
3	/	165.02
4	7.06(^{1}H, d, *J*2.1)	126.4
5	7.43(^{1}H, d, *J*1)	134.8
6	/	132.6
7	7.02 (^{1}H, d, *J*2.5)	161.6
8	/	166.5
9	/	190.4
10	/	181.31
4a	/	126.1
8a	/	126.6
9a	/	121.5
10a	/	121.3
1-OH	11.38	/
3-OH	11.97	/
8-OH	12.01	/
6-Me	2.36	/

The spectrum of Compound 1 (Table 2) confirmed that Compound 1 was rhein and the molecular formula of Compound 1 was assigned as $C_{15}H_8O_6$.

There were two aromatic hydroxyl signals at δ_H 11.89 (with two hydrogen atoms), a carboxyl signal at δ_H 13.59, and five methyl signals in this spectrum: signals at δ_H 7.41, δ_H 7.72, δ_H 7.82, δ_H 7.75, and δ_H 8.16. The structure of -CH-CH-CH- was present in Compound 1. Proton signals at δ_H 7.41 and δ_H 7.75 respectively coupled with the adjacent protons. The signal at δ_H 7.82 coupled with two protons. The signals at δ_H 7.75 and δ_H 8.16 respectively indicated only one proton without coupling [21]. The ^{13}C-NMR of Compound 1 showed that there were four types of carbons (α, β, carbonyl, and quaternary) in the skeleton of AQS. The signal at δ182.5 was the carbonyl carbon and the signal at δ178 was the α carbon substituted by a hydroxyl. Two signals at δ190.82 and δ182.13 belonged to carbonyl carbons. A signal at δ190.82 was an α carbon connected to two hydroxyl groups, and the signal at δ182.13 was not connected with any hydroxyl group. The chemical shift of the α carbon was 126.6 without any substituent group. Connected with a hydroxyl group, the chemical shift increased to 162. Signals at δ161.52 and δ161.62 indicated two hydroxyl groups were connected with the α carbon. Two hydroxyl groups were respectively located on either side of the carbonyl carbon21.

In conclusion, H-7 (δ7.41), H-5 (δ7.72), H-6 (δ7.82), H-2 (δ7.75), and H-4 (δ8.16) were found in the ^{1}H-NMR of Compound 1. H-7 was connected with an aromatic hydroxyl group at position 1, and H-2 was connected with an aromatic hydroxyl group at position 7. The signal at δ13.59 suggested that a carboxyl group existed at H-3. The signal at δ11.89 indicated that two aromatic hydroxyl groups were presented. The ^{13}C-NMR of Compound 1 suggested that there were two carbonyl carbons in the structure of Fraction 1 and two hydroxyl groups were substituted at H-1 and H-8.

The structure of Compound 1 (Figure 3) corresponded to the structure of rhein (1,8-2-hydroxyl-3-carbonyl-AQS).

The spectrum results of Compound 2 were listed in Table 3.

There was a methyl signal at δ2.36 and a aromatic hydroxyl signal at δ11.38. Two aromatic hydroxyl signals were also observed at δ11.97 and δ12.01 in the ^{1}H-NMR of Compound 2.There were four methyl groups. Signals at δ6.52 and at δ7.02 were coupled with meta-position protons. The other two signals at δ7.06 and δ7.43 were split. These signals indicated that four methyl groups were not adjacent and the signal of the proton at δ6.52 was surrounded by strong electronic groups, such as aromatic hydroxyl and aromatic methyl groups, and the other signals at δ7.02, δ7.06, and δ7.43.Two carbonyl carbons were found at δ190.4 and δ181.31 in the ^{13}C-NMR of Compound 2. Two positions of the α carbon were connected with hydroxyl groups at δ190.4, but the carbonyl carbon signal at δ181.31 was not coupled. The chemical shift was 126.6 without substituent groups for the α carbon, but the chemical shift was approximately 162 with hydroxyl groups on the α carbon. For the β carbon, the chemical shift was 134.3 without substituent groups, and the chemical shift was approximately 163 with hydroxyl groups. There were three carbons at δ162.1, δ165.02, and δ166.5 with hydroxyl groups, and the positions of the substituent groups could not be determined. Two hydroxyl groups were connected to the α position of the carbonyl carbon at δ190.4, and no hydroxyl groups were found with another carbonyl carbon at δ181.31. The carbon with the last hydroxyl group was at the β position.

Signals of protons at δ6.52 (s, ^{2}H), δ7.02 (s, ^{7}H), δ7.06 (s, ^{4}H), and δ7.43(s, ^{5}H) were found in the ^{1}H-NMR of Compound 2. The correlation from the proton (δ2.37) to H-6 indicated the location of the methyl group. Two hydroxyl groups were substituted with two protons at δ11.38 (s, ^{1}H), δ11.97 (s, ^{3}H) and δ12.01(s, ^{8}H). Two carbonyl carbons were observed in the ^{13}C-NMR of Compound 2. The locations of H-1 and H-8 at the α-position of the carbonyl carbon were occupied by two hydroxyl groups. Another hydroxyl group was located at H-3. The ^{1}H-NMR analysis showed that it was coincident with the ^{13}C-NMR analysis.

The configuration of emodin was identified as 1, 3, 8-trihydroxy-6-methyl-AQS. Two aromatic hydroxyl groups were found at H-2, and one aromatic hydroxyl group and one aromatic methyl group were found at H-7. Additionally, an aromatic hydroxyl group was observed at H-4 and an aromatic methyl group was found at H-5. According to the strength of the electronic groups, the sequence of the four protons in the magnetic field was H-2>H-7>H-4>H-5.

The structure of Compound 2 (Figure 4) corresponded to the structure of emodin.

4. Conclusions

AQS belongs to a large group of substances that includes substitutions at the β position and the 1 and 2 positions, but most have three or four substitutions, such as emodin, rhein, and chrysophanol [22]. In this study, we found that contents of AQS increased after browning reaction in Chinese chestnut kernels indicating AQS could be produced during browning process. Rhein and emodin were the main components of AQS in the browned kernels. Additional information about other components of AQS in browned kernels should be researched in the future.

This was the first time to find AQS can be produced during both enzymatic and non-enzymatic browning reaction of Chinese chestnut kernels. In the future, other browned materials should be used to confirm this result. The formation and safety of AQS in browning reactions should be studied in detail.

Acknowledgments

This work was financially supported by the Beijing Key Construction of the Discipline of Processing and Storage of Agriculture Products. (PXM2009-014207-078172).

References

[1] Cai R, Chi J H and Qi C J 2007 Current status and problems of chestnut industry *Chin. Fruit* **1**1-3

[2] Kang M L and Mu D H 2003 Study on browning Chinese chestnut during the processing and its inhibition *Hebei Univ. Sci. Tech.* **24** 72-6

[3] Tao Y L, Qiu J Z, Lin H, Zhang S F, Ye M Z, Huang P H and Guo X Z 2001 Research on characteristics of peroxidase and polyphenol oxidase of chestnut *Food Sci.* **22** 64-7

[4] Qi J H, Wang F, Pang M X and Huang M Q 2009 Active phenolics compounds extraction from chestnut and their antioxidative activities *Chin. Cereals Oils Assoc.* **24** 141-68

[5] Prota G 1988 Progress in the chemistry of melanins and related metabolites *Med. Res. Rev.* **8** 525-56

[6] Lopez-Serrano M and Ros-Barcelo A 1995 Peroxidase in unripe and processing-ripe strawberries *Food Chem.* **52** 157- 60

[7] Sapers G M, Hicks K B and Miller R L 2001 Antibrowning Agents. *In Food Additives*, 2nd ed. Branen A L, et al. (New York: Marcel Dekker Inc.) p 543

[8] Gacche R N, Warangkar S C and Ghole V S 2004 Glutathione and cinnamic acid: natural dietary components used in preventing the process of browning by inhibition of polyphenol oxidase in apple juice *Enzyme Inhib. Med. Chem.* **19** 175-9

[9] Fennema, O R 1996 Food chemistry (New York: Marcel Dekker Inc) pp 492-4

[10] Madinez M V and Whitaker J R 1995 The biochemistry and control of enzymatic browning *Trends in Food Science & Technology* **6** 195-200

[11] Hansel R and Sticher O 2007 Pharmakognosie–Phytopharmazie (Germany: Springer: Heidelberg) pp 234-7

[12] Andersen D O, Weber N D, Wood S G, Hughes B G, Murray B K and North J A 1991 In vitro

virucidal activity of selected anthraquinones and anthraquinone derivatives *Antiviral. Res.* **16** 185-96

[13] Ifesan B O, Hamtasin C, Mahabusarakam W and Voravuthikunchai S P 2009 Assessment of antistaphylococcal activity of partially purified fractions and pure compounds from Eleutherine americana *J. Food Prot.* **72** 354-9

[14] Agarwal S K, Singh S S, Verma S and Kumar S 2000 Antifungal activity of anthraquinone derivatives from Rheum emodi *J. Ethnopharmacol* **72** 43-6

[15] Manojlovic´ N T, Solujic´ S, Sukdolak S and Milosev M 2005 Antifungal activity of *Rubia* tinctorum, Rhamnus frangula and Caloplaca cerina *Fitoterapia* **76** 244–6

[16] Ammara R B et al. 2009 Antioxidant and free radical-scavenging properties of three flavonoids isolated from the leaves of Rhamnus alaternus *L.* (Rhamnaceae): A structure–activity relationship study *Food Chem.* **116** 258–64

[17] Fenig E, Nordenberg J, Beery E, Sulkes J and Wasserman L 2004 Combined effect of aloe-emodin and chemotherapeutic agents on the proliferation of an adherent variant cell line of Merkel cell carcinoma *Oncol. Rep.* **11** 213-7

[18] Qi J H, Yang X Y and Huang M Q 2011 Identification of chemicals from enzymatic-browning reaction of chestnut (Castanea mollissima) *Scientia Agricultura Sinica* **44**(8) 1678-87

[19] Yuan Y and Chen W S 2000 Content determination of three kinds of anthraquinones in the root of Rumex by HPLC *Acad. J. Sec. Mil. Med. Univ.* **21**(10) 952-4

[20] Yang X Y, Qi J H, Wang J, Huang M Q and Wu Y H 2010 Study on the extraction and stability of the enzymatic browning product in chestnut kernel *Chinese Agricultural Science Bulletin* **26**(17) 103-7

[21] Zhang H, Guo Z J, Wu N, Xu W, Han L, Li N and Han Y 2012 Two novel naphthalene glucosides and an anthraquinone isolated from Rumex dentatus and their antiproliferation activities in four cell lines *Molecules* **17** 843-50

[22] Curreli N, Sollai F, Massa L, Comandini O, Rufo A, Sanjust E, Rinaldi A and Rinaldi A C 2001 Effect of pant-derived naphthoquinones on the growth of *Pleurotus sajor-caju* and degradation of the compounds by fungal cultures *J. Basic Microb* **41** 253-9

Synthesis and antifungal activities of 3-alkyl substituted thieno[2,3-d]pyrimidinones

H M Wang[1], S H Deng[1], A H Zheng[1], Q Y Zhang[1], X B Chen[1], X H Zeng[1,2] and Y G Hu[1,2]

[1]Laboratory of Chinese Herbal Pharmacology, Renmin Hospital, Hubei University of Medicine, Shiyan 442000, P.R. China

E-mail: zengken@126.com; ygy999@163.com

Abstract. The 3-aryl substituted thieno[2,3-d]pyrimidinones **3** by sequential reaction of iminophosphorane **1**, aromatic isocyanates and various nucleophiles (HY), found some compounds showed good antitumor and antibacterial activities. Meanwhile, aliphatic isocyanates were applied in the reaction to prepare 3-alkyl substituted thieno[2,3-d]pyrimidinones, but there are no reports of their antifungal activities. As a continuation of our research for new biologically active heterocycles, we herein wish to report a facile synthesis and antifungal activities of 3-alkyl substituted thieno[2,3-d]pyrimidinones **6** via easily accessible iminophosphorane **1**. The growth inhibitory effect of one concentration (50mg/L) of compounds **6** against five fungus(*Fusarium oxysporium, Rhizoctonia solani, Colletotrichum gossypii, Gibberella zeae and Dothiorella gregaria*) in vitro was tested by the method of toxic medium. Compound **6d** showed the best inhibition rate against Gibberella zeae with 85.68%.

1. Introduction

The derivatives of N-containing heterocycles containing the thieno[2,3-d]pyrimidinone scaffold, which are structurally related to purinones, obtains much scientific attenttion because they have remarkable biological properties. They proved to show significant antifungal, antitumor, anticonvulsant and angiotensin antagonistic activities.[1-3] Therefore, the reports of synthesis and biological activity research on thienopyrimidinones are numerous.[4-7]

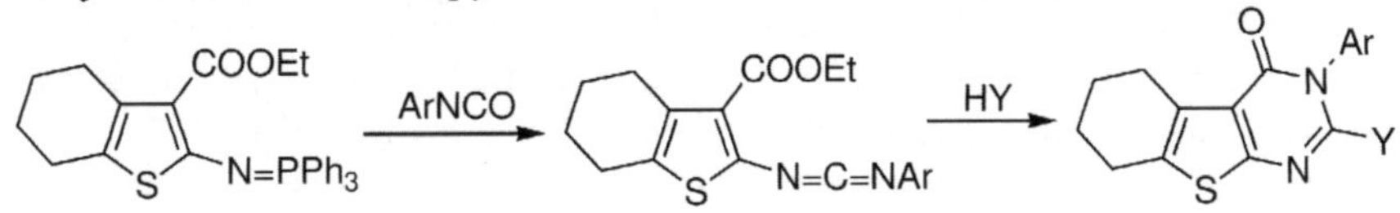

1 **2** **3**

We once provided an efficient synthesis of various 3-aryl substituted thieno[2,3-d]pyrimidinones 3 by sequential reaction of aryl substituted isocyanates, iminophosphorane, and various nucleophiles (HY), found some compounds showed good antitumor and antibacterial activities.[8-10] Meanwhile, aliphatic isocyanates were applied in the reaction to prepare 3-alkyl substituted thieno[2,3-d]pyrimidinones,[11] but there are no reports of its antifungal activities. To continue in this research for finding new antifungal compouds containing the 3-alkyl substituted thieno[2,3-d]pyrimidinone scaffold, herein we wish to report antifungual activities of 3-alkyl substituted thieno[2,3-d]pyrimidinones **6** synthetized via easily sequential reaction of aryl substituted isocyanates, iminophosphorane, and various nucleophiles (HY).

2. Results and discussion

We have reported that the reaction of iminophosphorane **1** with aromatic isocyanates carried out smoothly at 0-5 °C to give carbodiimides **2**,[10] however, as aliphatic isocyanates were used, the mixture took place very slow at room temperature. Good yields of carbodiimide **4** was obtained when the mixture was refluxing for 24 hours. Generally the obtained carbodiimide **4** unpurified should be directly utilized in further reaction. The mixture of carbodiimide **4** with phenol or alcohol can give 2-alkoxy or 2-aryloxy 3-alkyl substituted thieno[2,3-d]pyrimidinones **6** by using catalytic amount of sodium ethoxide or K_2CO_3 at room temperature in moderate to good yields (**6a-6k** in Table 1).

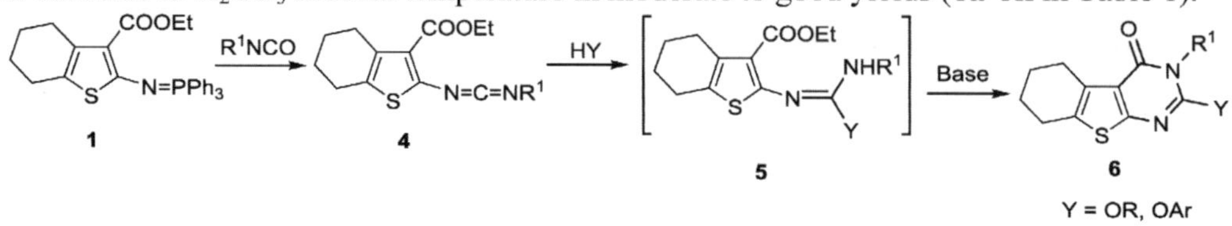

The structure of the synthesized compounds **6** was confirmed by their spectral data and elemental analyses. For example, the ^{1}H NMR spectral data of **6c** show the signal of NCH at 5.62 ppm as broad absorption, and signal of CH_3 at 1.57 ppm as doublets. The cyclohexyl ring's signals appeared at 3.00-2.67 and 1.87-1.82 ppm as mutiple absorptions. The phenyl signals appeared at 7.37-7.23 ppm. The MS spectrum of **6f** shows molecule ion peak (M^+) at m/z 354 with 100% abundance.

Table 1. Preparation of Compounds **6.**

No.	R^1	Y	Conditions	Yield (%)
6a	Bu	MeO	r.t./24 h	65
6b	i-Pr	MeO	r.t./24 h	70
6c	i-Pr	4-BrC_6H_4O	r. t./18h	80
6d	i-Pr	2-Me C_6H_4O	r. t./15h	67
6e	i-Pr	3-Me C_6H_4O	r. t./18h	67
6f	i-Pr	4-Me C_6H_4O	r. t./15h	65
6g	i-Pr	4-Cl C_6H_4O	r. t./24h	75
6h	i-Pr	2,4-2Cl-6-Me C_6H_2O	r.t./18 h	73[a]
6i	Bu	2-Me C_6H_4O	r.t./15 h	65
6j	Bu	2,4-2Cl-6-Me C_6H_2O	r. t./24h	67
6k	Bu	4-Br C_6H_4O	r. t./18h	80

a. Known Compound[11]

The growth inhibitory effect of one concentration (50mg/L) of compounds **6** against five fungus(*Fusarium oxysporium, Rhizoctonia solani, Colletotrichum gossypii, Gibberella zeae and Dothiorella gregaria*) in vitro was tested by the method of toxic medium (Table 2).

The results showed that the inhibitory effects of most compounds 6 are not obvious, but compounds 6c, 6i, 6g and 6d showed moderate to good inhibitory effects, especially Compound **6d** showed the best inhibition rate against *Rhizoctonia solani* with 85.68% (Table 2).

Analysis of the structure-activity relationship showed that groups R^1 and Y have influence on its antifungul activities. When the R^1 is isopropyl, compounds 6 showed good activity relatively. Meanwhile, the Y is aryloxy, especially the scaffold containing methyl in phenyl, compounds 9 have good activity relatively.

Table 2. In vitro growth inhibitory effect of compounds **6** (50mg/L).

No.	Inhibition rate (%)[b]				
	Fusarium oxysporium	*Rhizoctonia solani*	*Colletotrichum gossypii*	*Gibberella zeae*	*Dothiorella gregaria*
6a	9.09	54.17	33.33	31.03	19.05
6b	4.55	50.00	14.29	17.24	9.52
6c	42.86	56.76	34.78	36.00	40.00
6d	62.38	85.68	70.87	78.37	73.34
6e	33.33	67.57	52.17	44.00	46.67
6f	33.33	67.57	52.17	44.00	46.67
6g	62.86	52.43	40.43	42.00	43.33
6h	38.10	48.65	43.48	32.00	26.67
6i	52.38	75.97	60.87	72.00	66.67
6j	38.10	48.65	43.48	32.00	26.67
6k	42.86	32.43	30.43	32.00	33.33

Inhibition rate (%) = (Controlled colony diameter - Processing colony diameter) / (Contrast colonies diameter - Inoculating colony diameter) * 100.

3. Experimental

IR spectra (absorption in cm-1, with KBr pellets, recorded by a PE-983 infrared spectrometer apparatus); MS were measured on a Finnigan Trace MS spectrometer. 1H NMR spectra (chemical shifts (δ) in ppm, (CH3)4Si (δ = 0) as an internal reference, recorded in CDCl3 or DMSO on a Varian Mercury Plus 400 (400 Hz) or Varian Mercury Plus 600 (600 Hz) spectrometer apparatus); Melting points (determined by a X-4 model melting points apparatus, uncorrected); Elementary analyses (determined by a Vario EL III elementary analysis apparatus). The growth inhibitory effect of one concentration (50mg/L) of compounds **6** against five fungus in vitro was tested by the method of toxic medium. Coefficients of variation were <10%.

*3.1. Preparation of 2-Alkoxy 3-alkyl substituted thieno[2,3-d]pyrimidinones **6a-6b***

The obtained carbodiimide 4 unpurified was mixed with alcohol(15 ml) including catalytic amoun of RO^-Na^+(5% in mole number). The mixture was stirred for 12-24 hours at room temperature, then the solvent was removed off under reduced pressure, the residue was recrystallized from the mixed solvent (ether: ethanol, 2:1) to give 2-Alkoxy 3-alkyl substituted thieno[2,3-d]pyrimidinone to give 2-alkoxy 3-alkyl substituted thieno[2,3-d]pyrimidinone in moderate to good yield.

6a 3-Butyl-2-methoxy-5,6,7,8-tetrahydrobenzothieno[2,3-d]pyrimidin-4(3H)-one

The compound is white solid with melt point 83-84 °C, ^1H NMR ($CDCl_3$, 400 MHz) δ 4.05-3.99 (m, 5H, NCH_2 and OCH_3), 2.98-2.69 (m, 4*H, 2*CH_2), 1.87-1.81 (m, 4*H, 2*CH_2), 1.64-1.37 (m, 4*H, 2*CH_2), 0.95 (t, J = 7.2 Hz, 3*H, CH_3); MS m/z (%), 292 (M^+, 100), 275 (50), 261 (30), 236 (87), 208 (78), 192 (12). Elemental Elemental Analysis for $C_{15}H_{20}N_2O_2S$: Calculated. C, 61.62; H, 6.89; N, 9.58. Found C, 61.46; H, 6.82; N, 9.53.

6b 2-Methoxy-3-(i-propyl)-5,6,7,8-tetrahydrobenzothieno[2,3-d]pyrimidin-4(3H)-one

The compound is white solid with melt point 129-130 °C, ^1H NMR ($CDCl_3$, 400 MHz) δ 5.48-5.40 (m, 1*H, NCH), 4.03 (s, 3*H, OCH_3), 2.98-2.69 (m, 4*H, 2*CH_2), 1.87-1.81 (m, 4*H, 2*CH_2), 1.45 (d, J = 6.8 Hz, 6*H, 2*CH_3); MS m/z (%), 278 (M^+, 97), 248 (56), 235 (99), 221 (82), 207 (100), 179 (55), 151 (41). Elemental Elemental Analysis for $C_{14}H_{18}N_2O_2S$: Calculated. C, 60.41; H, 6.52; N, 10.06. Found C, 60.54; H, 6.37; N, 10.01.

*3.2. Preparation of 2-Aryloxy 3-alkyl substituted thieno[2,3-d]pyrimidinones **6c-6k***

The obtained carbodiimide 4 unpurified was mixed with phenol (1.5 mmol) in CH_3CN (15 ml) including catalytic amoun of K_2CO_3 (10% in mole number). The mixture was stirred for 12-24 hours and filtered, then the solvent was removed off under reduced pressure, the residue was recrystallized from the mixed solvent (ether: ethanol, 2:1) to give 2-Aryloxy 3-alkyl substituted thieno[2,3-d]pyrimidinone.

6c 2-(4-Bromo-phenoxy)-3-isopropyl-5,6,7,8-tetrahydro-benzothieno[2,3-d]pyrimidin-4(3H)-one

The compound is white solid with melt point 181-182 °C; ^{1}H NMR ($CDCl_3$, 400 MHz) δ 7.31-6.97(m, 4*H, Ar-H), 5.62(bs, 1*H, NCH), 2.97-2.69 (m, 4*H, 2*CH_2), 1.85-1.83 (m, 4*H, 2*CH_2), 1.57 (d, J=6.4Hz, 6*H, 2*CH_3); MS (*m/z*, %), 420(74), 418(M^+, 66), 376(45), 350(19), 333(11), 296(10), 254(100), 226 (27), 205(23), 177(24), 161(9), 43(23); Elemental Analysis for $C_{19}H_{19}BrN_2O_2S$: Calculated. C, 54.42; H, 4.57; N, 6.68. Found. C, 54.56; H, 4.45; N, 6.61.

6d 3-Isopropyl-2-o-tolyloxy-5,6,7,8-tetrahydro-benzothieno[2,3-d]pyrimidin-4(3H)–one

The compound is white solid with melt point 168-169 °C; ^{1}H NMR ($CDCl_3$, 400 MHz) δ 7.30-7.05(m, 4*H, Ar-H), 5.66 (bs, 1*H, NCH), 3.00-2.66 (m, 4*H, 2*CH_2), 2.21(s, 3*H, Ar-CH_3), 1.83-1.82 (m, 4*H, 2*CH_2), 1.61 (d, J=6.8Hz, 6*H, 2*CH_3); MS (*m/z*, %), 354 (M^+, 100), 311(31), 268(51), 179(19); Elemental Analysis for $C_{20}H_{22}N_2O_2S$: Calculated. C, 67.77; H, 6.26; N, 7.90. Found. C, 67.61; H, 6.13; N, 7.79.

6e 3-Isopropyl-2-m-tolyloxy-5,6,7,8-tetrahydro-benzothieno[2,3-d]pyrimidin-4(3H)–one

The compound is white solid with melt point 127 ℃; ^{1}H NMR ($CDCl_3$, 400 MHz),δ7.31-6.97(m, 4H, Ar-H), 5.63(bs, 1H, NCH), 2.98-2.68 (m, 4*H, 2*CH_2), 2.40 (s, 3*H, Ar-CH_3), 1.83-1.65 (m, 4*H, 2*CH_2), 1.58 (d, J=6.8Hz, 6*H, 2*CH_3); MS (*m/z*, %), 354 (M^+, 100), 311(53), 284(37), 269(96), 254 (53), 241(67), 221(16), 205(28), 177(21), 91(24), 43(15); Elemental Analysis for $C_{20}H_{22}N_2O_2S$: Calculated. C, 67.77; H, 6.26; N, 7.90. Found. C, 67.72; H, 6.35; N, 7.99.

6f 3-Isopropyl-2-p-tolyloxy-5,6,7,8-tetrahydro-benzothieno[2,3-d]pyrimidin-4(3H)-one

The compound is white solid with melt point 163-164 °C; ^{1}H NMR ($CDCl_3$, 400 MHz) δ 7.26-7.05(m, 4*H, Ar-H), 5.63(bs, 1*H, NCH), 2.98-2.67 (m, 4*H, 2*CH_2), 2.39 (s, 3*H, Ar-CH_3), 1.83-1.65 (m, 4*H, 2*CH_2), 1.58 (d, J=6.8Hz, 6*H, 2*CH_3); MS (*m/z*, %), 354 (M^+, 100), 311(32), 268(100), 241 (65), 204(25), 176(20), 90(27), 77(15), 43(20); Elemental Analysis for $C_{20}H_{22}N_2O_2S$: Calculated. C, 67.77; H, 6.26; N, 7.90. Found. C, 67.71; H, 6.35; N, 7.81.

6g 2-(4-Chloro-phenoxy)-3-isopropyl-5,6,7,8-tetrahydro-benzothieno[2,3-d]pyrimidin-4(3H)–one

The compound is white solid with melt point 168 °C; ^{1}HNMR ($CDCl_3$, 400 MHz) δ 7.41-7.11(m, 4H, Ar-H), 5.62(bs, 1*H, NCH), 2.99-2.67 (m, 4*H, 2*CH_2), 1.84-1.82 (m, 4*H, 2*CH_2), 1.57 (d, *J* =6.4Hz, 6*H, 2*CH_3); MS (*m/z*, %), 377(22), 374 (M^+, 91), 331(36), 296(10), 304(29), 261(22), 254(100), 226(14), 205(33), 176(23), 43(26); Elemental Analysis for $C_{19}H_{19}ClN_2O_2S$: Calculated. C, 60.87; H, 5.11; N, 7.47. Found. C, 60.66; H,5.21; N, 7.41.

6h 2-(2,4-dichloro-6-methylphenoxy)-3-isopropyl-5,6,7,8-tetrahydrobenzothieno[2,3-d]pyrimidin-4(3H)-one

See reference [11]

6i 3-Butyl-2-(2-methylphenoxy)-5,6,7,8-tetrahydrobenzothieno[2,3-d]pyrimidin-4(3H)-one

The compound is white solid with melt point 114-115 °C, ^{1}H NMR ($CDCl_3$, 400 MHz) δ 7.30-7.10 (m, 4*H, Ar-H), 4.23 (t, *J* = 7.6 Hz, 2*H, NCH_2), 3.00-2.68 (m, 4*H, 2*CH_2), 2.25 (s, 3*H, CH_3), 1.84-1.46 (m, 8*H, 4*CH_2), 0.99 (t, *J* = 7.2 Hz, 3*H, CH_3); MS m/z (%), 368 (M^+, 100), 351 (45), 311 (21), 268 (24), 90 (11). Elemental Elemental Analysis for $C_{21}H_{24}N_2O_2S$: Calculated. C, 68.45; H, 6.56; N, 7.60. Found. C, 68.35; H, 6.62; N, 7.50.

6j 3-Butyl-2-(2,4-dichloro-6-methylphenoxy)-5,6,7,8-tetrahydrobenzothieno[2,3-d]pyrimidin-4(3H)-one

The compound is white solid with melt point 181-183 °C, ^{1}H NMR ($CDCl_3$, 400 MHz) δ 7.35 (s, 1*H, Ar-H), 7.18 (s, 1*H, Ar-H), 4.26 (t, *J* = 7.6 Hz, 2*H, NCH_2), 2.97-2.68 (m, 4*H, 2*CH_2), 2.18 (s, 3*H, CH_3), 1.84-1.45 (m, 8*H, 4*CH_2), 0.99 (t, *J* = 7.2 Hz, 3*H, CH_3); MS m/z (%), 436 (M^+, 100), 419 (42), 380 (79), 345 (85), 317 (40), 162 (25). Elemental Elemental Analysis for $C_{21}H_{22}Cl_2N_2O_2S$: Calculated. C, 57.67; H, 5.07; N, 6.40. Found. C, 57.57; H, 5.11; N, 6.35.

6k 2-(4-Bromo-phenoxy)-3-butyl-5,6,7,8-tetrahydro-benzothieno[2,3-d]pyrimidin-4(3H)-one

The compound is white solid with melt point 144-145 °C; ^{1}HNMR ($CDCl_3$, 400 MHz) δ 7.57-7.08 (m, 4H, Ar-H), 4.19 (t, *J* =7.6Hz, 2*H, NCH_2), 3.00-2.68 (m, 4*H, 2*CH_2), 1.84-1.74 (m, 6*H, 3*CH_2), 1.47-1.42 (m, 2*H, CH_2), 0.99 (t, *J* =7.2Hz, 3*H, CH_3); MS (*m/z*, %), 433(M^+, 100), 415(28),

375(22), 348(13), 333(14), 297(19), 261(57), 253(97), 204(45), 177(31), 161(14), 57(11); Elemental Analysis for $C_{20}H_{21}BrN_2O_2S$: Calculated. C, 55.43; H, 4.88; N, 6.46. Found. C, 55.56; H, 4.79; N, 6.53.

Acknowledgments
We gratefully acknowledge financial support of this work by the Open Project of Hubei Key Laboratory of Wudang Local Chinese Medicine Research (Hubei University of Medicine (Grant No. WDCM009), the Hubei Province health and family planning scientific research project (Grant No. WJ2015Z113), the of Shiyan Municipal Science and Technology Bureau Science and technology project (Grant No. 15Y12), and the scientific research innovation team of Hubei University of Medicine (grant Nos. 2014QDJZR10, 2014CXZ01). H.M. Wang and S.H. Deng are first coauthors and contributed equally to this work.

References
[1] Kikuchi H, Yamamoto K, Horoiwa S, Hirai S, Kasahara R, Hariguchi N, Matsumoto M and Oshima Y 2006 Exploration of a new type of antimalarial compounds based on febrifugine *J. Med. Chem.* 49 4698
[2] Modica M, Santagati M, Russo F, Sevaggini C, Cagnotto A and Mennini T 2000 High affinity and selectivity of [[(arylpiperazinyl) alkyl] thio] thieno [2, 3-d] pyrimidinone derivatives for the 5-HT 1A receptor. Synthesis and structure–affinity relationships *Eur. J. Med. Chem.* **35** 677
[3] Wang Y D, Johnson S, Powell D, McGinnis J P, Miranda M and Rabindran K 2005 Inhibition of tumor cell proliferation by thieno [2, 3-d] pyrimidin-4 (1H)-one-based analogs *Bioorg. Med. Chem. Lett.* **15** 3763
[4] Duval E, Case A, Stein R L and Cuny G D 2005 Structure–activity relationship study of novel tissue transglutaminase inhibitors *Bioorg. Med. Chem. Lett.* **15** 1885
[5] Al-Omran F and El-Khair A A 2004 Novel synthesis of thieno[2, 3-b]pyridine and substituted 2-thienylthiourea derivatives as antibiotic agents *J. Heterocycl. Chem.* **41** 909
[6] Perspicace E, et al. 2013 Design, synthesis and biological evaluation of new classes of thieno[3, 2-d]pyrimidinone and thieno[1, 2, 3]triazine as inhibitor of vascular endothelial growth factor receptor-2 (VEGFR-2) *J. Med. Chem.* **63** 765
[7] Wu C H, et al. 2010 Design and synthesis of tetrahydropyridothieno[2, 3-d]pyrimidine scaffold based epidermal growth factor receptor (EGFR) kinase inhibitors: the role of side chain chirality and Michael acceptor group for maximal potency *J. Med. Chem.* 53 7316
[8] Ding M W, Chen Y F and Huang N Y 2004 New Efficient Synthesis of 1, 2, 4-Triazolo[5, 1 - b] quinazolin-9(3H)-ones via a Tandem Aza-Wittig/Heterocumulene-Mediated Annulation Eur. *J. Org. Chem.* **2004** 3872
[9] Liu M G, Hu Y G and Ding M W 2008 New iminophosphorane-mediated synthesis of thieno [3′, 2′: 4, 5]thieno[3, 2-d]pyrimidin-4(3H)-ones and 5H-2, 3-dithia-5, 7-diaza-cyclopenta[c, d] indenes *Tetrahedron* **64** 9052
[10] Wang H M, Guo S B, Hu Y G, Zeng X H and Yang G Y 2015 Synthesis and Antitumor Activity of Some Novel 5,6,7,8-Tetrahydrobenzo-thieno[2,3-d]pyrimidin-4(3H)-one Derivatives Chin. *J. Org. Chem.* **35** 1075
[11] Zheng A H, Long J Y, Wang H M and Zeng X H 2007 2-(2, 4-Dichloro-6-methylphenoxy)-3-isopropyl-5, 6, 7, 8-tetrahydrobenzothieno[2, 3-d] pyrimidin-4 (3H)-one *Acta Cryst.* **E63** 1142

Research of the diurnal soil respiration dynamic in two typical vegetation communities in Tianjin estuarine wetland

Q Zhang[1], W Q Meng[2] and H Y Li[1, 3]

[1]College of Environmental Science and Engineering, Nankai University, Tianjin, China

[2]College of Urban and Environment Science, Tianjin Normal University, Tianjin, China

Email: eialee@nankai.edu.cn

Abstract. Understanding the differences and diurnal variations of soil respiration in different vegetation communities in coastal wetland is to provide basic reliable scientific evidence for the carbon "source" function of wetland ecosystems in Tianjin.Measured soil respiration rate which changed during a day between two typical vegetation communities (*Phragmites australis*, *Suaeda salsa*) in coastal wetland in October, 2015. Soil temperature and moisture were measured at the same time. Each of the diurnal curves of soil temperature in two communities had a single peak value, and the diurnal variations of soil moisture showed a "two peak-one valley" trend. The diurnal dynamic of soil respiration under the two communities had obvious volatility which showed a single peak form with its maximum between 12:00-14:00 and minimum during 18:00. The diurnal average of soil respiration rate in *Phragmites australis* communities was 3.37 times of that in *Suaeda salsa* communities. Significant relationships were found by regression analysis among soil temperature, soil moisture and soil respiration rate in *Suaeda salsa* communities. There could be well described by exponential models which was $y = 0.245e^{0.105t}$ between soil respiration rate and soil temperature, by quadratic models which was $y = -0.276x^2 + 15.277x - 209.566$ between soil respiration rate and soil moisture. But the results of this study showed that there were no significant correlations between soil respiration and soil temperature and soil moisture in *Phragmites australis* communities (P>0.05). Therefore, under the specific wetland environment conditions in Tianjin, soil temperature and moisture were not main factors influencing the diurnal variations of soil respiration rate in *Phragmites australis* communities.

Keywords: soil respiration; *Phragmites australis* communities; *Suaeda salsa* communities; regression analysis

1. Introduction

Soil, considered as the largest and the slowest turnaround time carbon pool in the terrestrial ecosystem [1], is one of the important source or sink of atmospheric CO_2. Soil respiration is a major way of CO_2 emission from the land into the atmosphere [2-3], and is considered to be one of the largest flux in the global carbon cycle [4-5].The amount of carbon released from soil respiration is about 50-76 Pg every year [6], about 10 times larger than the carbon released by fossil fuel combustion [7]. In recent years, studies on soil carbon flux of forest ecosystems [8-9], farmland ecosystems [10-11], grassland ecosystems [12] and wetland ecosystems [13-14] are relatively abundant but researches in Tianjin coastal wetland are still less. Wetland known as the "kidney of the Earth" is the transition zone between terrestrial ecosystems and aquatic ecosystems, with abundant biodiversity and more complex habitat, is one of the most productive ecosystems in the world. The area of wetlands is about 5.14 million square kilometers, only accounting for 4% to 6% of the Earth's land area. However, it is the largest carbon pool in the world with its carbon stocks of about 770×108t which account for 35% of the terrestrial biosphere carbon [15], and plays an important role in global carbon cycle. Therefore, the impact of soil respiration to the wetland carbon cycle should not be overlooked.

In this paper, the typical vegetation communities of estuary wetland in Tianjin were chosen as the research objects. In October, the diurnal dynamics of soil carbon flux of two typical vegetation communities (*Phragmites australis, Suaeda salsa*)surrounding the wetland were separately measured by LI-8100A (LI-COR, Lincoln, NE, USA), and the differences in soil respiration in different vegetation types and the diurnal variation of soil carbon flux were analyzed in order to provide basic reliable scientific evidence for the carbon "source" function of wetland ecosystems in Tianjin.

2. Study area and research method

2.1. Study area

Yongding River, located in the north of Tianjin, was built in 1971 with the total length of 66km. Yongding River is situated in the coastal plain with the flat terrain which tilts from west to east. Elevation above sea level is smaller and the ground elevation is about 2.5m. The regional climate is warm temperate semi-humid monsoon with the characteristics of significant monsoon and four distinctive seasons. The average annual precipitation for the area is 566.1 mm, mainly concentrated in June to September. The average air temperature is 12.6℃ for many years [16]. The regional soil is salinization soil, which gives priority to with saline damp soil. Halophytes of estuary area are abundant in Yongding River, such as *Phragmites australis* and *Suaeda salsa*, which are in zonal distribution in general.

2.2. Research method

2.2.1. Experimental design. Two kinds of typical vegetation communities(*Phragmites australis, Suaeda salsa*) which were distributed widely in the surrounding wetland, were chosen as research objects through a preliminary comparison and analysis of field survey data. Five sample plots in size of 20m x 20m were laid in each vegetation community along the wetland from far and near, of which plot one to four for *Suaeda salsa* community and plot five for *Phragmites australis* community. The study was conducted to do measure in mid-October 2015. At each plot three monitoring points were randomly set, and the distance between adjacent monitoring points was not less than 5m. Each monitoring point soil set a PVC soil isolation ring whose front end inserted into the soil surface. In order to reduce the effects of the PVC ring placed on soil respiration rate, it should be balanced after 24-48 hours before measurement. The soil surface litter and fresh plant seedings were removed before measuring so as to reduce the impact of CO_2 released by woody debris decomposition and aerial parts of plants on measurement results [15]. The diurnal variations of soil respiration at three monitoring sites in each sample plots were measured during one day, every two hours from 6:00 to 18:00 by LI-8100A (LI-COR, Lincoln, NE, USA). Each monitoring point measured continuously three times during the same period. Soil moisture and soil temperature were measured at the same time. In addition, the topsoil of each monitoring point was collected for measuring the total nitrogen and total carbon contents after being air drying and sieving.

2.2.2. Assay method of soil properties. Soil total nitrogen content was determined by the Kjeldahl method (K-370).Soil total carbon content was determined by the potassium dichromate external heating method.

2.2.3. Statistical analysis. All the data was sorted by EXCEL, and the regression analysis was conducted by SPSS 22.0.

3. Results

3.1. Comparative analysis of environmental factors characteristics
Different vegetation types in the wetland had different soil physical and chemical properties. This study analyzed the soil physical and chemical properties of two vegetation types (*Phragmites australis*, *Suaeda salsa*) (Table1). Soil temperature varied widely in two researched. The ranges in *Suaeda salsa* communities and *Phragmites australis* communities were 9.842℃(minimum of 9.308℃, maximum of 19.15℃) and 8.7℃(minimum of 10.15℃, maximum of 18.85℃) respectively. The diurnal average soil temperature (13.714℃) in *Phragmites australis* communities was higher than that in *Suaeda salsa* communities (12.907℃). Compared with the *Phragmites australis* communities, the range of soil moisture in *Suaeda salsa* communities was smaller. But the soil moisture in *Phragmites australis* communities was significantly greater than that in *Suaeda salsa* communities. The soil total carbon content was similar in two vegetation communities, but the soil total nitrogen content was higher in *Phragmites australis* communities, 2.332 mg/g.

Table 1. Soil environmental factors of two vegetation communities.

Environmental Factor	Statistics	Type of Community	
		Suaeda salsa	*Phragmites australis*
Soil Temperature/℃	Mean	12.907	13.714
	Range	9.308—19.15	10.15—18.85
Soil Moisture/%	Mean	27.792	32.2
	Range	26.01—29.68	25.65—37.65
Total Carbon(mg/g)	Mean	1.349	1.423
Total Nitrogen(mg/g)	Mean	1.569	2.332

Table 2. Statistics of soil respiration rate in two vegetation communities.

Type of Community	N	Maximum /[umol/(m^2 · s)]	Minimum /[umol/(m^2 · s)]	Range /[umol/(m^2 · s)]	Mean /[umol/(m^2 · s)]	Standard Deviation /[umol/(m^2 · s)]	Standard Error /[umol/(m^2 · s)]	Coefficient of Variation
Suaeda salsa	7	1.711	0.409	1.302	1.036	0.436	0.165	0.421
Phragmites australis	7	4.645	2.075	2.570	3.493	1.040	0.393	0.298

3.2. Diurnal dynamic variation of environmental factors
The major environmental factors that affected the soil respiration were atmospheric temperature, atmospheric humidity, soil temperature, soil moisture, etc [14]. The diurnal dynamic variation of soil temperature in two vegetation communities between 6:00 and 18:00 was shown in Figure 1. Each of the diurnal curves of soil temperature in two communities had a single peak value. Both of the two communities reached a minimum at 6:00, increased gradually between 8:00-10:00, and the maximum at 12:00, then gradually decreased between 14:00-16:00 to about 10℃ at 18:00.

The diurnal dynamic variation of soil moisture in two vegetation communities between 6:00-18:00 was shown in Figure 2. The diurnal variations of soil moisture showed a "two peak-one valley" trend in both of two communities, but the *Phragmites australis* communities were significant. Soil moisture reached a minimum at 6:00, and the maximum at 16:00 which in *Phragmites australis* communities

IOP Conf. Series: Earth and Environmental Science **41** (2016) 012016 doi:10.1088/1755-1315/41/1/012016

was higher of 37.65 %. Soil moisture increased slowly between 6:00-10:00 in *Suaeda salsa* communities, probably because *Suaeda salsa* communities distributed intensively with a higher canopy density and limited light of the surface made soil temperature vary slightly and moisture evaporate slowly.

3.3. Diurnal variation of soil respiration rate

Differences of the factors such as vegetation type and soil property influenced on diurnal variation pattern of soil respiration rate in sample plots [17]. The diurnal dynamic of soil respiration under the two communities both had obvious wave which showed a single peak form with the maximum between 12:00-14:00 and the minimum at 18:00(Figure 3). Different vegetation communities had different diurnal average, maximum, minimum and variation range of soil respiration(Table 2). The maximum and minimum soil respiration rate in *Phragmites australis* communities were higher than that in *Suaeda salsa* communities. The diurnal variation range of soil respiration in *Phragmites australis* communities was larger. But the smaller coefficient of variation explained that soil respiration rate on diurnal variation scale had a high variation degree in *Phragmites australis* communities. The diurnal average of soil respiration rate in *Phragmites australis* communities was 3.37 times higher than that of *Suaeda salsa* communities. The diurnal variation trend of soil respiration rates in two communities were consistent with the soil temperature. Since the temperature difference was larger in autumn, and soil temperature in vegetation communities is closely linked to changes in regional temperature, the peak-valley value of soil respiration rate had a significant relationship with soil temperature. But soil respiration rates in two communities had begun to decline before soil moisture reaching the peak(Figure 4).

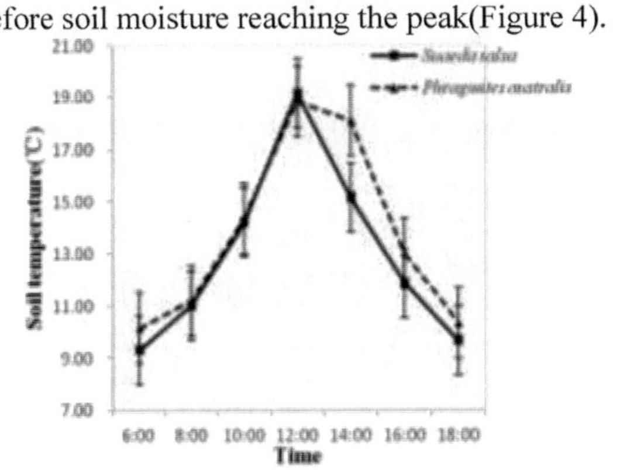

Figure 1. The diurnal variations of soil temperature in two vegetation communities.

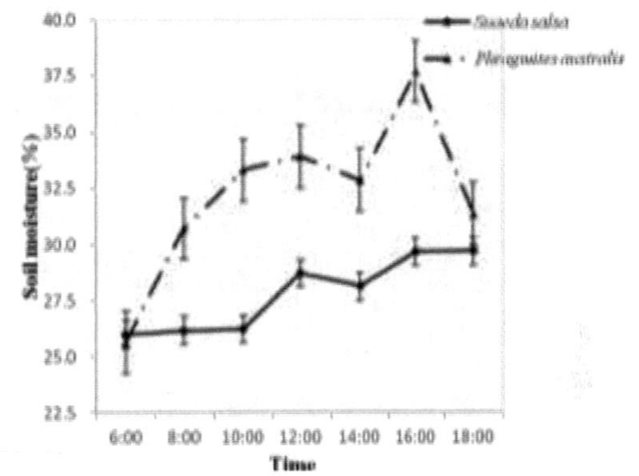

Figure 2. The diurnal variations of soil moisture in two vegetation communities.

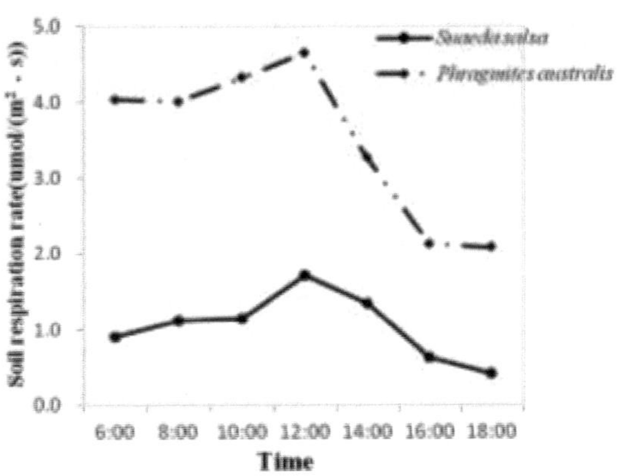

Figure 3. The diurnal variations of soil respiration in two vegetation communities.

IOP Conf. Series: Earth and Environmental Science **41** (2016) 012016 doi:10.1088/1755-1315/41/1/012016

3.4. Correlation between soil respiration rate and environmental factors

3.4.1. Effect of soil temperature on soil respiration. Soil temperature and soil moisture were respectively the control factors and limiting factors of soil respiration, and whose combined effect could explain most variations of soil respiration [18]. Different region probably had different sensitivity of soil respiration to soil temperature. The results of regression analysis for soil respiration rate and soil temperature in Suaeda salsa communities indicated that the soil respiration showed significant exponential correlation with soil temperature($R=0.760, R^2=0.577, P<0.05$)(Figure 5).The model was $y = 0.245e^{0.105t}$.

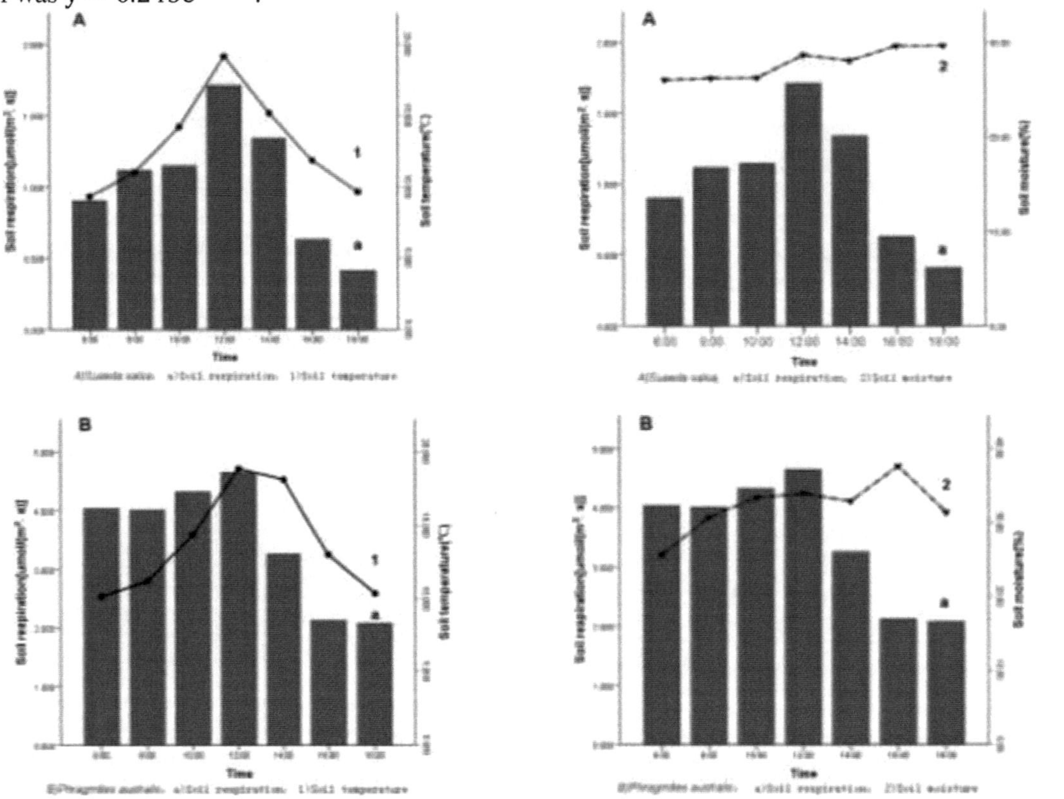

Figure 4. The diurnal variation tendencies of soil respiration and soil temperature and soil moisture in two vegetation communities.

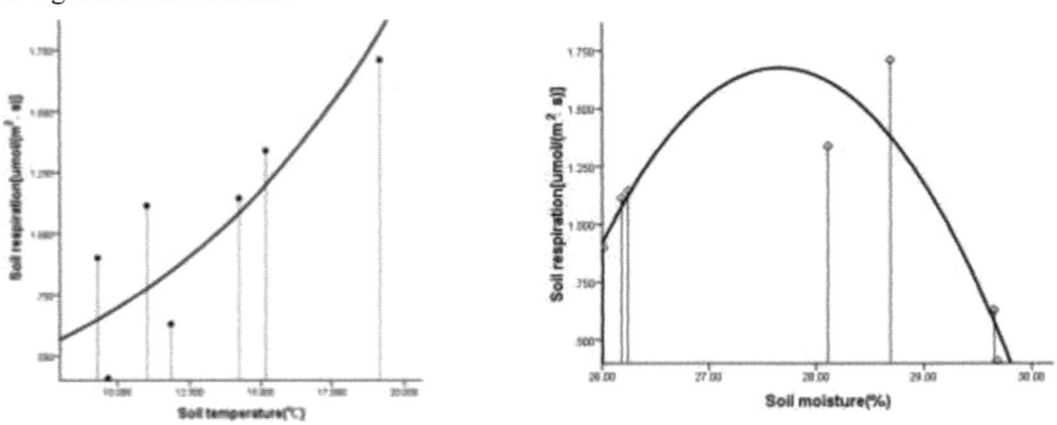

Figure 5. Relationship between soil respiration and soil temperature in *Suaeda salsa* communities.

Figure 6. Relationship between soil respiration and soil moisture in *Suaeda salsa* communities.

3.4.2. Effect of soil moisture on soil respiration. Soil respiration rate response to soil moisture was more sensitive due to large fluctuations of water content in the topsoil. To discuss the impact of soil moisture on soil respiration rate in Suaeda salsa communities, soil carbon flux and soil moisture were fitted by nonlinear regression analysis. The results showed that there was significant correlation between soil respiration rate and soil moisture described by quadratic models which was $y = -0.276x^2 + 15.277x - 209.566$ (R=0.903, R^2=0.815, P<0.05) (Figure 6).

Likewise, to discuss the impact of soil temperature and soil moisture on soil respiration rate in *Phragmites australis* communities, the soil carbon flux, soil temperature and soil moisture were fitted by linear regression analysis respectively. From the results, the larger significance probability(R=0.691, R^2=0.478, P>0.05) cannot explain there was a significant correlation between soil respiration rate and soil temperature and moisture in *Phragmites australis* communities.

4. Discussion

Soil carbon fluxes in wetland ecosystem had spatial and temporal variability. Soil respiration heterogeneity was determined by many factors. Different soil environmental factors, vegetation community types and environmental conditions could lead to different soil respiration rates [19]. Mingfeng Li has shown that SOC and TN contents directly or indirectly decided the variation of soil respiration in the relatively stable environmental factors condition [20]. *Phragmites australis* could absorb the nitrogen in soil and water, and its residues and absorbed nitrogen settled into the soil together after a period of time, hence soil total nitrogen content was higher in *Phragmites australis* communities. Xueyang Yu *et al.* have reported that the total carbon accumulation rates of *Suaeda salsa* marsh from the Liaohe Delta was mainly controlled by the accretion rate [13]. The vegetation coverage in situ plants determined organic carbon concentration of soil sediments which were in the sediments surface layer above 10 cm in the coastal wetland. The soil organic carbon was decomposed into CO_2, and microbial activity was relatively high in rich SOC soil to promote the soil respiration. However, the total carbon contents were similar in two communities in this study. Therefore the research concluded that the soil total carbon was not the dominant factor affecting the soil respiration.

Different vegetation types not only affected the soil organic matter content, but also had direct impacts on soil temperature and soil moisture [21]. In this study, as a result of the wide distribution of *Suaeda salsa* communities possibly, the vegetation canopy density was high. In contrast to *Phragmites australis* communities whose lower canopy density made the sunlight pass through branches and leaves to the soil easily, and the tightness was relatively poor. So the diurnal average of soil temperature in *Phragmites australis* communities was higher than that in *Suaeda salsa* communities, and the two variation ranges of soil temperature were larger. At the same time, due to *Phragmites australis* which were aquatic plants and their special habitat conditions, soil moisture was significantly greater than that in *Suaeda salsa* communities.

The behavior in diurnal dynamic of soil respiration was accorded with most research conclusions. Yanbing Xie has demonstrated that the daily pattern of soil respiration rate was obvious in *Phragmites australis* communities wetland in Panjin, Liaoning province, showing a single peak curve which was consistent with the temperature curve [22]. Changchun Song and Yiyong Wang pointed out that CO_2 emissions in marsh wetland plants-soil system were associated with the root layer soil temperature, presented a significant exponential correlation [23]. The relation between soil respiration rate and soil temperature in *Suaeda salsa* communities in the wetland of Yongding River was consistent with distribution models fitted by other wetland ecosystems [24, 25, 26].

In addition to environmental factors, biological factors were also important to affect soil respiration. The temporal dynamic rules of the soil respiration could be divided into diurnal and seasonal variation. Biological factors were smaller change variables on a circadian scale which was the reason why the diurnal variation rules of soil respiration could not be explained by biological factors [27]. Numerous studies have shown that, temperature was the closest environmental factors to carbon exchanged in ecosystems and soil carbon fluxes [28,29].And in this study, there were no significant correlations between soil respiration and soil temperature and soil moisture in *Phragmites australis* communities by the linear regression analysis($P>0.05$). Some researchers considered that there were weak correlations between temperature and soil carbon fluxes, which was greatly influenced by light radiation [30]. Zixian Du has shown that, for *Phragmites australis* communities in Minjiang riverside wetlands, soil temperature was not the main impact factors in diurnal respiration, which was consistent with this study [31].

5. Conclusions

(1)Each of the diurnal curves of soil temperature in two communities had a single peak value. Both of the two communities reached a minimum at 6:00, increased gradually between 8:00-10:00, and the maximum at 12:00, then gradually decreased between 14:00-16:00 to about 10℃ at 18:00. And the diurnal variations of soil moisture showed a "two peak-one valley" trend. Soil moisture reached a minimum at 6:00, and the maximum at 16:00 , which in *Phragmites australis* communities was higher.

(2)The diurnal dynamic of soil respiration in two communities had obvious volatility which showed a single peak form with its maximum between 12:00—14:00 and minimum at 18:00. The diurnal average of soil respiration rate in *Phragmites australis* communities was 3.37 times of the *Suaeda salsa* communities.

(3) Significant relationships were found by regression analysis among soil temperature, soil moisture and soil respiration rate in *Suaeda salsa* communities. There could be best described by exponential models which were $y = 0.245e^{0.105t}$ between soil respiration rate and soil temperature, by quadratic models which was $y = -0.276x^2 + 15.277x - 209.566$ between soil respiration rate and soil moisture. But the results of this study showed that there were no significant correlations between soil respiration and soil temperature and soil moisture in *Phragmites australis* communities ($P>0.05$). Therefore, in the specific wetland environment conditions in Tianjin, soil temperature and moisture were not main factors influencing the diurnal variations of soil respiration rate in *Phragmites australis* communities.

Acknowledgments

This work was financially supported by the National Natural Science Foundation (31370700), National Natural Science Youth Fund Project (41301096).

References

[1] Schlesinger W H 1982 Carbon storage in the calishe of arid soils: a case study from Arizona *Soil Sci.* **133** 247-55
[2] Schimel D S, Braswell B H, Holland E A , McKeown R, Ojima D S, Painter T H, Parton W J and Townsend A R 1994 Climatic, edaphic, and biotic controls over storage and turnover of carbon in soils *Glob. Biogeochem. Cycles.* **8** 279–93
[3] Schlesinger W and Andrews J 2000 Soil respiration and the global carbon cycle *Bio-geochem.* **48** 7–20
[4] Kirschbaum M U F 1995 The temperature dependence of soil organic matter decomposition ,and the effect of global warming on soil organic C storage *Soil Biol. Biochem.* **27** 753–60
[5] Raich J W and Schlesinger W H 1992 The global carbon dioxide flux in soil respiration and its relationship to vegetation and climate *Tellus B.* **44** 81–99
[6] Post W M *et al* 1982 Soil carbon pools and world life zones *Nature* **298** 156-9
[7] Schlesinger W H 1977 Carbon balance in terrestrial detritus *Annu. Revi. Ecol. and System.* **8** 51-81
[8] Yang J Y and Wang C K 2006 Effects of soil temperature and moisture on soil surface CO_2 flux

of forest in Northeastern China *J. Plant Ecol.* **30** 286-94

[9] Liu Y, Han S J and Hu Y L 2005 Effects of soil temperature and humidity on soil respiration rate under Pinus sylvestriformis forest *Chin. J. Appl. Ecol.* **16** 1581-5

[10] Huang Y, Liu S L and Shen Q R 2001 Model establishment for simulating soil organic carbon dynamics *Scie. Agric. Sini.* **34** 465-8

[11] Wang H S, Zhang J S and Meng P 2007 Changes and mechanism of soil respiration on winter wheat field in low mountain and hill regions of North China *Chin. J Agrometeorol.* **28** 21-4

[12] Chen N N, Guan D X and Jin C J 2011 Characteristics of soil respiration on Horqin Meadow *Chin. J. Gras.* **33** 82-7

[13] Yu X Y, Ye S Y and Nicholas L Y 2014 Carbon sequestration along vegetation coverage gradient in the Suaeda salsa marsh from the Liaohe Delta *Geol. Chin.* **41** 648-57

[14] Wang F C, Liu J Z and Chen Y J 2013 Yellow River delta wetland soil respiration and the environmental factor Analysis *Yell. River* **35** 81-4

[15] Ju M T, Wang Y X and Meng W Q 2009 *Protection and Evaluation of Wetland Ecosystem* (Beijing: Chemical Industry Press) p 1

[16] Liu W, Guo Y Z and Feng Y N 2006 Ecological environment impact assessment of regulation project in Yongding New River *Haihe. Water Reso.* **5** 9-11

[17] Feng C Y, Lv S H and Gao J X 2008 Soil respiration characteristics of different vegetation types in the mountain areas of North China *J. Beijing Fore.Univ.* **30** 20-6

[18] Mielnick P C and Dugas W A 2000 Soil CO_2 flux in a tallgrass prairie *Soil Biol. and Biochem.* **32** 221-8

[19] Sha C Y, Tan J, Wang Q and Wang M 2015 Methane and carbon dioxide emissions from different types of Riparian wetland *Ecolo. and Environ. Sci.* **24** 1182-90

[20] Li M F, Dong Y S, Geng Y B and Qi Y C 2004 Analyses of the correlation between the fluxes of CO_2 and the distribution of C&N in grassland soils *Environ. Sci.* **25** 7-11

[21] Liu S H and Fang J Y 1997 Effect factors of soil respiration and the temperature's effects on soil respiration in the global scale *Acta Ecolo. Sini.* **17** 469-76

[22] Xie Y B, Jia Q Y, Zhou L, Li R P and Lv G H 2006 Soil respiration and its controlling factors at phragmites communis wetland in Panjin *J. Meteorol. and Environ.***22** 53-8

[23] Song C C and Wang Y Y 2006 Responses of soil temperature in wetland ecosystem to air temperature and their effects on CO_2 emission *Chin. J. Appl. Ecol.* **17** 625-9

[24] Yang J J, Lv G H and Zhang Y 2009 Soil respiration of different plant communities in Ebinur Lake watershed *Res. Environ. Sci.* **22** 362-70

[25] Qiao M Y, Shen C C, Shi H H, Ding D W and Guo Z 2015 Diurnal change of soil carbon flux of island forests: A case study in Beichangshan Island *Mari. Environ. Sci.* **34** 377-83

[26] Lu Y K, Liu J Z, Chen Y J, Lu M and Xiao S W 2012 The research on carbon dioxide flux of the three communities in the Dongpinghu riparian zone *Chin. Agric. Sci. Bulletin* **28** 8-14

[27] Xie Y B, Jia Q Y, Li R P and Lv G H 2009 Research on the influences of biological factors on the soil respiration of reed ecosystem in the wetland of Panjin *J. Anhui Agric. Sci.* **37** 18070-72

[28] Boone R D *et al* 1998 Roots exert a strong influence on the temperature sensitivity of soil respiration *Nature* **396** 570-2

[29] Craine J M and Wedin D A 2002 Determinants of growing season soil CO_2 flux in a Minnesota grassland *Biogeochem.* **59** 303-13

[30] Raich J W and Tufekcioglu A 2000 Vegetation and soil respiration: Correlations and controls *Biogeochem.* **48** 71-90

[31] Li Z X, Zeng H D and Huang X H 2010 Soil respiration and controlling factors at Phragmites communis community in riverside wetlands *J. Subt. Reso. Environ.* **5** 49-55

Antifungal activities of (E)-4-benzylidene-5-oxopyrrolidine-2-carboxamides and 6-oxo-1,2,3,6-tetrahydropyridin-2-carboxamides synthezed via ugi reaction from baylis-hillman bromides

X H Zeng[1], L C Xiang[1], H L Li[1], H M Wang[1,2] and X B Wang[1,2]

[1]Laboratory of Chinese Herbal Pharmacology, Renmin Hospital, Hubei University of Medicine, Shiyan 442000, P.R. China

E-mail: meirwang@126.com; 459560483@qq.com

Abstract. The derivatives of N-containing heterocycles containing the pyrrolidone or pyridinone moiety are of high importance because they have extensive biological properties including antifungal, anti-muscarinic, anti-cancer, anti-viral and anti-HIV activities. In our previous work, we described an one-pot transition-metal-free, base-mediated synthesis of pyrrolidinones ((E)-4-benzylidene-5-oxopyrrolidine-2-carboxamides **6**) and an one-pot regioselective synthesis of previously seldom reported pyridinones (6-oxo-1,2,3,6-tetrahydropyridin-2-carboxamides **9**) via Ugi reaction from Baylis-Hillman Bromides. Herein, the growth inhibitory effect of one concentration (50mg/L) of compounds **6** and **9** against fungi (*Gibberella zeae*) in vitro was tested by the method of toxic medium. Ihe results showed that the inhibitory effects of compounds **6** against *Gibberella zeae* are not obvious, but compounds **9** showed moderate to good inhibitory effects against *Gibberella zeae*. Compound **9h** showed the best inhibition rate against *Gibberella zeae* with 95%.

1. Introduction

The derivatives of N-containing heterocycles containing the pyrrolidone or pyridinone moiety are of high importance because they have extensive biological properties including antifungal, anti-muscarinic, anti-cancer, anti-viral and anti-HIV activities. [1-8]

Constructing pyrrolidone scaffolds by condensation between the carboxylic acid or ester with different primary amines primarily is the classic synthetic methodology.[9] In addition, the rearrangement of β-lactam, the cycloaddition reaction, Michael addition reaction, and intramolecular allylic alkylation reaction catalysed by transition metal have been reported.[10-13] Synthesis the construction of 1,2,3,6-tetrahydropyridin-6-one scaffold have been payed attention to intramolecular ammonolysis, reduction of piperadinedione and RCM reaction. However, these strategies typically require lengthy reaction steps, yield products with low regioselectivities, and are compatible with few functional groups.[14-16]

The Baylis–Hillman (B-H) reaction, as an effective synthetic tool for building C–C bonds by a straightforward one-pot reaction, can obtain molecules with complex functional groups. The adducts constructed by B-H reaction, have been used to prepared extensive heterocyclic compounds, such as quinolinones, pyrroles, 1,3-thiazinones and pyrrolopyrazines.[17-19]

The Ugi reaction as one important reaction of the isocyanide-based multicomponent reactions shows advantages to traditional two components reactions with effective and atom-economical character. And as a superior tool for multifarious synthesis for heterocycles similar to pharmaceuticals or natural products, It have received increased attention, because it can generate a complex α-acylamino- substituted carboxamide to be used for synthesis of nitrogen heterocyclic compounds widely.[20, 21]

In our previous work, we described an one-pot base-mediated synthesis of pyrrolidinones. The reaction of 2-(Bromomethyl)-3-aryl-2-propenoic acid (Baylis-Hillman bromides) **1**, obtained from Baylis-Hillman adducts, with primary amines **2**, isocyanides **3** and arylglyoxals **4** to give the (E)-4-benzylidene-5-oxopyrrolidine-2-carboxamides **6** with 5-oxo-4,5-dihydro-1,4-oxazepine-3-carboxamides **5** via a sequential Ugi and intramolecular cyclization reaction in good yield, and Cs_2CO_3 was used to adjust the pH value of the solution to 6 during the reaction process.[18]

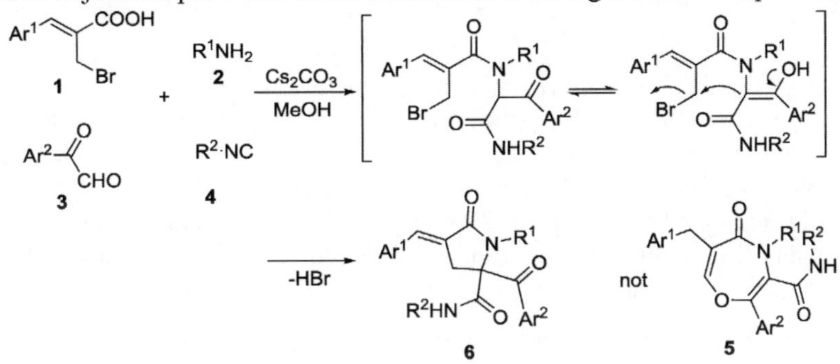

Then we disclose an one-pot regioselective synthesis of previously seldom reported pyridinones (6-oxo-1,2,3,6-tetrahydropyridin-2-carboxamides **9**), Which has been prepared by a linear domino reaction based on I-MCR and the hydrolysis of ylide intermediates from from Phosphonium salts **7** of Baylis-Hillman bromides **1**, primary amines **2**, isocyanides **3** and arylglyoxals **4** in one pot in the presence of K_2CO_3 in methanol， and no 6-oxo-1,6-dihydropyridine-2-carboxamide **8** was found.[19]

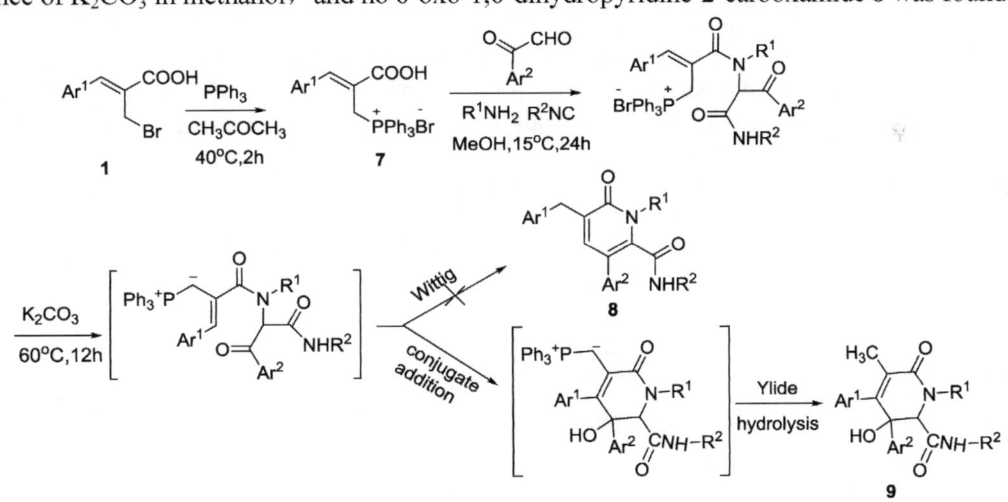

Herein， We have described the antifungal activities of (E)-4-benzylidene-5-oxopyrrolidine-2-carboxamides **6** and 6-oxo-1,2,3,6-tetrahydropyridin-2-carboxamides **9** Synthezed via Ugi reaction from Baylis-Hillman Bromides.

2. Material and methods

IR spectra (absorption in cm^{-1}, with KBr pellets, recorded by a PE-983 infrared spectrometer apparatus); MS were measured on a Finnigan Trace MS spectrometer. ^1H NMR spectra (chemical shifts (δ) in ppm, $(CH_3)_4Si$ (δ = 0) as an internal reference, recorded in $CDCl_3$ or DMSO on a Varian Mercury Plus 400 (400 Hz) or Varian Mercury Plus 600 (600 Hz) spectrometer apparatus); Melting points (determined by a X-4 model melting points apparatus, uncorrected); Elementary analyses (determined by a Vario EL III elementary analysis apparatus). The growth inhibitory effect of one concentration (50mg/L) of compounds **6** against *Gibberella zeae* in vitro was tested by the method of toxic medium. Coefficients of variation were <10%.

2.1. Preparation of (E)-4-benzylidene-5-oxopyrrolidine-2-carboxamides **6**
In the mixture of primary amine **2** (1.5 mmol) and the Baylis-Hillman bromomethyl cinnamic acid [17] **1** (1.5 mmol) in MeOH (10 mL) at room temperature, was putted the aryl substituted glyoxal **4** (1.5 mmol) and the alkyl substituted isocyanide **3** (1.5 mmol) consecutively. Cs_2CO_3 was used to adjust the pH value of the solution to 6 during the reaction process. After stirring for 12 hours at room temperature, the mixture was chilled to precipitate the solid. Then the resulted solid was filtered and recrystallized from the mixed solvent (ether: ethanol, 2:1) to give (E)-4-benzylidene-5-oxopyrrolidine-2-carboxamide.

2.2. Preparation of 6-oxo-1,2,3,6-tetrahydropyridin-2-carboxamides **9**
In the mixture of the Baylis-Hillman phosphonium salt **7** (1.5 mmol) and the aryl substituted glyoxal **4** (1.5 mmol) in methanol (15 mL), was putted primary amines **2** (1.5 mmol), and the alkyl substituted isocyanide **3** (1.5 mmol) successively. The mixture was elevated the temperature to 60 °C and K_2CO_3 (0.42 g, 3 mmol) was added, after stirring for 24 hours at room temperature. The mixture was stirred at 60 °C for 12 hours and filtered, then the solvent was removed off under reduced pressure, the residue was purified by silica gel chromatography (ethyl acetate: hexanes, 5:1) to give 6-oxo-1,2,3,6-tetrahydropyridin-2-carboxamide.

3. Results and discussion
The growth inhibitory effect of one concentration (50mg/L) of (E)-4-benzylidene-5-oxopyrrolidine-2-carboxamides **6a-6t** (Table 1) and 6-oxo-1,2,3,6-tetrahydropyridin-2-carboxamides **9a-9k** (Table 2) against *Gibberella zeae* in vitro was tested by the method of toxic medium with commercial diniconazole as positive control.

The results showed that the inhibitory effects of compounds **6** against *Gibberella zeae* are not obvious (Table 1), but compounds **9** showed moderate to good inhibitory effects against *Gibberella zeae*, especially compounds **9e**, **9f**, **9h** and **9j**. Compound **9h** showed the best inhibition rate against *Gibberella zeae* with 95% (Table 2). The inhibition rate of the commercialization diniconazole was 96% under the same control conditions.

Analysis of the structure-activity relationship showed that group Ar^1 has little effect on the bactericidal activity of compound **9**, but groups R^1 and Ar^2 have great influence on its bactericidal activity. When the R^1 is alkyl, especially isopropyl, compounds **9** showed good activity relatively. However, the Ar^2 is 4-bromophenyl, compounds **9** have no bactericidal activity.

Table 1. In vitro growth inhibitory effect of compounds **6** against *Gibberella zeae* (50mg/L).

No.	Ar^1	Ar^2	R^1	R^2	Inhibition rate[a]
6a	phenyl	4-bromophenyl	*n*-propyl	*t*-butyl	17%
6b	phenyl	4-chlorphenyl	benzyl	*t*-butyl	29%
6c	phenyl	4-chlorphenyl	*n*-propylopyl	*c*-hexyl	32%
6d	phenyl	4-chlorphenyl	*n*-propyl	*t*-butyl	37%
6e	phenyl	4-chlorphenyl	benzyl	*n*-butyl	29%
6f	phenyl	4-chlorphenyl	*n*-propyl	*n*-butyl	27%
6g	phenyl	4-chlorphenyl	*i*-propyl	*c*-hexyl	33%
6h	phenyl	4-chlorphenyl	*i*-propyl	*t*-butyl	34%
6i	phenyl	4-chlorphenyl	*n*-bu	*t*-butyl	21%
6j	phenyl	4-chlorphenyl	methyl	*t*-butyl	31%
6k	phenyl	4-chlorphenyl	methyl	*c*-hexyl	32%
6l	phenyl	4-chlorphenyl	ethyl	*t*-butyl	39%
6m	phenyl	4-chlorphenyl	ethyl	*c*-hexyl	28%
6n	phenyl	4-chlorphenyl	benzyl	*c*-hexyl	41%

6o	4-chlorphenyl	4-bromophenyl	*n*-propyl	*t*-butyl	35%
6p	phenyl	4-bromophenyl	benzyl	*t*-butyl	32%
6q	4-chlorphenyl	4-bromophenyl	benzyl	*t*-butyl	37%
6r	4-chlorphenyl	4-bromophenyl	*i*-propyl	*t*-butyl	29%
6s	4-chlorphenyl	4-chlorphenyl	4-bromophenyl	*t*-butyl	41%
6t	4-chlorphenyl	4-chlorphenyl	phenyl	*t*-butyl	41%

Inhibition rate = (Controlled colony diameter - Processing colony diameter)/(Contrast colonies diameter - Inoculating colony diameter) * 100%.

Table 2. In vitro growth inhibitory effect of compounds **9** against *Gibberella zeae* (50mg/L).

No.	Ar^1	Ar^2	R^1	R^2	Inhibition rate
9a	phenyl	4-bromophenyl	*n*-propyl	*t*-butyl	27%
9b	phenyl	4-bromophenyl	*i*-propyl	*t*-butyl	24%
9c	phenyl	4-chlorphenyl	benzyl	*t*-butyl	32%
9d	phenyl	4-bromophenyl	benzyl	*t*-butyl	32%
9e	phenyl	4-chlorphenyl	*i*-propyl	*t*-butyl	67%
9f	phenyl	4-chlorphenyl	*n*-propyl	*t*-butyl	61%
9g	phenyl	4-bromophenyl	*i*-propyl	*c*-hexyl	20%
9h	phenyl	4-chlorphenyl	*i*-propyl	*c*-hexyl	95%
9i	phenyl	4-chlorphenyl	benzyl	*c*-hexyl	22%
9j	4-chlorphenyl	4-chlorphenyl	*n*-propyl	*t*-butyl	51%
9k	4-chlorphenyl	4-chlorphenyl	*n*-propyl	*c*-hexyl	10%

Inhibition rate = (Controlled colony diameter - Processing colony diameter)/(Contrast colonies diameter - Inoculating colony diameter) * 100%.

Acknowledgments

We gratefully acknowledge financial support of this work by the Open Project of Hubei Key Laboratory of Wudang Local Chinese Medicine Research (Hubei University of Medicine (Grant No. WDCM009), the Hubei Province health and family planning scientific research project (Grant No. WJ2015Z113), the of Shiyan Municipal Science and Technology Bureau Science and technology project (Grant No. 15Y12), and the scientific research innovation team of Hubei University of Medicine (grant Nos. 2014QDJZR10, 2014CXZ01).

References

[1] Kenda B M, Matagne A C, Talaga P E, Pasau P M, Differding E, Lallemand B I and Fuks B 2004 Discovery of 4-substituted pyrrolidone butanamides as new agents with significant antiepileptic activity *J. Med. Chem.* **47**(3) 530-49

[2] Kraus J L, Bouygues M, Courcambeck J and Chermann J C 2000 Use of proline bioisosteres in potential HIV protease inhibitors: phenylalanine-2-thiophenoxy-3-pyrrolidinone: synthesis and anti-HIV evaluation *Bioorg. Med. Chem. Lett.* **10**(17) 2023-6

[3] Jayasinghe L, Abbas H K, Jacob M R, Herath W H and Nanayakkara N D 2006 N-Methyl-4-hydroxy-2-pyridinone Analogues from Fusarium oxysporum *J. nat. prod.* **69**(3) 439-42

[4] Sanderson P E, Cutrona K J, Dorsey B D, Dyer D L, McDonough C M, Naylor-Olsen A M and Krueger J A 1998 L-374,087, an efficacious, orally bioavailable, pyridinone acetamide thrombin inhibitor *Bioorg. Med. Chem. Lett.* **8**(7) 817-22

[5] Goldman M E, Nunberg J H, O'Brien J A, Quintero J C, Schleif W A, Freund K F and Hoffman J M 1991 Pyridinone derivatives: specific human immunodeficiency virus type 1 reverse transcriptase inhibitors with antiviral activity *Proc. Natl. Acad. Sci. USA* **88**(15) 6863-7

[6] Nunberg J H, Schleif W A, Boots E J, O'Brien J A, Quintero J C, Hoffman J M and Goldman M E 1991 Viral resistance to human immunodeficiency virus type 1-specific pyridinone reverse transcriptase inhibitors *J. virl.* **65**(9) 4887-92

[7] Saari W S, Hoffman J M, Wai J S, Fisher T E, Rooney C S, Smith A M and O'Brien J A 1991 2-

Pyridinone derivatives: a new class of nonnucleoside, HIV-1-specific reverse transcriptase inhibitors *J. Med. Chem.* **34**(9) 2922-5

[8] Himmel D M, Das K, Clark A D, Hughes S H, Benjahad A, Oumouch S and Meyer C 2005 Crystal structures for HIV-1 reverse transcriptase in complexes with three pyridinone derivatives: a new class of non-nucleoside inhibitors effective against a broad range of drug-resistant strains *J. Med. Chem.* **48**(24) 7582-91

[9] Wu Y H and Feldkamp R F 1961 Pyrrolidines. I. 1-Substituted 3-Pyrrolidinylmethyl Alcohols and Chlorides1 *J. Org. Chem.* **26**(5) 1519-24

[10] Alcaide B, Almendros P, Cabrero G and Ruiz M P 2005 Organocatalytic ring expansion of β-lactams to γ-lactams through a novel N1-C4 bond cleavage. Direct synthesis of enantiopure succinimide derivatives *Org. lett.* **7**(18) 3981-4

[11] Miyabe H, Ueda M, Fujii K, Nishimura A and Naito T 2003 Tandem carbon-carbon bond-forming radical addition-cyclization reaction of oxime ether and hydrazone *J. Org. Chem.* **68**(14) 5618-26

[12] Tan D Q, Atherton A L, Smith A J, Soldi C, Hurley K A, Fettinger J C and Shaw J T 2012 Synthesis of a γ-Lactam Library via Formal Cycloaddition of Imines and Substituted Succinic Anhydrides *J. Comb. Chem.* **14**(3) 218-23

[13] Vogel S, Bantreil X, Maitro G, Prestat G, Madec D and Poli G 2010 Palladium-catalyzed intramolecular allylic alkylation of α-sulfinyl carbanions: a new asymmetric route to enantiopure γ-lactams *Tetrahedron Lett.* **51**(11) 1459-61

[14] Huang C G, Chang B R and Chang N C 2002 An efficient synthesis of γ-substituted α, β-unsaturated-δ-lactams. Formal synthesis of (±)-protoemetinol *Tetrahedron Lett.* **43**(15) 2721-3

[15] Grison C, Genève S and Coutrot P 2001 Enantioselective synthesis of α, β-unsaturated γ-and δ-lactams *Tetrahedron Lett.* **42**(23) 3831-4

[16] De Matteis V, van Delft F L, Tiebes J and Rutjes F P 2006 A Ring-Closing Metathesis Pathway to Fluoroviny-Containing Nitrogen Heterocycles *Eur. J. Org. Chem.* **2006**(5) 1166-76.

[17] Basavaiah D, Reddy B S and Badsara S S 2010 Recent contributions from the Baylis− Hillman reaction to organic chemistry *Chem. Rev.* **110**(9) 5447-674

[18] Zeng X H, Wang H M, Wu L and Ding M W 2013 One-pot synthesis of 5-oxopyrrolidine-2-carboxamides via a tandem Ugi 4CC/S N cyclization starting from Baylis–Hillman bromides *Tetrahedron* **69**(19) 3823-8

[19] Zeng X H, Wang H M and Ding M W 2015 Unexpected Synthesis of 5, 6-Dihydropyridin-2 (1 H)-ones by a Domino Ugi/Aldol/Hydrolysis Reaction Starting from Baylis–Hillman Phosphonium Salts *Org. lett.* **17**(9) 2234-7

[20] Gulevich A V, Zhdanko A G, Orru R V and Nenajdenko V G 2010 Isocyanoacetate derivatives: synthesis, reactivity, and application *Chem. Rev.* **110**(9) 5235-331

[21] Zeng X H, Wang H M, Yan Y M, Wu L and Ding M W 2014 One-pot regioselective synthesis of β-lactams by a tandem Ugi 4CC/S N cyclization *Tetrahedron* **70**(23) 3647-52

The influence of substrates rates on the germination characteristic of a soil seed bank

N Zhao[1], M X He[1], H Y Li[1, 3]and W Q Meng[2]

[1]College of Environmental Science and Engineering, Nankai University, Tianjin 300350, China

[2]College of Urban and Environmental Science,Tianjin Normal University, Tianjin 300387, China)

E-mail: eialee@nankai.edu.cn

Abstract. Soil seed bank (SSB) is considered as an important way of vegetation restoration, it can fleetly achieved vegetation diversification and the course of succession when the topsoil mixed with planting substrates. In this paper, a greenhouse germination method was used to explore the effect on germination characteristic of soil seed bank by adding different inorganic substrates, such as activated carbon, perlite and vermiculite. The results showed that perlite and vermiculite can effectively promote the germination of soil seed bank, but also significantly promote Shannon-Wiener diversity index. When vermiculite mixed with the topsoil in 40%, the germination effect of soil seed bank was more obviously than other groups ; at this time, the density of soil seed bank reached 6×10^5 plants/m^2, Shannon Wiener diversity index reached 1.4354. Therefore, it was more conducive to improve the soil seed bank density and species diversity by adding 40% vermiculite in the topsoil.

1. Introduction

Soil seed bank (SSB) is a collection of viable seeds present in the soil and on its surface with regeneration potential. As the "genetic potential library"[1] in the ground vegetation, soil seed bank plays an extremely important role in the process of vegetation restoration and vegetation potential updates[2]. In recent years, with the international status of the degraded damaged land and ecological system's restoration and reconstruction rising, the research of vegetation degradation's soil seed bank were studied extensively and deeply in our country (such as grassland, wetland and abandoned land's degradation [3-5]). However, soil seed bank's research which can be applied to the actual vegetation restoration is still few.

The research of soil seed bank on foreign country started earlier and the application in engineering practice also had a breakthrough, especially there were lots of soil seed bank's application case in Japan, such as using soil seed bank applied to the vegetation restoration for reservoir bare land, waste dump and high-speed road surrounding [6-7]. In addition, the Japanese scholar have researched on spray seeding afforestation methods in topsoil mixed with planting substrates, which mainly reflected in the forest vegetation restoration[8] and slope greening[9]. Based on the above reasons , a greenhouse germination method was used in this study, in order to explore the influence of different rates of inorganic substrates on the germination characteristics of soil seed bank, and provide valuable reference on soil seed bank for the future application in vegetation restoration project.

2. Materials and methods

2.1. Topsoil sampling site

Sampling site was located in Tianjin Wuqing District, China (E116° 43'46"-E117° 19'59", N39° 07'05"-N39° 42' 20"), which belonged to the typical suburban junction area, and also had relatively rich flora and soil seed bank diversity. The matter accumulation quantity of soil organic was not much, and the soil was slightly alkaline. There were abundant plant species, such as *Ulmus pumila*, *Fraxinus velutina*, *Tamarix chinensis*, *Robinia pseudoacacia,Morus alba,* et al.

2.2. Sampling method

The research started in March 2014, 30 samples were randomly selected from the sampling site (the sample area was 1 m×1 m), and collected 0-15 cm of topsoil as a soil sample. The samples were mixed evenly and removed the impurities (sand, stone, leaves, etc.), then arranged samples in snakeskin bag and brought them to the laboratory. Meanwhile, the vegetation condition was recorded, including species composition, quantity and vegetation coverage.

2.3. Experimental design

In the practical seedling applications, such as agriculture, horticulture and flower cultivation, the use of substrate is a very common phenomenon [10-11]. In this study, three kinds of inorganic substrate were selected, in which included activated carbon (A), perlite(P) and vermiculite(V). The reason for choosing these substrates were that these substrates were widely used and easily obtained in Tianjin.

And then, a greenhouse germination method was used to explore the germination characteristic of soil seed bank, in which include the germination density, Shannon-Wiener diversity and the dynamic of seed germination. The germination greenhouse was located at the Tianjin University of Technology, where germination experiments were implemented. After removal of gravel or dead leaves, the soil samples were mixed with three kinds of inorganic substrates in 10%, 20%, 30%, 40% respectively. The mixture of substrate and topsoil were laid in the seedbed to a thickness of 5cm (the size of germination disk is 50 cm×20 cm×5cm). In addition, this research provided a set of blank control group (without adding any substrates). The germination experiment began in May 5th and ended in May 30th. The experiment needed to keep the soil moist, and record the amount and type of soil seed germination and other data regularly.

2.4. Data analysis

The research data were all analyzed and disposed in Excel and R3.0.2 software. In the calculation of the community diversity index, the Shannon-Wiener diversity index was selected in this study, and the formula was as follows:

$$\text{Shannon-Wiener diversity index: } H = -\sum_{i=1}^{S}(P_i \times \ln P_i) \tag{1}$$

In the formula: P_i was the ratio of germination amount of plant i to the total germination amount of the soil seed bank.

3. Result and discussion

3.1. The germination density of soil seed bank

As shown in figure 1, relative to the blank control group (the dashed line), perlite and vermiculite effectively promoted the germination of the soil seed bank, but activated carbon had little effect on seed germination. When vermiculite mixed with topsoil in 40%, the density of soil seed bank was obvious, as much as 6×10^5 plants/m^2. Due to the inorganic substrates had larger porosity, which could promoting the air circulation and keeping moisture, and it also could promoting the growth of the plant roots and the germination of soil seed bank; in addition, when it compared with the other inorganic substrates, the fertilizer ability of activated carbon was much worse which caused the little effect on the density of soil seed bank.

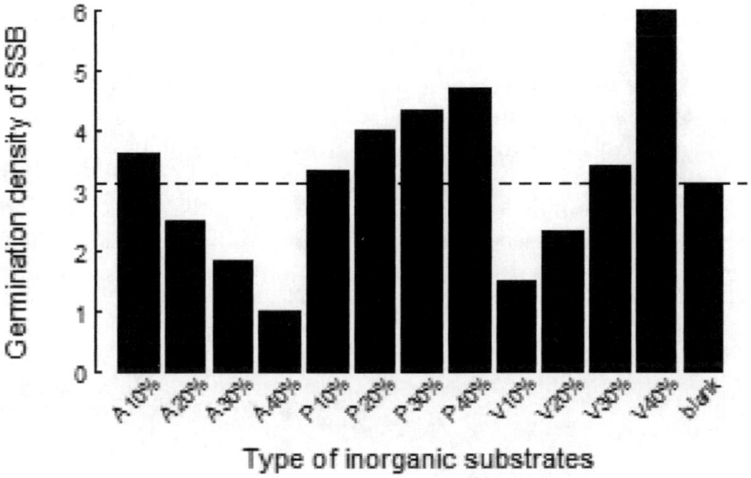

Figure 1. The germination density of soil seed bank by adding different inorganic substrates (units:10^5 plants/m^2).

3.2. *Shannon-Wiener diversity index*

As shown in figure 2, relative to the blank control group (the dashed line), most of the groups improved the Shannon-Wiener diversity of germination species, especially perlite and vermiculite had significant promoting effect on Shannon-Wiener diversity index, while activated carbon in10% to 30% has inhibitory effect on it. Through analyzed the type and the quantity of seed germination species, found that adding 10%-30% activated carbon in topsoil greatly promoted *Setaria viridis* growing (more than 80% of the total number of seedlings germination). Shannon-Wiener diversity index not only related to the species, but also related to the individual distribution uniformity, therefore it caused the phenomenon that Shannon-Wiener diversity index reduced by adding activated carbon.

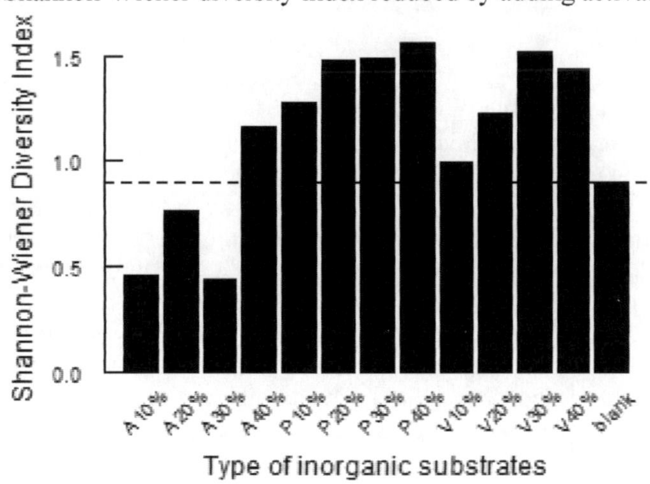

Figure 2. Shannon-Wiener diversity index of soil seed bank by adding different inorganic substrates.

3.3. *Dynamic curve of seed germination*

Several groups with high seed germination density were selected as the research objects, in order to explore the effect of different ratio of inorganic substrates (activated carbon, perlite and vermiculite) on seed germination. As shown in figure 3, it could be seen that the germination curve generally presented "S" shape, and with "smooth growing-rapid growing-stable" period, but the time for each period differed depended on different inorganic substrates. Compared to the others treatment, the topsoil mixed with 40% vermiculite had very significant effect on the late period of seed germination. In addition, the number of seed germination in blank control group was more efficient on germination density in the period of 3/16 till 3/26. On the one hand, it was because the blank control group didn't contain any inorganic substrates, and the content of topsoil was higher than other groups, thus the seeds in the topsoil was relatively higher; while the content of topsoil in other groups relatively less, so the number of seeds was reduced; In other hand, with the time of seed germination and seedling growth, the soil nutrients in germination disk decreased gradually, while the inorganic substrates played a multiple action on water conservation and mineral fertilizer. Therefore, the effect of inorganic substrates on the germination of soil seed bank was mainly reflected in the late stage of germination.

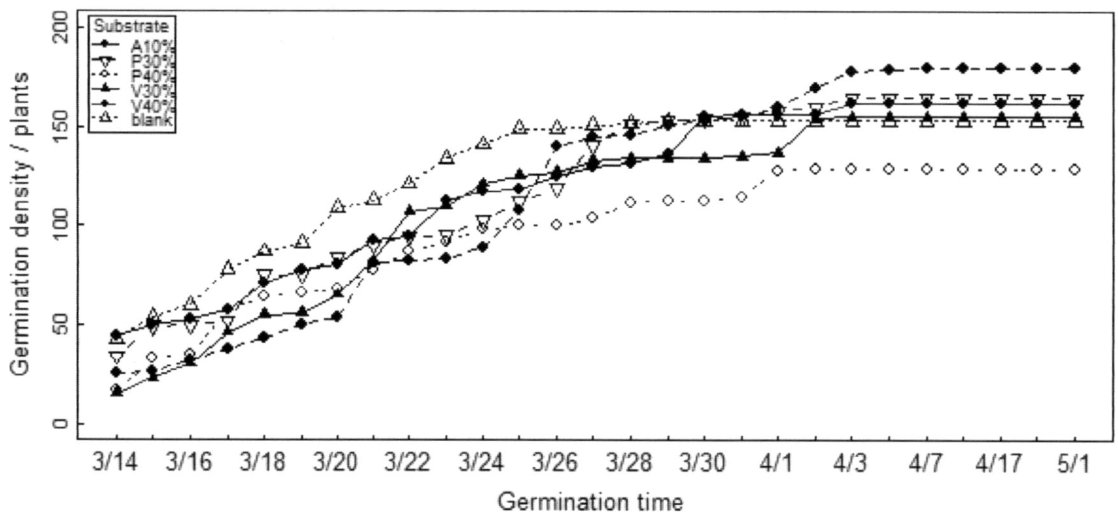

Figure 3. Germination curve of soil seed bank by adding different inorganic substrates.

4. Conclusion

Based on the analysis of the above, inorganic substrates promoted the soil seed bank density and species diversity in a certain extent. When vermiculite mixed with topsoil in 40%, its germination effect of soil seed bank was obvious, the density of soil seed bank reached 6×10^5 plants/m^2, Shannon Wiener diversity index reached 1.4354. Therefore, it was more conducive to improve the soil seed bank density and species diversity by adding 40% vermiculite in topsoil. But the promoting effect on seed germination by adding inorganic substrates was not highly significant. This may related to the specific nature of the inorganic substrates, although the chemical properties of inorganic substrates was relatively stable, but its fertilizer retention capacity and buffering capacity was relatively poor, so its effect on the promotion of seed germination was not particularly significant. Therefore, adding substrates can be used as one of artificial measures for soil seed bank application in vegetation restoration project, which needed deeper study in germination characteristic of soil seed bank by adding more other type substrates, so as to offer reference to engineering application in soil seed bank.

Acknowledgment

This work is financially supported by the National Natural Science Foundation (31370700) and National Natural Science Youth Foundation (41301096).

References

[1] Thompson K, Band S R and Hodgson J G 1993 Seed size and shape predict persistence in the soil *Funct. Ecol.* **7** 236-41

[2] Coffin D P and Lauenroth W R 1989 Spatial and temporal variation in the seed bank of semiarid grassland *Am. J. Bot.* **76**(1) 53-8

[3] Liu H L, Mao L C, Yao H Y and Chen C 2012 Analysis of soil seed banks for different degraded stages of *Seriphidium transillense* desert grassland *Acta Agrect. Sinica* **20**(3) 413-8

[4] Li C M, Long L, Xu C L,Wang M M and Zhang D G 2015 Characteristics of soil seed banks of alpine meadow under different degradation degrees in eastern Qilian mountains *Acta Agrect. Sinica* **23**(5) 957-62

[5] Chen Y, Jiao J Y, Wang N and Jia Y F 2012 Characteristic of soil seed banks under different erosion zones on abandoned land in the Hilly-Gullie loess plateau *Res. Soil Water Cons.* **19**(1)1-5

[6] Umehara T 1983 Restoration method of pioneer plant by scattering method topsoil in the forest *J. Jpn. Soc. Revegetational Technol.* **9** 1-8

[7] Kubo M, Hosogi D and Matsue M 2009 Artificial slope revegetation using forest soil seed bank in Akashi Kaikyo National Government Park *J. Jpn. Soc. Revegetational Technol.* **35**(3)473-8

[8] Hosogi D 2005 Soil seed bank in forest topsoil as planting material (<SPECI AL ISSUE>Living beyond the times: mechanisms and applications of dormancy) *J. Jpn. Soc. Revegetational Technol.* **30** 509-13

[9] Nakamura K, Ishizaka T and Taniguchi S 2005 A case study on the revegetation method using topsoil seed bank *J. Jpn. Soc. Revegetational Technol.* **31** 107-10

[10] Gruda N and Schnitzler W H 2004 Suitability of wood fiber substrates for production of vegetable transplants II *Sci. Hortic.* **100**(1-4) 333-40

[11] Kang H M, Zhang Q X and Tang J 2005 Research advances on growth media *Chin. J. Soil Sci.* **36**(1)124-7

Long-term meteorological trends in the Naiman County of northern China

S X Yao[1], C C Zhao[1, 3], S Y Wang[1], H F Wang[1] and T H Zhang[2]

[1] Lanzhou City University, Lanzhou 730070, China

[2] Cold and Arid Regions of Environmental and Engineering Research Institute, Chinese Academy of Sciences, 320 Donggang West Road, Lanzhou 730000,China

E-mail: zhao_chch1978@163.com

Abstract. The annual trends of six meteorological variables (precipitation, evaporation, relative humidity, sunshine hours, and wind-speed and air temperature) were analyzed by wavelet analyses and mutation analyses of Mann-Kendall method based on daily average data observed from Naiman National automatic meteorological station from 1970 to 2014. Results indicated: (1) The ranges of annual mean precipitation, evaporation, relative humidity, daily sunshine hours, wind-speed and air temperature were 213~567 mm, 1408~2192 mm, 46%~58%, 7~9 h, 2.7~3.8 m/s and 5.8~8.6°C,respectively. (2) Air temperature and daily sunshine hours had increased by 0.29°C /10a and 0.11h/10a ($P<0.01$) respectively over 45 years (1970~2014); (3) Wavelet analysis indicated that wind-speed had a long cycle (26 years); the cycle for daily sunshine hours was somewhat shorter (18 years); evaporation had no obviously cycle; and the other three variables had both shorter and longer cycles. (4) Mutation analysis determined six meteorological parameters that differed from each other, e.g., evaporation had a mutation from low to high occurred in 2001; relative humidity had a mutation from high to low occurred in 2000; daily sunshine hours had a mutation from low to high occurred in 1993.

1. Introduction

Long-term trends in meteorological variables are some of the most important issues of the present time. The analysis of long-term trends is essential for the assessment of the impacts of climate variability and change in a given area [1]. Climate change research developed rapidly over the last century from 1 article in 1907 to 862 articles in 2009. Moreover, China, from 8 to 22 articles during 1992–1998, showed great interest in climate change research, quickly catching up with developed countries in 2009, with 377 articles which ranked among the top 6 in the world [2].

Naiman County (42°55′ N, 120°42′ E, 360 m a.s.l.) is located in the eastern part of Inner Mongolia, China, which is one of four well-known sandy areas in northern China, and are generally thought to originate from sand and dust storms that occur frequently in the arid and semi-arid regions of northern China [3,4]. Temperature, precipitation and other are major factor of climate change. Analysis their changes and characteristics are important for region's environmental and ecological. Many studies are analyzed individual elements of change, rarely multi-factor analysis. The method based on non-parametric is first application in this region. The objective of this study was to provide a brief overview of annual climate in Naiman County, by means of trend, wavelet and mutation analyses on six meteorological parameters: air temperature, precipitation, evaporation, sunshine hours, relative humidity and wind-speed on data collected continuously during 1970 to 2014 from the Naiman National automatic meteorological station.

2. Materials and methods

2.1 Study area description

Naiman County has a temperate, continental, semi-arid monsoonal climate. Annual precipitation is 366 mm. Mean annual potential pan-evaporation is approximately 1935 mm. The mean annual temperature is 6.4°C. The topography is characterized by dunes and inter-dune lowlands [5].

2.2 Data

All data were extracted from the Naiman National automatic meteorological station. The data was collected continuously from 1970 to 2014 and the analysis was applied to daily data sets for the following six parameters: (1) annual precipitation (mm); (2) annual evaporation (mm); (3) annual mean relative humidity (%); (4) annual mean daily sunshine hours (h); (5) annual mean wind-speed (m/s); and (6) annual mean air temperature (°C). The annual data of each meteorological parameter was the average of monthly date, which was obtained by averaged daily date.

2.3 Methods

The trend analysis of each meteorological variable (precipitation, evaporation, relative humidity, sunshine hours, wind-speed and air temperature) changed with year was regressed by the linear equation: $y= at + b$, where: y denotes meteorological variable, t denotes year; A=a*10 was the trend rate of variable. The positive (negative) A represents linear increasing (decreasing) of the variable in the statistical time [6]. Trends of significance test chose F-test method.

Period and mutation features of the six variables were analyzed by wavelet and mutation analyses using the non-parametric Mann-Kendall (M-K) Method [7,8]in Origin 8.0 software.

3 Results

3.1 Statistical analysis

The 45-year means of the six meteorological variables were: annual mean precipitation -344 mm, evaporation -1810 mm, relative humidity -53%, daily sunshine hours -8 h, wind-speed - 3 m/s and air temperature -7°C (Table 1). The coefficient of variation (CVs) for precipitation was the largest (27%), followed by evaporation (11%) and air temperature (10%); the other three variables had small CVs (<10%). On the whole, variation was moderate (10% < CVs < 100%) for precipitation, evaporation and air temperature, and lower (CVs ≤10%) for relative humidity, daily sunshine hours and wind-speed. The skewness was negative for evaporation and air temperature, and positive for the other four variables. This indicated that evaporation and air temperature were left deviated and had mainly large values, while the other four variables were right deviated and the majority of the values were small. The kurtosis of precipitation and relative humidity were positive, but negative for the other four variables. Precipitation and relative humidity had a peak in their distribution, and the values were concentrated over a limited range, while the other four variables had flat distributions and were spread over a much larger range.

Table 1. Statistical characteristic of six meteorological variables.

Variables	Min.	Max.	Mean	SD	CVs(%)	Skewness	Kurtosis
Precipitation (mm)	213.10	567.10	343.88	91.68	27	0.88	0.07
Evaporation (mm)	1408.00	2192.00	1810.20	199.12	11	-0.24	-0.66
Relative humidity (%)	45.94	58.01	52.57	2.47	5	0.17	0.51
Daily sunshine hours (h)	7.33	9.10	8.14	0.42	5	0.23	-0.66
Wind-speed (m/s)	2.70	3.75	3.24	0.25	8	0.35	-0.54
Air temperature (°C)	5.75	8.62	7.11	0.71	10	-0.07	-0.98

3.2 Trend analysis

The time series of the six meteorological variables are presented (Fig.1). From the linear trends of the six variables over 45 years (1970-2014), the relative humidity and wind-speed had downward trends (R^2>0.1) in recent 45 years, and with trend rate 0.62%/10a and 0.06 (m/s)/10a (P<0.01), respectively. While trends for the other four variables were upward, and the air temperature increased most obviously (with highest R^2 (0.29)), and the trend rate was 0.29°C /10a (P<0.01), followed by daily sunshine hours (with R^2=0.11 and trend rate 0.11h/10a (P<0.01)). The upward trends of precipitation and evaporation was weak (R^2 <0.02), and with trend rate 1.02 mm/10a and 20.96 mm/10a ((P>0.05)), respectively.

For convenience, let (a), (b), (c), (d), (e) and (f) denote precipitation, evaporation, relative humidity, daily sunshine hours, wind-speed and air temperature, respectively behind.

IOP Conf. Series: Earth and Environmental Science **41** (2016) 012019 doi:10.1088/1755-1315/41/1/012019

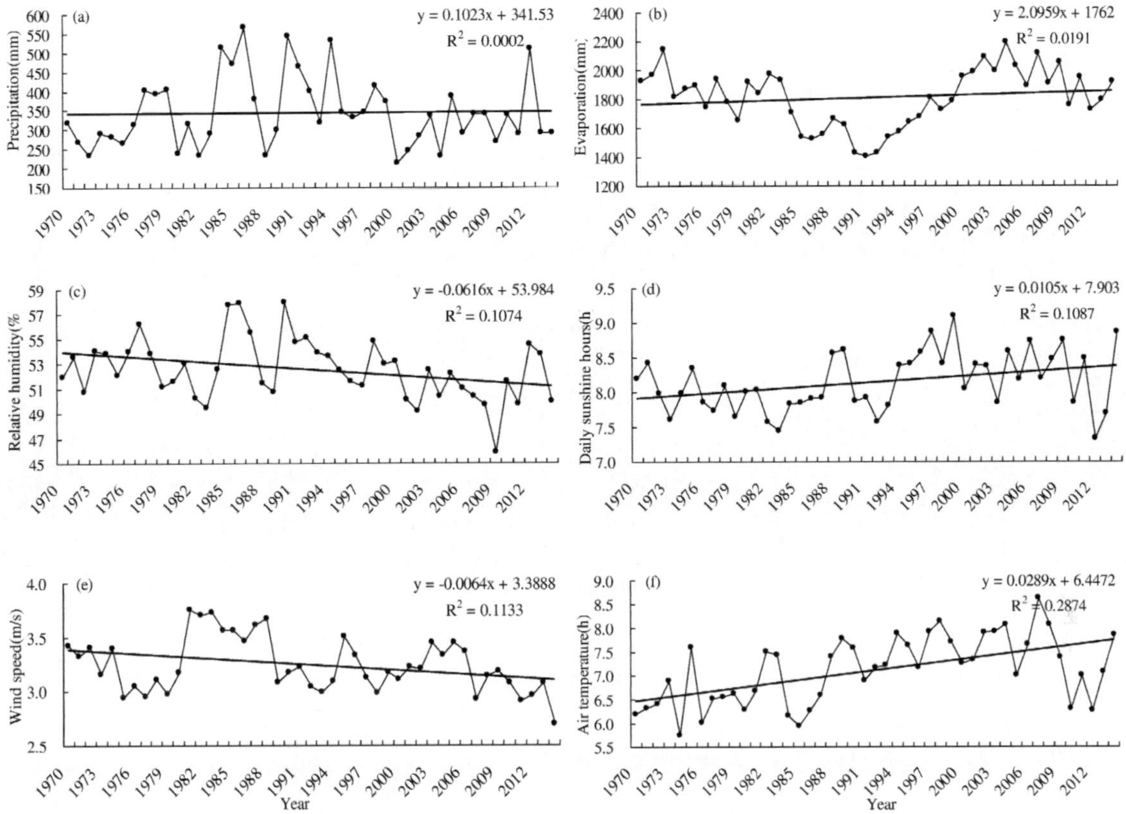

Figure1. Time series and trend analysis of six meteorological variables.

The trend of increased air temperature was consistent with trends reported for global temperatures [2, 9]. Increasing air temperature, number of daily sunshine hours and evaporation exacerbate drought, and thus exacerbate the deterioration of ecological environment of Naiman County [10].

3.3 Wavelet analysis

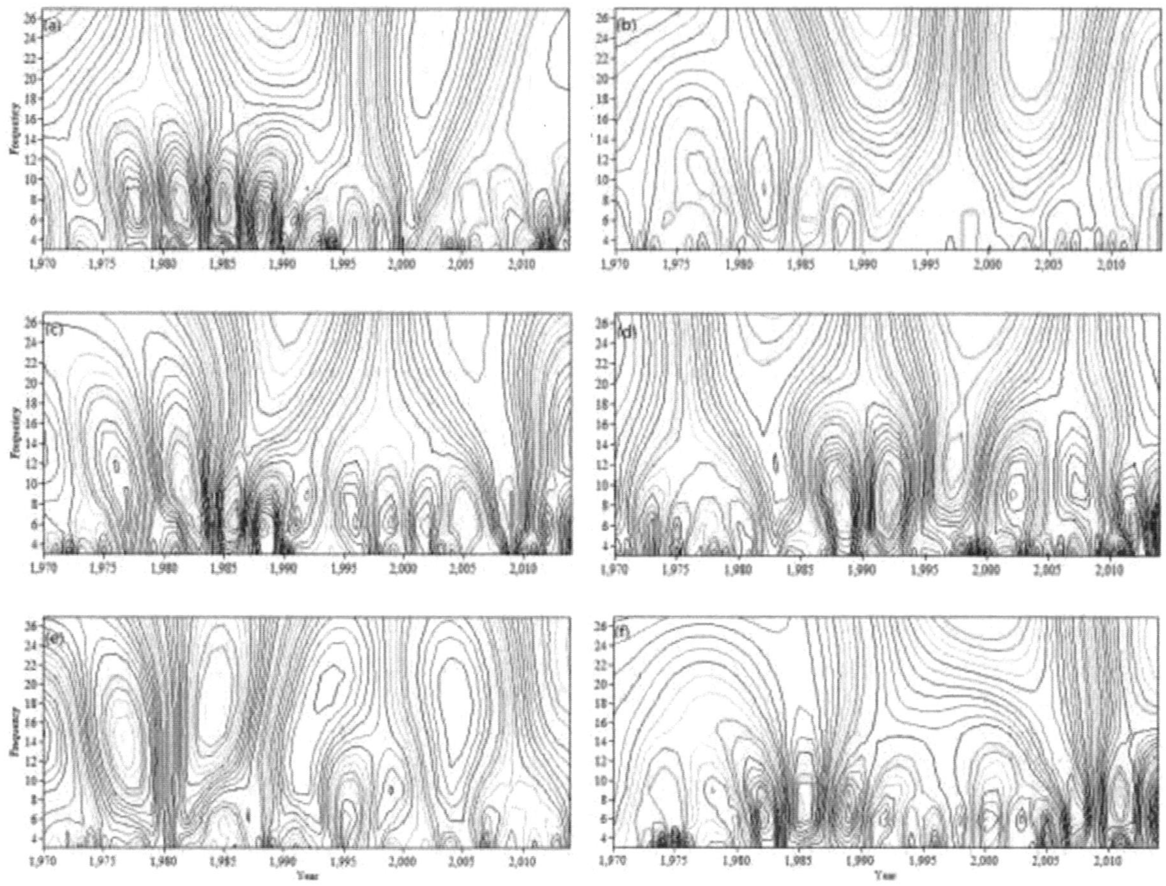

Figure 2. Wavelet analysis of six meteorological variables.

Wavelet analyses of the six meteorological variables suggest two characteristic cycles of annual precipitation, namely a 6-8 year short cycle (beginning after 1990) and a 16 year long cycle (from 1975 to 1990) (Figure 2a). This observation was consistent with the result of our previous study [11], and the period feature for annual precipitation was more obvious in 1975~1990. The period features of relative humidity were 13-year short cycles (beginning after 1985) and a 25-year long cycle (before 1985) (Figure 2c). Number of daily sunshine hours had an 18-year period feature (especially after 1985) (Figure 2d). The period features of wind-speed were 13-year short cycles and a 26-year long cycle (Figure 2e). Air temperature had 7-year short cycles and a 16-year long cycle at the same time (Figure 2f). There was no apparent period feature for evaporation (Figure 2b).On the whole, wind-speed had the longest cycle (26 years), followed by the number of daily sunshine hours (18 years), evaporation had no obviously period feature, and the other three variables had both short and long cycles.

3.4 Mutation analysis

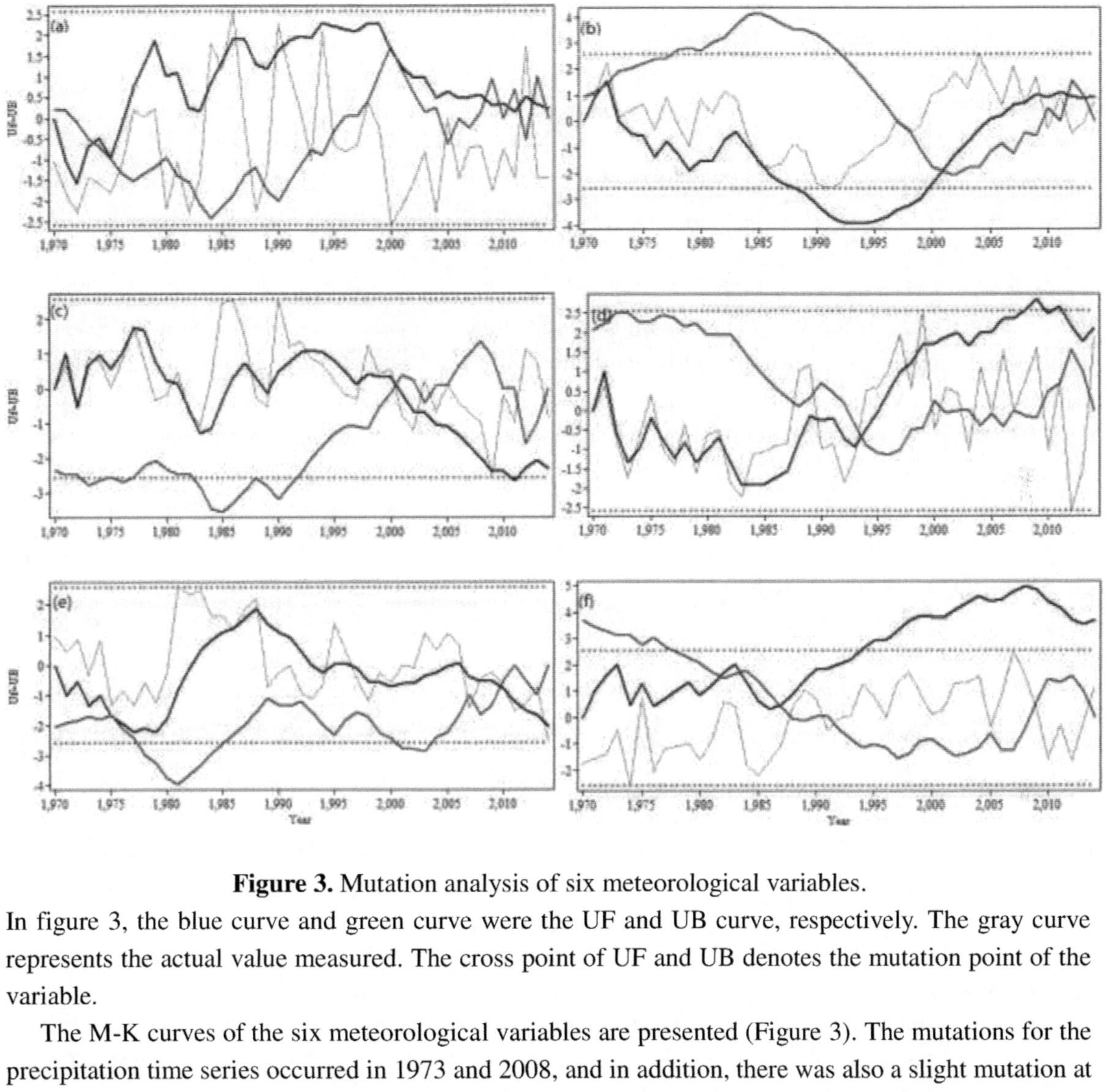

Figure 3. Mutation analysis of six meteorological variables.

In figure 3, the blue curve and green curve were the UF and UB curve, respectively. The gray curve represents the actual value measured. The cross point of UF and UB denotes the mutation point of the variable.

The M-K curves of the six meteorological variables are presented (Figure 3). The mutations for the precipitation time series occurred in 1973 and 2008, and in addition, there was also a slight mutation at 2000. That was, precipitation had a mutations from high to low occurred in 2008, and had a mutations from low to high occurred in 1973 (Figure 3a). The mutations in the evaporation time series occurred in 1972, 2001, 2011 and 2013, so, evaporation had a mutations from high to low occurred in 1972 and 2011, and had a mutations from low to high occurred in 2001 (Figure 3b). Relative humidity had only one mutation point, at 2000, and it increased before 2000, and then decreased (Figure 3c). A single mutation for the number of daily sunshine hours occurred in 1993, and this variable decreased before 1993, but increased after 1993 (Figure 3d). Wind-speed had two mutation points: 1975 and 2009; Wind-speed had a mutations from low to high occurred in 1975, and had a mutations from high to low occurred in 2009. The mutations for the time series of air temperature was complex, it had three mutation points: 1981, 1983 and 1986. The air temperature had a mutation from low to high occurred in 1986 and had a mutation from high to low occurred in after 1981. In summary, the mutation points of the six meteorological variables differed from each other. After 2000, precipitation and relative humidity decreased, but evaporation, daily sunshine hours and air temperature increased, this provides further evidence of warming air temperatures in Naiman County.

4. Discussion

The annual mean air temperature of Naiman County for the period 1970-2014 was 7.11°C, and there was significant increase in air temperature, and air temperature of Naiman County increased by 0.29°C /10a, which was consistent to the result of China for the period 1951–2004: the average annual mean Tmax and Tmin over China increased by 0.147°C /10a and 0.283°C /10a respectively [12]. The related research results also showed that there was obvious warming over the past century for many parts of the world, has been strongly associated with increases in air temperature [12-14], and the distributions of both daily maximum and minimum temperatures have significantly shifted towards higher values [15]. In fact, the climate on earth is changing rapidly, and the global surface temperature has increased by 0.2°C/10a over the last 30 years [16]. And the global average surface temperature of the Earth has increased by 0.6±0.2°C since 1900 and it is likely that the rate and duration of the warming are greater than at any time in the past 1000 years [17]. But for Australia, it has warmed ~0.8°C over the last century [18]. That was, there was a similar trend for climate system, but the rate of warming was differently among different areas.

The annual mean precipitation of Naiman County for the period 1970-2014 was 343.88 mm, and it has a weak upward trend, with trend rate 1.02 mm/10a. The research results of precipitation in Naiman County for the period 1961-2009 [19] and 1970-2010 [11] indicated that there was a decreased trend for annual precipitation. Because of different time-series of research data selected, the results of the analysis has a certain influence, the results vary. The trend of precipitation has showed a stable during this period during not pass the significance test. However, this trend is same as desertification in northern of China. On a continent-wide basis, the trend of precipitation is not statistically significant because of high interannual variability [18]. Because under the trend of warring for air temperature, there was a wide range of uncertainty in rainfall trends [18].

5. Conclusions

We have presented trend analyses of annual precipitation, evaporation, relative humidity, daily sunshine hours, wind-speed and air temperature of Naiman County in eastern Inner Mongolia, China, over the years 1970-2014. The main conclusions of the study are summarized as follows:

(1) The annual mean precipitation, evaporation, relative humidity, daily sunshine hours, wind-speed and air temperature were 344 mm, 1810 mm, 53%, 8h, 3 m/s and 7°C, respectively. Precipitation and relative humidity had a peak in their distribution range and values tended to be concentrated over a narrow range. Evaporation and air temperature were left deviated and the majority of the values were large, while for other variables the results were opposite (no peaks in the distribution and no concentration of values over the range).

(2) There is a general warming signal, with the air temperature increasing by 0.29 °C/10a during 1970-2014, as did daily sunshine hours (0.11h/10a), evaporation (20.96 mm/10a) and precipitation

(1.02 mm/10a). Relative humidity and wind-speed decreased over the period, with trend rate were 0.6%/10a and 0.06 (m/s)/10a, respectively.

(3) The period features of the six meteorological variables were that wind-speed had a long cycle (26 years), as did daily sunshine hours (18 years), evaporation had no obvious period feature, and the other three variables had both short and long cycles.

(4) The mutation points of the six variables differed from each other. On the whole, after 2000, precipitation and relative humidity decreased, but evaporation, daily sunshine hours and air temperature increased.

Acknowledgments
This study was supported by the National Natural Science Foundation of China (No. 31300388; No. 41361013); Dr. Start-up Science Research Foundation of Lanzhou City University (No. LZCU-BS2013-09; No. LZCU-BS2013-12) and Chancellor Research and Innovation Fund of Lanzhou City University (No. LZCU-XZ2014-12). Thank you to Dr. H.R. Kutcher, University of Saskatchewan, Canada for suggestions on the manuscript.

References

[1] Marin L, Birsan M V, Bojariu R, Dumitrescu A, Micu D M and Manea A 2014 An overview of annual climatic changes in Romania: trends in air temperature, precipitation, sunshine hours, cloud cover, relative humidity and wind-speed during the 1961–2013 period *Carpathian Journal of Earth and Environmental Sciences* **9**(4) 253-8

[2] Li J F, Wang M H and Ho Y S 2011 Trends in research on global climate change: a science citation index expanded-based analysis *Global and Planetary Change* **77** 13-20

[3] Wang T 2000 Land use and sandy desertification in the North China *Journal of Desert Research* **20**(2) 103-7

[4] Li F R, Zhang A S, Duan S S 2005 Patterns of reproductive allocation in Artemisia halodendron inhabiting two contrasting habitats *Acta. Oecologica* **28** 57-64

[5] Zhao H L, Zhou R L, Su Y Z, Hua Z, Zhao L Y and Drake S 2007 Shrub facilitation of desert land restoration in the Horqin sand land of Inner Mongolia *Ecological Engineering* **3** 1-8

[6] Qu C P, Guan D X, Wang A Z, Jin C J, Yuan F H, WU J B and Ni P 2009 Characteristics of climate change in Horqin Sandy Land in past 56 years *Chinese Journal of Ecology* **28**(11) 2326-32

[7] Mann H B 1945 Nonparametric tests against trend *Econometrica* **13** 245-59

[8] Kendall M G 1975 Rank correlation methods *Charles Griffin. London* **1** 202

[9] Solomon S 2007 *Climate Change Cambridge* (New York: Cambridge University Press)

[10] Sun X M and Zhao X Y 2009 Assessment of Ecological Risk to Climate Change of the Farming –Pastoral Zigzag Zone in Northern China *Acta Scientiarum Naturalium Universitatis Pekinensis* **45**(4) 713-20

[11] Yao S X, Zhang T H and Zhao C C 2014 Multiple-time-scale analysis on precipitation in Naiman Banner of Horqin Sandy Land in 1970-2010 *Journal of Desert Research* **34**(2) 542-9

[12] Hua W, Samuel S P S and Wang H J 2014 Analysis of sampling error uncertainties and trends in

maximum and minimum temperatures in China *Advances in atmospheric sciences* **31** 263–72

[13] Bonsal B R, Zhang X, Vincent L A and Hogg W D 2001 Characteristics of daily and extreme temperatures over Canada *J. Climate* **14**(9) 1959–76

[14] Meehl G, Tebaldi C, Walton G, Easterling D and McDaniel L 2009 Relative increase of record high maximum temperatures compared to record low minimum temperatures in the U.S *Geophys. Res. Lett.* **36** L23701

[15] Markus G D and Lisa V A 2012 The shifting probability distribution of global daytime and night-time temperatures *Geophys. Res. Lett.* **39** L14707

[16] Houghton J T, Ding Y, Griggs D J, Noguer M, Linden P J and Dat X 2001 Climate change 2001: the scientific basis, contribution of working group I to the third assessment report of the international panel on climate change (New York :Cambridge University Press)

[17] Hughes L 2003 Climate change and Australia: trends, projections and impacts *Austral Ecology* **28** 423–43

[18] Liu X P, He Y H, Zhao X Y，Zhang T H, Li Y L and Yun J Y 2011 Characteristics of precipitation in Naiman region of Horqin Sandy land *Research of Soil and Water Conservation* **18**(2) 155-8

Study on territorial risk assessment in Beibu Gulf of Guangxi

Y Wang[1, 4], L S Hou[2] and M S Zhang[3]

[1] Development Research Center of China Geological Survey, Beijing 100037, China

[2] China Geological Survey, Beijing 100037

[3] Xi'an Center of Geological Survey, China Geological Survey

E-mail: wangyaopku@pku.edu.cn

Abstract. This study is based on the theories of terrestrial exploration risk in the Beibu Gulf Economic Zone in the Guangxi Zhuang Autonomous Region. Using ArcGIS and modelling methods, the spatial distributions of resources and environmental carrying capacity, terrestrial exploration dynamics, and terrestrial exploration potential are studied and evaluated in an integrated way. The results of the study are as follows: Mashan county, Longan county, Luchuan county, et al. have some low-risk land that can be explored, and infrastructure investment should be increased in these locations. Whereas, low-risk terrestrial development is possible in Qinzhou city, Heng county, Longan county, et al. and should be carried out in a planned and gradual fashion. The medium risk regions are distributed all over the region, but are concentrated in the city area of Nanning. However, the middle and north parts of Yulin city, the south part of Chongzuo city, Qinnan district, et al. are also in the same category and priority can be given to improve the quality and efficiency of these regions. The high risk regions, mainly distribute in the Liangqin district, Yining district, Binyang county, et al. This region should focus on the development principles of priority protection, appropriate development, and point-like development. Whereas, the extremely-high-risk areas, of which more than 60% are in Tiandeng county, Daxin county, Shanglin county, et al., are unsuitable for further development. The three-dimensional risk matrix method can make up for the deficiencies of other methods and has broad prospects in regional terrestrial development risk assessment.

Keywords: The Beibu Gulf Economic Zone; resources and environmental carrying capacity; terrestrial development dynamics; terrestrial development potential; risk assessment

IOP Conf. Series: Earth and Environmental Science **41** (2016) 012020 doi:10.1088/1755-1315/41/1/012020

1. Introduction

The Beibu Gulf Economic Zone is the only gulf area in the South China. In 2008, the rapid industrialization and urbanization was included as the national general strategy for regional development. Therefore, Scientific evaluation of the spatial characteristics of territorial resources and identification of the risk factors related to the territorial development is important for the territorial planning of the Beibu Gulf Economic Zone.

Currently, the general trend of risk assessment is reflected by the transition from health risk assessment to eco-risk assessment as well as from the effects of individual pollutants to the inclusion of composite effects from more than one pollutant. Also from chemical pollutants to non-chemical pollution,local to regional studies even global studies and from qualitative to semi-quantitative and quantitative assessments.

There are several problems in recently conducted studies on the territorial development risks. (1) There is no uniformity in the concepts, such as environmental risk assessment, eco-risk assessment and urban development risk assessment. All are defined differently at present; none of these concepts provide an overall measurement technique for regional territorial development risks. (2) Limited effortshave been made to comprehensively study territorial development risks. While many authors have intensively studied various risks over the past few years [1-22], few of them have looked at these risks in a comprehensive fashion. (3) More intensive studies must be carried out on territorial development riskson a regional scale. However, due to limitations in methods and available data, the major risk assessment-related research units of China are limited to districts/counties, drainage basins or smaller administrative institutions, and few have attempted regional-scale research. (4) There are many methods for risk assessment, but methods for large-scale studies are not as operable. Facile methods that involve fewer indices are usually not sufficiently reliable. However, the results of some reliable models like the system kinematics model are frequently exposed to uncertainties in large regional studies as the model construction involves the establishment of complex feedback relations.

This study conducts a risk assessment on territorial development in the Beibu Gulf Economic Zone as a model for regional-scale research and explores the theory and practice of regional territorial development risks in order to provide new guidelines and referable research experienceon the territorial development risk assessment. In this study, territorial development risk assessment is defined to include three implications;

a) A comprehensive analysis of the resources and environmental carrying capacity, development, dynamics and development potential of different regions to establish the risk levels;

b) A division of different classes of risk regions according to spatial differentiation based on the natural environment, socioeconomic development level, eco-system characteristics and type of human activity;

c) The provision of risk-prevention measures after determining the risk type, risk distribution and risk level.

This study aims to establish the pattern, characteristics and mechanism of territorial development risks in coastal areas to provide both case support for regional-scale research on risk assessment in connection with territorial planning and a reference for the territorial planning of coastal economic zones.

IOP Conf. Series: Earth and Environmental Science **41** (2016) 012020 doi:10.1088/1755-1315/41/1/012020

2. Study area

Geographically the Beibu Gulf Economic Zone is located in the central south part of the Guangxi Zhuang Autonomous Region, it lies between 107° 22′ – 109° 05′ E, 21° 27′ – 24° 03′ N. Nanning, Beihai, Qinzhou, Fangchenggang, Yulin and Chongzuo are the important cities of this region. The economic zone has a population of approximately 22.12 million and covers about 7.34 million hectares of land. The region is densely occupied by platforms and large areas of hills and plains. The seafloor is host to rich oil and natural gas resources. There are many mineral resources typically include non-metallic minerals. Water resources are abundant in the region, mainly two major water systems; the Xijiang water system of the Pearl River basin and the river water systems along the coast of South Guangxi, which belong to the coastal basin. The mean annual water resource quantity is 34.96 billion cubic metres. The sea area of the economic zone is about 129,300km. Whereas, the coastline stretches from Gaoqiao Town on Baisha Peninsula at the Guangxi-Guangdong border in the east to Beilunhe estuary at the China-Vietnam border in the west. The total continental coastline is 1,628.59 km.

Territorial development in the Beibu Gulf Economic Zone is subject to the following risk groups:

1) The land resources development risk group composed of water and soil erosion risk, soil pollution risk, geologic disaster risk, arable land low productivity risk and land ‘three-ization’ (desertification, paludification and salinization) risk.

2) The water resources development risk group composed of flood disaster risk, drought risk, water pollution risk.

3) The marine resource development risk group composed of improper spatial arrangement of development and utilization, the need to adjust marine industry structures, low science and technology level, marine environmental pollution and marine eco-degradation.

4) And the mineral resources development risk group composed of low comprehensive utilization rate, insufficient deep processing capability, lack of a commercial mineral exploration mechanism, mine geologic disasters, land resource occupation and destruction, impact and destruction to groundwater systems and environmental impact from wastewater and waste residue.

Figure 1. Location of the Beibu Gulf Economic Zone.

3. Research data and methods

3.1. Research data

The data used inthis study include land utilization survey data, statistical data of mineral resources, geologic disaster survey data, eco-conservation data, pollutant survey data and water resource survey data of 2009. Also, 30-mresolution digital elevation model (DEM) and gradient data downloaded from the InternationalScientific Data Platform. The monthly weather observation data of 2000–2010 from 27 weather stations in and around the Beibu Gulf Economic Zone downloaded from the China Meteorological Data Sharing Service System (including air pressure, hours of sunshine (h), maximum/minimum temperature (°C), average temperature (°C), average relative humidity (%) and average wind velocity (m/s)). The soil map of China with a scale of 1:1000000 and transportation data downloaded from the Environmental and Ecological Science Data Center for West China (http://westdc.westgis.ac.cn) of the National Natural Science Foundation of China (including high-speed railways, general railways, expressways, national highways and provincial highways). Also, 1:250000 geological map of China, 1:200000 land utilization distribution map of the Beibu Gulf Economic Zone, and data of the planned urban development groups from 'Territorial Plan of Beibu Gulf Economic Zone 2011–2030'[23] used in this study.

3.2. Research methods

3.2.1. Calculation of resources and environmental carrying capacity. The assessment index system for this study is designed using analytic hierarchical process (AHP) and includes 'comprehensive assessment of resources and environmental carrying capacity' as the target layer and the indices measuring the resources and environmental carrying capacity as the index layer. A comprehensive assessment index system for resources and environmental carrying capacity is established, which covers four first-tier indices (land resources, water resources, geologic environment and eco-environment) and six second-tier indexes. The weights of these indices are established using the SaatyT. L1–9 scale.

Table 1. Index system of resources and environmental carrying capacity.

Target Layer	Criterion Layer	Index Layer
Comprehensive assessment index system for resource and environmental carrying capacity	Territorial resource carrying capacity (0.25)	Available land resource abundance (0.5)
		Land system susceptibility (0.5)
	Water resource carrying capacity (0.25)	Available water resource abundance (0.6)
		Water environmental carrying capacity index (0.4)
	Geologic environmental carrying capacity (0.25)	Geologic disaster hazard (1)
	Eco-environmental carrying capacity (0.25)	Eco-function importance (1)

A mathematical assessment model is established using the index weighted average method:

$$V = \sum_{i=1}^{n} W_i B_i$$

Here, V is the integrated weighted assessment score, W_i is the weight of each index, B_i is the assessment score of each index and n is the number of indices.

3.2.2. Calculation of the territorial development potential. The territorial development potential is defined as the theoretical territorial development potential minus the land available for development after the comprehensive limitations in territorial development. The higher the territorial development potential, the lower the territorial development risk. The theoretical territorial development potential is defined as the proportion of land available for development and utilization after deducting existing urban construction land, water conservancy land, special land and transportation land from the total land area of the administrative region from the total area of the region, as calculated by the following:

$$Q_S = (1 - P_b - P_c - P_d - P_e - Z_D - Z_H - Z_Q - Z_S - Z_L) \times 100$$

$$Q_l = (1 - P_b - P_c - P_d - P_e) \times 100$$

Here, Q_S is the territorial development potential, Q_l is the theoretical territorial development potential, P_b, P_c, P_d and P_e are the proportions of urban construction land, special land, transportation land and water conservancy land, respectively, in the total area, Z_D, Z_H, Z_Q, Z_S and Z_L are the landform, geologic disaster, climate, eco-safety and food safety limitations, respectively.

3.2.3. Calculation of the territorial development dynamics. (1) Accessibility. The distances of a region from the transport trunk lines such as high-speed railways, general railways, expressways, national highways and provincial highways are used to represent the accessibility of the region. First, distribution map with a 30-m spatial resolution representing the distances from high-speed railways, general railways, expressways, national highways and provincial highways is produced by spatial analysis. Then, the calculation results are linearly normalized between 1 and 100. The accessibility is calculated according to the formula:

$$T = \frac{\sum T_i}{}$$

Here, T is the accessibility of a region, and T_i is the linearly normalized distance from this region to traffic trunk i. Finally, the natural breaks (Jenks) classification method is used to classify the accessibility.

(2) Urban construction land. The distances of a region from the existing urban and rural construction, land are used to represent the influence of the land in this region. Specifically, a distribution map showing the shortest distances to the existing urban and rural construction lands calculated by spatial analysis is first drawn. The calculation results are then linearly normalized to between 0 and 100 to allow the calculation of territorial development dynamics. The larger the value of the influence of the existing urban and rural construction lands, the closer this region is to the existing urban and rural construction lands, and the greater the influence of the existing urban and rural construction lands of this region. Finally, the natural breaks (Jenks) classification method is used to categorize the influence of the existing urban and rural construction lands in the Beibu Gulf Economic Zone into five grades: low, sub-low, medium, sub-high and high.

(3) Influence of planned city groups. The distance of a region from the planned city group is used to represent the influence of this group in the region. Specifically, a distribution map showing the shortest distances from the planned city group calculated by spatial analysis is first drawn. The calculation results are then linearly normalized to between 0 and 100 to allow calculation of the land development potential. The larger the value of the influence of the city group, the closer this region is to the planned city group and the greater the influence of the city group in the region. Finally, the natural breaks (Jenks) classification method is used to classify the influence of the planned city group in the Beibu Gulf Economic Zone into five grades: low, sub-low, medium, sub-high and high.

(4)The territorial development dynamics. By integrating the accessibility, the influence of the existing urban and rural construction lands and the influence of the planned city group, the territorial development dynamics are calculated. Specifically, the averages of the accessibility, the influence of the existing urban and rural construction lands, and influence of the city group are first calculated. The calculation results are then linearly normalized to between 0 and 100. Finally, the territorial development dynamics are derived. And finally, the natural breaks (Jenks) method is used to classify the territorial development dynamics of the Beibu Gulf Economic Zone into five grades: low, sub-low, medium, sub-high and high.

3.2.4. Calculation of the Territorial Development Risk. Three-dimensional risk matrix refers to the 3D magic cube that incorporates resources and environmental carrying capacity, territorial development potential and existing development potential into one single framework and forms a 3 ×

3 × 3 matrix in a three-dimensional space (Figure 2). There are 27 types of combinations, each representing a region with a different development risk. Specifically, the scores of individual cells representing resources and environmental carrying capacity, whereas, the existing development dynamics and regional development potential are prioritized by value and classified according to some prominent mutation points in the data sequence. The overall scores of resources and environmental carrying capacity, existing development dynamics and regional development potential are clustered into three grades: low, medium and high. The highest score of the cluster is recorded as class 1 and coordinated at 3. The medium score is recorded as class 2 and coordinated at 2. The lowest is recorded as class 3 and coordinated at 1. Thus, the coordinates of the cells are expressed by the combinations of 1, 2 and 3, and each of these combinations represents one magic cube unit in the 3D magic cube. According to the risk level represented by each magic cube unit, the risk level of each cell (low, sub-low, medium and high risk grades), is obtained (Table 2). In this study, a region with no potential is defined as a very high-risk region and excluded from the three-dimensional risk matrix calculation.

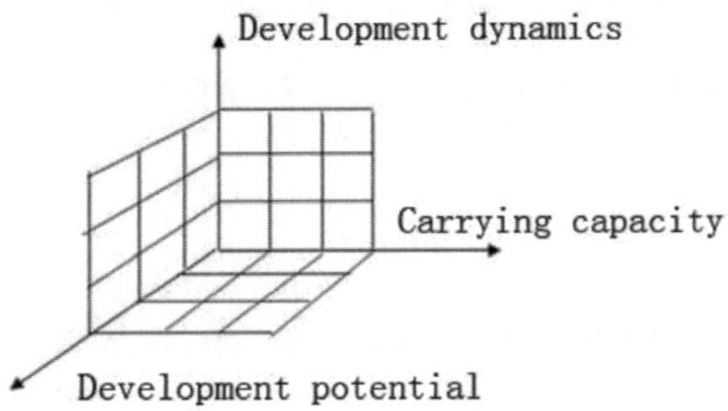

Figure 2. Three-dimensional risk matrix model representing terrestrial development risk.

Table 2. The units of the magic cube and orientation of major development function.

Cube Unit	**Risk Rating Region**
(3,1,3); (3,2,3); (3,1,2); (2,1,3)	Low
(3,2,2); (3,2,1); (3,1,1); (2,2,3); (2,1,2); (2,1,1)	Sub-low
(1,3,3); (1,3,2); (2,3,3); (2,3,2); (3,3,3); (3,3,2); (3,3,1); (1,1,2)	Medium
(1,2,1); (1,1,1); (1,2,2); (1,2,3); (1,1,3); (1,3,1); (2,3,1); (2,2,1); (2,2,2)	High
Regions with no development potential	Very high

Note: The 3D magic cube units are resources and environmental carrying capacity, existing development dynamics and regional development potential.

Low-risk regions are characterized by high resources and environmental carrying capacity, low development dynamics and great development potential. They are suitable for mass development, which can involve modest risk.

Sub-low-risk regions are characterized by high resources and environmental carrying capacity and average development dynamics and development potential; they can be taken as a key backup development region. The development of these regions must address the potential impact on the eco-environment and be implemented in a scheduled, progressive manner.

Medium-risk regions are characterized by high development potential, low resources and environmental carrying capacity and great development potential.

High-risk regions are characterized by low resources and environmental carrying capacity along with high or average development potential. The development of these regions must be implemented in a cautious manner so that the resources and the environment are conserved.

Very high-risk regions are defined as regions with no development potential. No development is recommended for these regions.

4. Analysis and results

4.1. Resources and Environmental Carrying Capacity

The single-factor analysis of resources and environmental carrying capacity indicates that more than half of the land system in the Beibu Gulf Economic Zone has higher susceptibility. Regions with susceptible land systems are typically found in the north-central and eastern areas of the Beibu Gulf Economic Zone. Nanning, Yulin, Beihai and Qinzhou cities have relatively susceptible land systems, whereas Fangchenggang and Chongzuo have more ideal land system conditions. The surface water environment capacity of the Beibu Gulf Economic Zone appears to be overload-free. The offshore water environment of Guangxi has quite good quality. Most waters are clean or relatively clean, and the water environmental carrying capacity is excellent. The geological disaster hazard in the Beibu Gulf Economic Zone as comparativelyto the entire area is quite low, although many unstable slopes exist that could cause geologic disasters. The major part of the Beibu Gulf Economic Zone is ecologically important, only the downtown areas of Chongzuo and Yulin, Tiandeng, Long'an and Binyang counties, and some of the coastal counties/cities have relatively low ecological importance.

Using the administrative map of the Beibu Gulf Economic Zone and the GIS spatial graphic expressions, each district/county is assigned a spatial cluster type and identified on the map to derive an integrated distribution map of the resources and environmental carrying capacity of the Beibu Gulf Economic Zone (Figure 3).

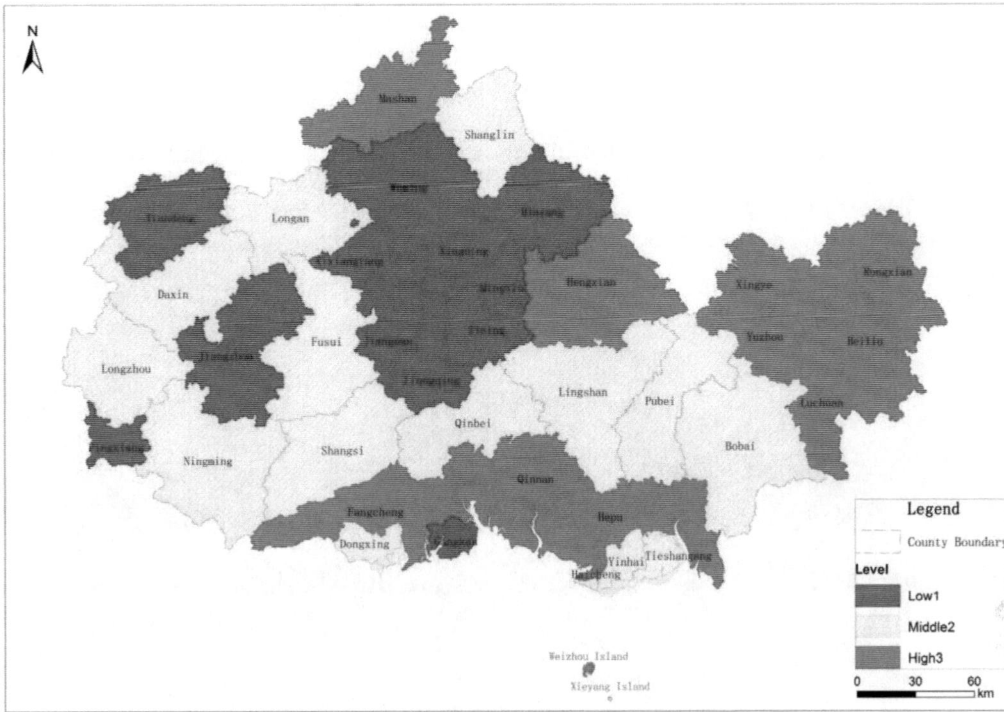

Figure 3. Cluster analysis of resource and environmental carrying capacity of the Beibu Gulf Economic Zone.

The high carrying capacity regions are characterized by great overall resources and environmental carrying capacity, including 11 districts/counties (Mashan county, Fangcheng district, Qinnan district, Hepu county, Haicheng district and most of the districts/counties in Yulin city), which all together make up 30% of the total area of the economic zone. These regions have relatively abundant land and water resources, excellent geological and water environment, low eco-susceptibility,modest eco-importance and relatively high eco-environment support capability.

Medium carrying capacity regions are characterized by medium resources and environmental carrying capacity, including 11 districts/counties (Long'an and Shanglin counties of Nanning city, Dongxing city, Yinhai and Tieshangang districts of Beihai city, Chongzuo city, and most of the counties/districts of Qinzhou city), which collectively contribute 45% of the total area of the economic zone. Most of these districts/counties have limited available land resources, whereas some of them have insufficient water resources, medium or better geological safety, good water and atmospheric environments and average eco-susceptibility and eco-importance.

Low carrying capacity regions are characterized by relatively poor resources and environmental carrying capacity, they include 12 districts/counties (Jiangzhou district, Pingxiang city and Tiandeng county of Chongzuo, Gangkou district of Fangchenggang and most parts of Nanning city), which mutually contribute 25% of the total area of the economic zone. These areas have limited available land resources, relatively abundant available water resources, relatively safe geological environment, good water and atmospheric environments, slightly susceptible ecosystems, and quite high eco-importance. In general, the eco-environmental carrying capacity is superior to the resource carrying capacity in these areas.

4.2. Territorial development potential

Figure 4 shows the cluster analysis result of the territorial development potential of the Beibu Gulf Economic Zone.

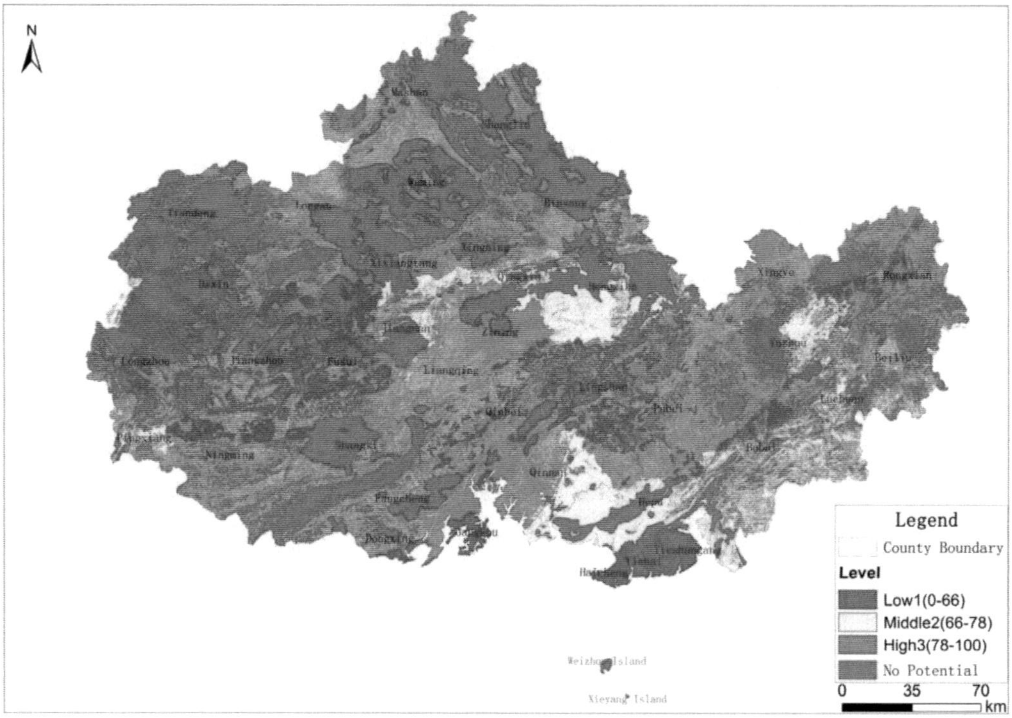

Figure 4. Cluster analysis of terrestrial exploration potential in the Beibu Gulf Economic Zone.

The above map shows the territorial development potential of the Beibu Gulf Economic Zone is generally great, with significant spatial differences between the east and west. Regions with greater theoretical potential are typically found in the plateaus and mountains in the west, while the coastal plain areas in the south have smaller potentials. By and large, Qinzhou city possesses a high overall territorial development potential, whereas, the south-central part of Chongzuo and the northern part of Nanning have relatively high territorial development potentials. Yulin and Beihai cities have lower territorial development potentials. The low development potential of Tiandeng County in northern Chongzuo is attributable to the heavy soil rocky causing the major part of the area unsuitable for development potential.

4.3. Territorial development dynamics

Figure 5 shows the territorial development dynamics of the Beibu Gulf Economic Zone.

IOP Conf. Series: Earth and Environmental Science **41** (2016) 012020 doi:10.1088/1755-1315/41/1/012020

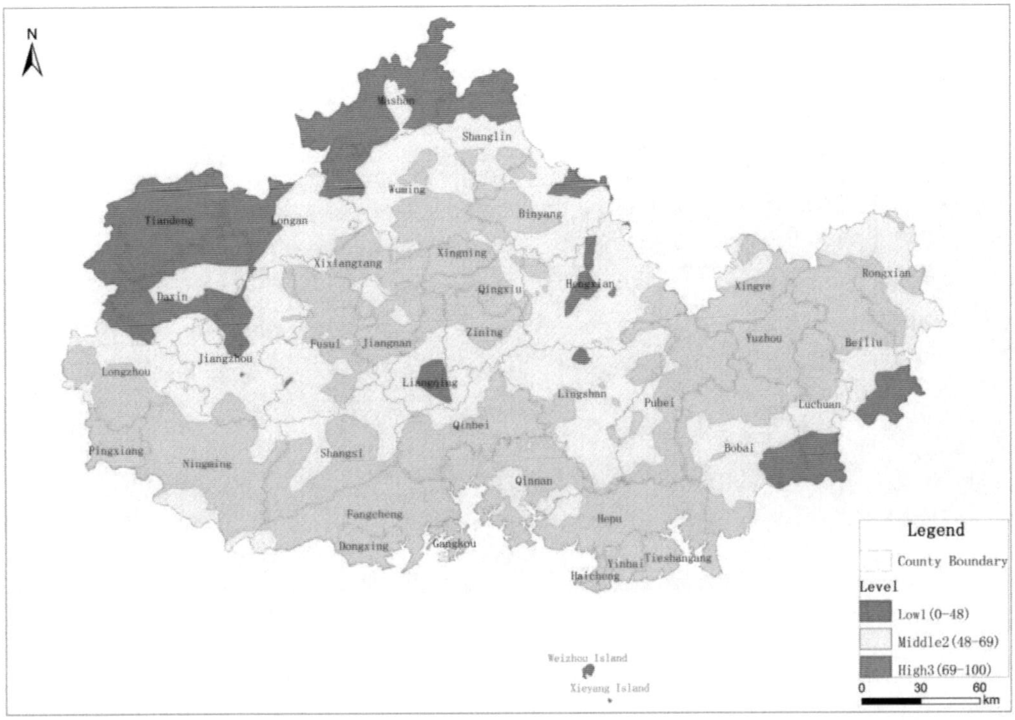

Figure 5. Cluster analysis of terrestrial development dynamics in the Beibu Gulf Economic Zone.

The above map shows that the territorial development dynamics of the coastal and border areas of the Beibu Gulf Economic Zoneis noticeably higher than that of the inland areas. The development dynamics of the central south of Beibu Gulf Economic Zone are higher than that of the north. Beihai and Fangchenggang cities have high territorial development dynamics grades across the cities. The downtown area of Nanning and the downtown and neighboring areas of Yulin also have high development dynamics.

4.4. The territorial development risk

On the basis of the 3D Magic Cube units, the main types of the resulting Magic Cube units of the districts/counties are analyzed in three types. A territorial development risk division plan of the Beibu Gulf Economic Zone is worked out (Figure 6).

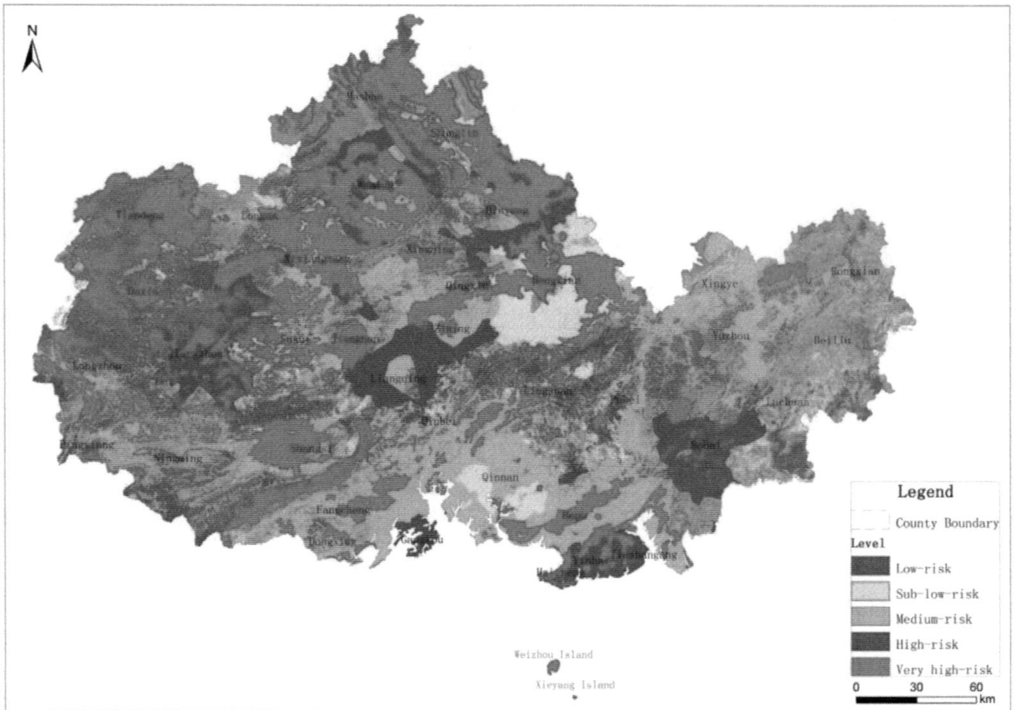

Figure 6. Planning map of the terrestrial development risk in the Beibu Gulf Economic Zone.

Low-risk regions: These regions share the characteristics of high resources and environmental carrying capacity, relatively high eco-safety, relatively low existing development density and sufficient future development potential. Future more, these low-risk regions should fully utilize their resource advantages and speed up industrialization and urbanization. In addition, they should also receive industrial transfers from optimized development regions and population transfers from restricted and prohibited development regions and boost industrial clustering. They should also address the protection of the local eco-environment.

Sub-low-risk regions: These regions share the characteristics of relatively high resources and environmental carrying capacity along with average existing development density and regional development potential. As backup key development regions, it is important that development takes account of potential impacts of the eco-environment, utilize more labor and funds for resource regeneration, try to improve the resource and environmental carrying capacity to the extent that exceeds the total resources and environmental consumption needed for the economic–social development, and proceed in a scheduled, progressive manner. To develop livable, pioneering, regional, modern, and central cities poised to have manufacturing and modern services. It is important for these areas to build themselves into livable modern central cities, important bases to receive industrial transfers, famous minor business cities and demonstration areas for unified urban–rural development. Further, to increase agricultural resource conservation and industrial upgrading, conserve and efficiently utilize resources, and strike a balance between the protection of agricultural production lands and urbanization and industrialization.

Medium-risk regions: These regions share the characteristics of large population density, highland development ratio, relatively high energy or water consumption per unit industrial output value and high existing development density. Also, supported by their physical locations, technological potential and the economic dynamics. Because of the financial strength accumulated from previous economic development, the regional development potentials are still quite high. In addition, medium-risk regions should prioritize quality and efficiency and strive to develop land and energy efficient industries.

High-risk regions: These regions share the characteristics of weak resources and environmental carrying capacity, low existing development dynamics, low population density, bad economic concentration conditions, average regional development potential, and large areas of natural conservation zones within the territory. In future development, high-risk regions should first follow the principle of conservation, moderate development and point development, control the development intensity, gradually shrink the land spaces of towns and rural resident settlements, further implement incentive development within the existing layout, and reserve more spaces for safeguarding the virtuous cycle of the ecosystem.

Very high-risk regions: These regions are not recommended for further development. Instead, mandatory protection should be implemented according to the law and relevant plans. Any human interference with the natural ecology should be controlled, and tourist and eco industries should be moderately developed. In Tiandeng, Mashang and Shanglin counties, where the land is heavily rocky desertified, and the eco-environment is relatively susceptible, it is important that more effort be made to protect ecologically important spaces like woodlands and eco-functions such as water-soil conservation and water source conservation, strengthen the protection of important water source conserving functions like Liuwanshan of Yulin city, develop green landscape resources on three-dimensional bases, and strongly explore traditional Chinese medicine and tourism businesses characterized by local cultures, sightseeing, and eco-resorts.

5. Conclusions

Based on GIS technology, a risk assessment index system and a 3D magic cube risk assessment model for territorial development are established to assess the risks involved in the territorial development ofthe Beibu Gulf Economic Zone. Some recommendations for territorial development in this study area are provided. Following are the main conclusions:

(1) It is important to define the right direction for future territorial development of different types of risk regions. Low-risk regions should put more investment on infrastructure to prepare themselves to receive industrial and population transfers from other regions, speed up industrialization and urbanization, make more effort to protect local eco-environments, and avoid predatory development. Sub-low-risk regions should take account of potential eco-environmental impacts to avoid the mass exhaustion of regional resources and environmental carrying capacity and implement scheduled, progressive development. Medium-risk regions should prioritize quality and efficiency, optimize their economic structures, transform their growth modes, and strive to develop land- and energy-efficient industries. High-risk regions should first follow the principle of conservation, moderate development and point development, strictly control the development intensity, explore green, pollution-free industries by adjusting measures to the local particularities, and become important eco functions of the Beibu Gulf Economic Zone. Very high-risk regions are not recommended for further development.

They should control any human interference to the natural ecology and moderately develop tourism andeco industries.

(2) In view of the existing risks involved in the development and utilization of territorial, water and mineral resources in the Beibu Gulf Economic Zone, measures must be taken to prevent or control these risks by continuing to stabilize the arable land size, moderately increase the construction land size, improve the intensive utilization level of construction land, refine the seismic disaster control mechanism, improve the water supply capability, establish strict water-conservation standards, strengthen water quality supervision, intensify geologic exploration to boost ore-prospecting breakthroughs, adjust the energy structure to extend industrial chains and safeguard the demand–supply balance of mineral resources.

(3) The assessment results indicate that the three-dimensional risk matrix assessment method offers a comprehensive assessment approach based on a number of territorial development system indices. It can overcome the shortcomings of other assessment methods and has vast potential when used for the comprehensive assessment of risks involved in regional territorial development. Further in-depth research on performing comparative result analysis by using three-dimensional risk matrix assessment method, principal component analysis method and k-means clustering method shall be conducted.

References

[1] Mortberg U M, Balfors B and Knol W C 2007 Landscape Ecological Assessment: A Tool for Integrating Biodiversity Issues in Strategic Environmental Assessment and Planning *Journal of Environmental Management* **82** (4) 457-70

[2] Ayalew L and Yamagishi H 2005 The Application of GIS Based Logistic Regression for Landslide Susceptibility Mapping in the Kakuda Yahiko Mountains *Central Japan Geomorphology* **65**(1/2) 15-31

[3] Biswajeet P and Saro L 2007 Utilization of Optieal Remote Sensing Data and GIS Tools for Regional Landslide Hazard Analysis Using an Artificial Neural Network Model *Earth Science Frontiers* **14**(6) 143-51

[4] Du J C and Teng H C 2007 3D Laser Scanning and GPS Technology for Landslide Earthwork Volume Estimation *Automation in Construction* **16** (5) 657-63

[5] Noferini L, Pieraccini M, Mecatti D, Macalusoa G, Atzenia C, Mantovanib M, Marcatob G, Pasutob A , Silvanob S and Tagliavinib F 2007 Using GB-SAR Technique to Monitor Slow Moving Landslide *Engineering Geology* **95**(3/4) 55-95

[6] Zhou J B and Zou J 2010 Vulnerability Assessment of Water Re-sources to Climate Change in Chinese Cities *Ecological Economy* **6**(2) 106-14

[7] Sullivan C A 2010 Quantifying Water Vulnerability: A Multi-di-mensional Approach *Stochastic Environmental Re-search and Risk Assessment* **25**(4) 627-40

[8] Hope B K 2006 An Examination of Ecological Risk Assessment and Management Practices *Environment International* **32**(8) 983 -95

[9] de Paz J M, Sanchez J and Visconti F 2006 Combined Use of GIS and Environmental Indices for Assessment of Chemical, Physical and Biological Soil Degradation in a Spanish Mediterranean Region *Journal of Environmental Management* **79**(2) 150-62

[10] Lu W Z and Wang W J 2005 Potential Assessment of the " Support Vector Machine" Method in Forecasting Ambient Air Pollutant Trends *Chemosphere* **59** (5) 693-701

[11] Wu C F and Pan W C 2003 Territorial Planning Theory and Method (Beijing: Science Press)

[12] Jing Y N 2010 Natural Disaster Risk Assessment (Shanghai: Shanghai Normal University)

[13] Liu Y 2008 Land Ecological Risk Assessment Study Based on Spatial Information Technology (Shanghai: Shanghai University)

[14] Zhao L 2011 Mineral Resource Development Environment Risk Assessment Based on the Information Diffusion and the Fuzzy Comprehensive Evaluation Method (Chengdu: Chengdu University of Technology)

[15] Xu L Y, Wang H M, Ma X Y, Chen J F and Ma J F 2013 Yunnan Drought Risk Management Framework and the Countermeasure Research *Areal Research and Development* **32**(2) 103-8

[16] Liu Y L 2011 Multi-Scale Natural Disaster Scenarios Assessment and Zoning (Shanghai: East China Normal University)

[17] Pan A D, Liu H P, Yang M Z, Huang T W and Li X 2010 Guangzhou Flood Risk Assessment Study *Areal Research and Development* **29** (5) 104-9

[18] Wang Y F, Gai M and Geng Y D 2013 Liaoning Coastal Economic Belt Water Shortage Risk Assessment *Areal Research and Development* **32** (2) 96-102

[19] Fu L H 2012 Ecological Risk Control of Land Use in Core Areas in Changsha–Zhuzhou-Xiangtan Regions Based on the Landscape Structure (Changsha: Hunan Normal University)

[20] Yan L and Xu X G 2010 Progress of the Research on Regional Ecological Risk Assessment *Areal Research and Development* **29** (1) 113-8

[21] Duan H L, Qian H S and Du Y 2011 Climate Change Related Risk for Citrus in Subtropical Zone in China under Future Climate Scenarios *Journal of Natural Resources* **26**(6) 971-80

[22] Yao H J, Meng X G and Qiang Z 2011 Territorial Planning in Guangxi Beibu Gulf Economic Zone (2011-2030) (Beijing: Chinese Academy of Land and Resource Economics)

[23] Saaty T L 1978 Modeling Unstructured Decision Problems: The Theory of Analytical Hierarchies *Mathematics and Computers in Simulation* **20**(3) 147 - 58

[24] Zheng Y, Wang J, Li X and Yu C 2014 Driving risk assessment using cluster analysis based on naturalistic driving data *IEEE, International Conference on Intelligent Transportation Systems IEEE* 2584-2589

Study on the structure of topsoil aggregates for Mu Us sandy land in China

Z H Li[1, 3], Q H Zhang[1], H W Tian[1] and Y H Xie[2, 3]

[1] Resource and Environment Department of Baotou Teachers' College, Baotou 014030,China

[2] Inner Mongolia Zoning Research Institute, Inner Mongolia Academy of Agricultural & Animal Husbandry Sciences, Hohhot 010031, China

E-mail: lizh360@163.com;xieyhsd@126.com

Abstract. It is significantly important to study the topsoil aggregates occurrence for land use, recovery at regional scale and protecting from topsoil sandification. Based on land sampling through experimental analysis and correlation statistics, this paper studies on the characteristics and the correlation of different habitat topsoil aggregates in Mu Us region. The results show: The soil aggregates are mainly micro aggregates in the range of 0.5~0.25mm in this area. The contents of soil aggregates are different in different depths which are 0cm>10cm>20cm respectively. The contents of soil aggregates are also different in different habitats, and the contents of >1mm soil aggregates are highest 4.27% and 3.55% in poplar woodland and *Calamagrostisepigeios* grassland sampled at the 0cm depth. After analyzing the stability of soil aggregates, the development of soil aggregates is not sufficient, and the stability of soil aggregates is poor. The soil structure is gradually optimized in the better area of vegetation protection or restoration.

Key words: Mu Us sandy land; Aggregates; Stability

1. Introduction

Mu Us sandy land is a region belonged to the tail-end of monsoon dry and little rain, the ecosystem environment was weak, the sandy land was large. And it is one of the much more serious areas of wind erosion and sandification in China. With the background of wind erosion and sandification, the wind erosion process by the action of wind force is also the deterioration process of topsoil structure. Therefore, studying on the structure of soil aggregates and the effects related factors, it has the important significance not only in theory and practice, but on the very definition of wind erosion, the influence in soil quality caused by soil degradation, and the related control.

Soil aggregates are glued by inorganic and organic matter, and their quantity and quality determine the soil characteristics and fertility. Soil aggregates and organic matter is the base for keeping the soil structure and fertility, they are integral while interacting, the former is the existence places for the later, and meanwhile, the latter is also the exiting cementation substance of the former. The soil aggregate stability and organic carbon characteristics have been paid much attention to studying in plantation [1].For the study of soil aggregate characteristics, Hessen has calculated the fractal dimension of soil aggregates by the weight distribution and the relationship between the average particle size[2]. O 'Brien has thought that the accumulation speed of soil organic carbon is faster than the soil nitrogen accumulation one. Soil C/N trend was not only going up with the increasing time but also the condition of the same vegetation [3]; The soil development is relatively slow because of the harsh natural environment in arid and semi-arid regions. Studying on soil aggregates is mainly concentrated on the forest soils which remain comparatively in good conditions. At present, studying on the formation and the stability of soil aggregates is less than wind erosion and desertification. Studying on soil aggregates is mainly qualitative with structural destruction of soil by the wind erosion. The reformative and rehabilitative structure has been put into effect by the measures of protecting desertification soil[4,5]. The change of soil structure has been done after adding external modified material in a certain sense[6,7,8]. And so has the effects of soil aggregates on different land-use type[9,10,11]. In addition, the soil aggregates of often studying for desertification are mainly focused on the study on soil crusts[12-15].Fang Shi-bo has thought that the development of biological crust can enhance the soil anti-scour ability and increase the time of soil crust anti-scour ability[16]. Gaohong-qian has also thought that the soil shows a trend of tapering with the biological crust development in loess hilly-gully region, the amount of sand is reduced, and the silt is increased [17]. Some scholars have researched the distribution of soil aggregates in the area for different land use [18-23]. Zhou Ping has analyzed soil granular structure with seven kinds of different ways of land use in loess hilly-gully region, and the results show that the natural grass, shrubs and artificial terrace of soil aggregate fractal dimension were lower than those of other land use types. The soil structure has good aggregation condition, high stability, small damage rate, strong ability against resist erosion. Improving soil fertility function is strong, and artificial grassland is the worst [19]. This paper studies the characteristics and interdependency of different habitat topsoil aggregates in Mu Us sandy land based on land sampling through experimental analysis and correlation statistics. Looking forward to that it is helpful for the study on the mechanism of resistance against the wind erosion of soil aggregates in Mu Us sandy land.

2. Overview of the studied area

The studied area is located in the center of Mu Us sandy land, Wu Shen Banner Erdos Inner Mongolia. The Geographical position is 38°24′～39°02′N, 108°41′～109°42′E. The average altitude is about 1300 m. The climate is semi-arid and temperate continental. Annual mean temperature is 6.5^{0}C, the average annual cumulative temperature is 890^{0}C, the annual precipitation is 360mm, mainly concentrated in June-September, the average annual evaporation 2300mm.The average annual wind speed is 3.3m/s, the annual sand windy and sand stormy days are 40～50, mainly concentrated in

Spring and Winter. Most of Mu Us sandy land is in chestnut soil steppe region, from northwest transition to brown soil semi-desert areas, from southeast transition to the loess plateau with warm Dark loessial soils area, it is just in a few transition regions of natural areas. In the studied area, the terrain is mainly the composition of flowing sand dunes, fixed sand dunes, semi fixed ones and dune bottomland in-between.

3. Materials and methods

3.1. Land investigation and sampling

As for Mu Us sandy land, nine samplings were got in different habitat soil. That's exactly in different landscape areas of fixed sand dunes, semi-fixed ones and dune bottom land in-between, also in willow woodland and *Calamagrostis epigeios* grassland. Digging the soil profile to the 20cm point, sampling in top-down turn every 10 cm vertically. Keeping a soil sample after eliminating debris according to fourfold division, and taking back for lab test with fresh soil.

3.2. Experimental measurement and calculation

The analysis of soil aggregates is through the method of dry sieving and wet sieving [24], weighting each soil sample 50g, testing the content of soil aggregates. For soil organic carbon testing, adopted potassium dichromate volumetric method---the external heating method [25].And processed through the particular ways as follow:

Wet sieving: under the natural structure indoor, first, breaking into small lumps about 1 cm in diameter , and removing plant residues, small stones and other impurities. Then carrying on the processing of soil sample with wet sieving method: put the sample 50g into the vibrating scalper, deionized in water for 5 mins, moved up and down in the water carefully, and then, got the results from the sieves 1.0 mm,0.5mm and 0.25mm respectively. 4 kinds of soil samples were got as the result: particle size > 1.0 mm, 0.5 ~ 1.0 mm, 0.25 ~ 0.5 mm, < 0.25 mm, then tested the weight and calculated the weight percentage respectively. Made sure that the amplitude of the movements was 3 cm, the movement speed was about 25 times per minute, so was the time for 2 mins. Also took much care for the soil sample blending adequately and the drying after wards. And the four-point method is used to divide the sample preparation.

Dry sieving: drying the soil samples in laboratory air. When the soil moisture content is 22% - 25%, the sieve on shaking table of soil sample needed processing under the speed 270 r/min, the shock 2 mins (keep the soil particle size of aggregate separative enough), repeating dry sieveing twice, then isolated soil aggregate are > 1.0 mm, 0.5 ~ 1.0 mm, 0.25 ~ 0.5 mm, and < 0.25 mm.

3.3. Data processing

Stability of soil aggregates is expressed with Mean Weight Diameter [26](MWD). The calculating formula of MWD:

$$MWD = \frac{\sum_{i=1}^{n}(X_i W_i)}{\sum_{i=1}^{n} W_i} \qquad 2.1$$

In the formula, X_i expresses the average diameter of soil aggregates; W_i is for the mass fraction of the corresponding X_i.

The calculating formula of the disruption rate on soil aggregates:

$$\text{Disruption rate on soil aggregates} = \frac{R-1}{R} \times 100\% \qquad 2.2$$

In the formula, R and 1 indicate the mass fraction of which is not only for the mechanical stability of soil aggregates but for water stability on their particle size >0.25mm respectively.All the statistical analyses were performed using SPSS13.0 software.

4. Results and analysis

4.1. The composition characteristics of soil aggregates

4.1.1. The composition characteristics of soil aggregates in general. The Table 1 indicates the content of different soil aggregates by both dry sieving and wet sieving. As shown in Table 1, by dry sieving, the content of soil aggregates is obviously different, of which the particles size is <0.25mm sampled from the different studied areas. After calculation, the average content of soil aggregates is 44.74% using dry sieving with <0.25mm particles size. Then under the dry sieving, the proportion of soil aggregates is highest in the range of 0.5～0.25mm, and the average content is 40.8%. Also by dry sieving, the average content of soil aggregates is 3.43% in the range of 1～0.5mm, and the lowest average content is 0.49% when the particle size of soil aggregates is >1mm. Next even some soil samples do not contain soil aggregates with >1mm particle size. However, for the soil samples through the wet sieving, the content of the soil aggregates of every soil layers is 43.67%, that is not obviously reduced. The average content of soil aggregates is 0.13% when the particle size is >1mm, some are disappeared thoroughly. Next as for the soil aggregates with 1～0.5mm particle size, the content is reduced to 2.24% sharply while for 0.5～0.25mm particle size, the content of soil aggregates is 41.29%,and is slightly increased. In general, it shows that the particle size of the soil aggregates in Mu Us sandy land is mainly 0.5～0.25mm, it is mainly the soil aggregates of small particles size, that is the soil aggregates of small particle size are more than the large ones.

4.1.2. The difference of soil aggregates content. Testing the content of soil aggregates in different soil profiles of 0cm,10cm and 20cm respectively, for the soil aggregates with the particle size >0.25mm,the average content of soil aggregates is highest, while for the particle size 1～0.5mm at the soil profile 0cm, the average content of soil aggregates is also highest. Moreover, for the particle size 0.5～0.25mm at the soil profile 20cm, the content of topsoil aggregates was the highest in Mu Us sandy land. It showed that the nearer the topsoil got, the higher the content of larger aggregates was; also that the content of soil aggregates with big particle size was different when the surface characteristics were different. For example, by dry sieving, the content of >0.25mm soil aggregates was up to 4.27% and 3.55% sampled at the depth 0cm in the poplar woodland and *Calamagrostis epigeios* grassland. But above all, the effect was very great on the content of large soil aggregates in topsoil, if different habitats, and the average content of >1mm soil aggregates is 0.62% at the 0cm for other sampling lands.

4.1.3. The difference of soil aggregates stability. Compared dry sieving to wet sieving, analyse Mean Weight Diameter(MWD) of soil aggregates(as Table 2). Firstly, by wet sieving, the data of MWD is not significantly different, and it is among 0.38-0.5mm whatever different depths are for a single land or for different lands at the same depth. Secondly, by dry sieving, the data of MWD is always bigger in the top of the soil profile for the same land although there is a big difference with different soil lands. Especially in the windward slope of sand dunes, poplar woodland and *Calamagrostis epigeios* grassland, the data of MWD is higher than in other lands. Finally, compared to the data of MWD through dry sieving with wet sieving, discover that there is some differences with the MWD data of dry sieving soil aggregates occurred both in the top-down soil and the different types of sampling lands, however, for the wet sieving soil aggregates there is almost no differences with the MWD data. Therefore, it is that the development of soil aggregates is not fully sufficient, only the primary micro aggregates development exists based on fine sand level, so the stability of the soil aggregates is poorer.

Table 1. The analysis of aggregates in profiles.

Sample Area Type	Depth /cm	Diameter (mm) Dry/Wet Sieving Aggregates Contents			
		>1	1～0.5	0.5～0.25	<0.25
Windward Slope of Sand Dune	0cm	2.07/0.29	2.96/0.77	46.28/46.88	51.3/47.94
	10cm	0.32/0.08	1.77/0.66	45.27/45.88	46.62/46.62
	20cm	0.04/0.01	1.41/0.57	42.97/43.29	44.42/43.86
Leeward Slope of Sand Dune	0cm	0/0	0.32/0.1	35.3/35.07	35.62/35.17
	10cm	0/0	0.54/0.09	45.94/45.93	46.48/46.02
	20cm	0.02/0	0.25/0.07	30.42/30.3	30.69/30.37
Inter-Sand-Dune Area	0cm	1.45/0	2.09/0.44	54.87/56.38	58.41/56.82
	10cm	0.02/0	1.03/0.55	67.95/67.29	69/67.84
	20cm	0.01/0	1.21/0.61	74.08/74.2	75.3/74.8
Sand Dune Top Area	0cm	0.03/0	0.09/0.02	27.51/27.28	27.63/27.3
	10cm	0.07/0.03	0.72/0.31	36.53/36.6	37.31/36.94
	20cm	0.04/0	0.82/0.3	46.26/46.5	47.12/46.8
Willow Area	0cm	3.55/0.93	5.23/2.51	46.59/46.92	55.37/50.36
	10cm	0.11/0.04	2.28/1.38	48.63/48.26	51.03/49.68
	20cm	0.07/0.02	3.74/2.69	58.06/57.97	61.87/60.67
Sand Willow Root Area	0cm	0.2/0.08	0.68/0.42	22.61/22.63	23.49/23.13
	10cm	0.02/0	0.12/0.04	18.76/18.48	18.89/18.52
	20cm	0.01/0	1.57/0.8	30.72/31.04	32.3/31.84
HedysarumLaeve Root Area	0cm	0.49/0.15	7.8/6.41	35.53/35.8	43.83/42.37
	10cm	0.56/0.33	11.87/10.59	32.91/33.39	45.33/44.31
	20cm	0.02/0	2.47/1.91	35.87/35.68	38.37/37.59
Grassland Area	0cm	0.09/0.01	2.28/1.65	35.51/35.6	37.88/37.25
	10cm	0/0	1.73/1.26	23.05/23.28	24.78/24.54

	20cm	0.01/0	2.58/1.78	39.61/40.11	42.19/41.89
calamagrostis Land	0cm	4.27/1.68	17.78/14.17	32.45/35.49	54.49/51.34
	10cm	0/0	9.63/4.41	44.75/48.91	54.38/53.32
	20cm	0/0	9.79/6	43.23/45.82	53.03/51.83

Note: the denominator is the weight of all levels of water stable aggregates, and the nominator is the weight of dry sieving aggregates.

As shown in Table 1, the content of soil aggregates is obviously decreased, of which the size is >1mm or 1～0.5mm for every soil profile after wet sieving. The calculated destruction rate of soil aggregates in different depths (as shown in Table 2), the results show that the destruction rate is the biggest at 0cm, and then it is smaller and smaller at 10cm and 20cm respectively. It shows that in the topsoil of different surface characteristics in Mu Us sandy land, the content of large particle size soil aggregates is reduced with the depth increased. Viewing the data of soil aggregates destruction rate as a whole shows that the destruction rate is small. And shows that micro aggregates take the main part, and the surface damage rate is maximum from the vertical variation. In addition, the difference of the vertical variation is not obvious with the part sampling environment from the top to the down, and that is the protection and the recovery get well with vegetation. Therefore, it indicates that the soil structure is gradually optimized in both areas of the good vegetation protection and restoration.

Table 2. The analysis data of aggregates in different depths.

Sample Area Type	Depth(cm)	Wet Sieving Averge Diameter MWD(mm)	Dry Sieving Averge Diameter MWD(mm)	Aggregates Damaging Rate (%)	Organic Matter Content (g/kg)
Windward Slope of Sand Dune	0cm	0.38	1.61	6.60	6.60
	10cm	0.38	0.35	1.60	1.90
	20cm	0.38	0.66	1.30	1.01
Leeward Slope of Sand Dune	0cm	0.38	0.35	1.30	0.79
	10cm	0.38	0.34	1.00	0.61
	20cm	0.38	0.37	1.10	0.59
Inter-Sand-Dune Area	0cm	0.38	0.44	2.70	1.32
	10cm	0.38	0.34	1.70	0.91
	20cm	0.38	0.34	0.70	0.87
Sand Dune Top Area	0cm	0.38	0.65	1.20	0.99
	10cm	0.38	0.35	1.00	0.89
	20cm	0.38	0.34	0.70	0.62
Willow Area	0cm	0.41	1.94	9.00	12.62
	10cm	0.39	0.86	2.60	1.73
	20cm	0.39	0.60	1.90	1.50
Sand Willow Root Area	0cm	0.38	0.46	1.50	1.07
	10cm	0.38	0.59	2.00	0.53

	20cm	0.38	0.33	1.40	0.84
HedysarumLaeve Root Area	0cm	0.43	0.83	3.30	2.68
	10cm	0.47	0.26	2.30	1.84
	20cm	0.39	0.32	2.00	0.76
Grassland Area	0cm	0.39	0.57	1.70	2.16
	10cm	0.39	0.32	1.00	0.99
	20cm	0.39	0.33	0.70	1.14
calamagrostis Land	0cm	0.50	1.18	5.80	3.95
	10cm	0.41	0.29	1.90	1.03
	20cm	0.42	0.28	2.30	0.92

4.2. The difference on the content of soil organic matter

With a comparative analysis of every land at 0 cm, 10 cm and the 20 cm profile of the soil organic carbon content respectively (shown Table 2).Organic carbon was highest at the 0cm sampling depth, and the contents of organic matter are all 3.57g/kg. Then for 9 lands sampling at the 10cm, the content of organic matter is 1.16g/kg, next at the 20cm, is 0.92g/kg. Therefore, shown as Table 2, the content of soil organic carbon is decreased successively for 9 lands at the 0cm, 10cm and 20cm respectively. Above all, the content of topsoil organic carbon is obviously different at the 0cm, 10cm and 20cm under the same environment of Mu Us sandy land, especially the content of organic carbon is the most obvious at the 0cm in the soil profile. As sampling soil from 9 different areas, their surface characteristics are different (shown as table 2), and the content of organic carbon is the highest in poplar woodland, and the content of organic carbon is all lower in other lands.

4.3. The correlation analysis among the characteristics value of soil aggregates

After taking the soil samples at different depth in MU Us sandy land,the soil aggregates are got through dry sieving and wet sieving (as shown in Table3). The correlation of the MWD,the damage rate and the content of soil organic carbon for the soil aggregates were analyzed by using SPSS 13.0 software. Firstly sampled at the 0cm by dry sieving aggregates, the result shows that the correlation coefficient is 0.914, and it is for MWD in0.01 confidence interval and the content of soil organic carbon, that the correlation coefficient is 0.923 got with soil destruction rate. In a word, they are always significant correlation. However, by wet sieving aggregates, the correlation coefficient is 0.264 for MWD in 0.05 confidence interval and the content of soil organic carbon, and then the correlation coefficient is 0.415 with soil damage rate. The result is not correlative. Secondly sampled at the10cm by dry sieving aggregates, the correlation coefficient is 0.123 for MWD in0.05 confidence interval and the content of soil organic carbon while it is 0.926 for the content of organic matter and the damage rate of soil aggregates in 0.01 confidence interval. So it is not correlative. Also by wet sieving aggregates, the correlation coefficient is 0.493 for MWD and the content of soil organic carbon. It is not correlative either. Finally sampled at the 20cm by dry sieving aggregates, it is 0.572 for MWD and the content of soil organic carbon, and it is 0.274 for the damage rate and the content of soil organic carbon. Meanwhile, sampled at the 20cm by wet sieving aggregates, it is 0.228 for MWD and the content of soil organic carbon. Still it is not correlative. In conclusion, when using dry sieving on the one hand, the content of soil organic matter is significant correlation with both soil aggregates and the damage rate. That is that soil organic matter is the important factor which affects the stability and aggregates occurrence of soil aggregates. On the other hand, the content of organic matter is not significant correlation with MWD of wet sieving aggregates, also means that the development of soil aggregates is in low level for this area, the content of organic matter is low. And the humus with high water stability and cohesion has not been differentiated; it has great relationship with the regional drought environment.

Table3. The correlation on the characteristic value of soil aggregates in different depths and the content of soil organic matter.

	0cm of the content of soil organic matter	10cm of the content of soil organic matter	20cm of the content of soil organic matter
MWD（mm)of wet sieving aggregates	0.206	0.493	0.228
MWD（mm)of dry sieving aggregates	.945**	0.123	0.572
the damage rate of dry sieving aggregates （%）	.940**	0.534	0.274

Note: here, ** represent significance level at $P < 0.01$ (99% confidence interval)

5. Discussion and conclusions

(1) Wind erosion and arid environment are important constraint conditions and background for topsoil aggregates occurrence and evolution in the area. According to the study, the soil aggregates are mainly micro aggregates in the range of 0.5~0.25mm. The contents of soil aggregates are different in different depths. The contents of topsoil aggregates get smaller as 0cm>10cm>20cm respectively,this result is identical with which of SHenfangyu and Wang yongdong[27].

(2) The contents of soil aggregates are also different in different habitats as followed: the contents of more than 1mm soil aggregates are up to 4.27% and 3.55% in poplar woodland and *Calamagrostis epigeios* grassland at the 0cm.

(3) Moreover, after analyzing the soil aggregates stability, the development of soil aggregates is not sufficient, then the stability of soil aggregates is poor; the damage rate of aggregates is low. The data of soil aggregates destructive rate shows that the content of soil aggregates is small in the soil of this area. Viewing from the vertical variation, surface damagerate is maximum. It means that the soil structure is gradually optimized in the good vegetation protection and restoration, this result is identical with which of Gao xiaofei, Liu heping, Six J and Elliott E T, the content of > 0.25 mm air drying soil aggregates and water-stable aggregates,the content is higher, the greater the stability of the soil structure[28,29].

(4) There is the synchronization between the content of soil organic matter and the change rule of structural system. Again analyzing the correlation on the characteristics value of soil aggregates, at the 0cm in 0.01 confidence interval, the result shows that the content of soil organic matter has significant

correlation not only with dry sieving aggregates but also the damage rate of soil aggregates. It indicates that soil organic matter is the important factor that affects the stability and aggregates occurrence of soil aggregates. Meanwhile, the content of organic matter does not have significant correlation with MWD by wet sieving aggregates, it shows clearly that the development of soil aggregates is lower in this area, it has great relationship with the regional drought environment. Yu Haiyan and Cha tonggang also thought that the soil organic carbon content is a downward trend with particle size decreasing.

Acknowledgement
This work was financially supported by the scientific research project of the Inner Mongolia Autonomous Region colleges and Universities (NJZY07118).

References

[1] Briedis C, Caires E F, Moraes Sa J C, Navarro, J D, Inagaki T M, Boer A, Neto C Q, Ferreira A D, Canalli L B and Santos J B 2012 Soil organic matter pools and carbon-protection mechanisms in aggregate classes influenced by surface liming in a no-till system *Geoderma* **170** 80-8

[2] Hessen D O, Agren G I,Anerson T R, Elser J J and Ruiter P C 2004 Carbon sequestration in ecosystems: the role of stoichiometry *Ecology* **85**(5) 1179-92

[3] O'Brien S L,Jastrow J D and Grimley D A 2010 Moisture and vegetation of soil orgnic carbon and total nitrogen in restored grassland *Global Change Biology* **16**(9) 2573-88

[4] Guo Y L and Xiang B 2005 Effects of Conversion of grass on aggregates in belt of criss cross *Inner Mongolia Prataculture* **17**(4) 53-5

[5] Chang X H 2004 Analysis of conservation tillage technology efficiency and application prospect *Farming and Cultivation* **1** 1-4

[6] Dong Z,Li H L and Zuo H J 2004 Wind protection and sand fixation effects of soil coagulant sand barrier *Science of Soil and Water Conservation* **2**(2) 45-9

[7] Cao L H and Zhao S W 2007 Study on improvements of modifiers on soil water-stable aggregates and its mechanisms in aeolian sandy soil *Journal of Soil and Water Conservation* **21**(2) 65-8

[8] Wei Z M and Gu S Y 2003 Effect of organic manure on main physical properties in windblown soil *Journal of Jilin Agricultural Sciences***28** (3) 16-8

[9] Wang Z F,Gao B and Li X G 2006 Effeets of different land use types on carbohydrate stability in arid grassland *Ournal of Gansu Agricaltural University* **41**(3) 91-5

[10] Zhao S W,Su J and Yang Y H 2005 Influence of the soil structure in loess hilly region of southern Ningxia under different man-made vegetation *Research of Soil and Water Conservation* **12**(3) 27-8

[11] Zhang Y S,Xie G D and Xiao Y 2005 Research on soil fractal feature of abandoned farmland A case study from Yijinhuoluo County,Inner Mongolia Au-tonomous Region *Chinese Journal of Eco. Agriculture* **13**(3) 42-4

[12] Cui Y,Lv Y Z and Li B G 2004 Physico-chemical properties of soil microbiotic crusts on erdos plateau *Soils* **36**(2) 197-202

[13] Zhang Y M 2005 Desert biological soil crustson the surface micro structure and its early developmental characteristics *Chinese Science Bulletin* **50**(1) 42-7

[14] Xue Y Y,Yan D R andLi T G2007 The research of desert m icrophytic crusts on erdos plateau *Journal of Inner Mongolia Agricultural University* **28**(2) 102-5

[15] Wu Y H,Gao Q and Yu X H 2003 Distribution influencing factors and monitoring of biological soil crusts *Chinese Journal of Ecology* **22**(3) 38-42

[16] Fang S B and Zhang X S 2011 Impact of Moss Soil Crust on Vegetation Indexes Interpretation *Spectroscopy and Spectral Analysis* **31**(3) 780-3

[17] Gao H Q, Zhao R G and Qin N Q 2012 Impact of biological soil crust on soil physical properties in the hilly loess plateau region *Journal of Nature Resources* **27**(8) 1316-26

[18] He S Q and Zhang Z C 2010 Organic carbon change and distribution of soil aggregates under different land uses *Bulletin of Soil and Water Conservation* **30**(1) 7-10

[19] Zhou P,Liu G B and Hou X L 2008 Fractal features of soil aggregate structure under differentl and use in the Hilly-gully region of Loess Plateau *Science of Soil and Water Conservation* **6**(2) 75-82

[20] Zhang M K, He Z L and Chen G C 1997 Formation of water-stable aggregates in red soil as affected by land use *Acta Pedologica Sinica* **34**(4) 359-66

[21] Zhang C, Liu G B and Xue S 2011 Fractal features of rhizosphere soil microaggregate and particle-size distribution under different vegetation types in the hilly-gully region of loess plateau *Scientia Agricultura Sinica* **44**(3) 507-15

[22] Zhang T 2009 Effect of continuous cropping of different vegetables on soil polysaccharides *Northern Horticulture* **8** 165-6

[23] Zhang Y M 2005 The soil crust micro structure and its early development characteristics on desert surface *Chinese Science Bulletin* **50**(1)42-7

[24] Istitute of Soil Science,Chinese Academy of Sciences 1978 Physical and chemical analysis of soil(Shanghai science and Technology Press) pp 514-8

[25] Istitute of Soil Science,Chinese Academy of Sciences 1978 Physical and chemical analysis of soil(Shanghai science and Technology Press) pp 83-5

[26] Yan B,Jia Z K and Han Q F 2010 Effects of different tillage on soil aggregates in the arid areas of South Ningxia *Agricultural Research in the Arid Areas* **28**(3) 58-63

[27] Shen F Y,Wang Y D and Li S Y 2015 Study on soil aggregates of the desert high way shelterbelt in tarim *Arid Zone Research* **32**(5) 910-17

[28] Wang X Y,Gao X F andLiu H P 2011 Review of analytical methods for aggregate size distribution and water-stability of soil macro-aggregates *Science of Soil and Water Conservation* **9**(3) 106-13

[29] Six J,Elliott E T and Paustian K 2000 Soil structure and soil organic matterII A normalized stability index and the effect of mineralogy *Soil Science Society of America Journal* **64**(3) 1042-9

[30] Yu H Y,Gong R N and Zhou Y 2015 Characteristics of soil aggregate stability and soil organic carbon under four typical artificial plantation in Beijing Badaling mountain area *Journal of Soil and Water Conservation* **29**(5) 162-6

IOP Conf. Series: Earth and Environmental Science **41** (2016) 012022 doi:10.1088/1755-1315/41/1/012022

Improvement of red pepper yield and soil environment by summer catch aquatic crops in greenhouses

X F Du[1], L Z Wang[2], J Peng[1], G L Wang[2], X S Guo[1], T G Wen[1], D L Gu[1], W Z Wang[1] and C W Wu[1, 3]

[1] Huai'an Institute of Agricultural Sciences, Huai'an 223001, P. R. China

[2] Department of Agricultural and Food Science, Huaiyin Institute of Technology, Huai'an 223002, P. R. China

E-mail: hacwwu@163.com

Abstract. To investigate effects of the rotation of summer catch crops on remediation retrogressed soils in continuous cropping, a field experiment was conducted. Rice, water spinach, or cress were selected as summer catch crops; bare fallow during summer fallow was used as the control group. Results showed that aquatic crops grown in summer fallow period could effectively reduce soil bulk density and pH, facilitate soil nutrient release, and improve soil physical and chemical properties compared with those grown in fallow period. Paddy-upland rotation could improve soil microbial members and increase bacterial and actinomycete populations; by contrast, paddy-upland rotation could reduce fungal populations and enhance bacterium-to-fungus ratio. Paddy-upland rotation could also actively promote activities of soil enzymes, such as urease, phosphatase, invertase, and catalase. The proposed paddy-upland rotation significantly affected the growth of red pepper; the yield and quality of the grown red pepper were enhanced. Summer catch crops, such as rice, water spinach, and cress significantly increased pepper yield in the following growing season by 15.4%, 10.2% and 14.0%, respectively, compared with those grown in fallow treatment. Therefore, the proposed paddy-upland crop rotation could be a useful method to alleviate continuous cropping problems involved in cultivating red pepper in greenhouses.

1. Introduction

Vegetable supply is a very important matter in China with the largest human population [1]. Greenhouse environments are suitable for vegetable growth throughout the year; as such, vegetable yield is usually higher in greenhouses than in open fields [2] As commercial demands for vegetables increase, growing vegetables in greenhouses has been recognized as a thriving industry in China. Moreover, greenhouse vegetable cropping has been increasingly popular in China; the total area allotted for greenhouse vegetable cropping reached 5×10^6 ha in 2012 [3]. Land areas for the greenhouse industry increase by 10% annually. Modern greenhouse facilities, planting modes, and technical systems have been designed and applied in particular regions for various crops. For instance, red pepper (*Capsicum annuum* L.) is considered as an important horticultural crop worldwide and commonly consumed in various ways in China. This crop is of economic and nutritional importance[4-7]. Furthermore, red pepper is one of the main crops in the greenhouse industry in China because this crop produces a high yield and economic profits. "Huai'an red pepper" has been considered as a national geographical trademark because this crop occupies an important position in the national red pepper market. The cultivation area of red pepper in the core region of Qingpu District in Huai'an city of Jiangsu province covers 2.7×10^4 hm^2; growing red peppers is a leading industry and source of income of local farmers [8]. However, continuous cropping has been associated with serious problems similar to those in other greenhouse industries; these problems severely cause soil retrogression, such as soil hardening, acidification, secondary salination, and nutrient imbalance, thereby leading to a decline in crop yield and quality [9-12]. Hence, remediation of retrogressed soils were one of the major goals of the greenhouse industry. Low-cost, highly efficient, and environmentally friendly remediation techniques should be developed to remediate retrogressed soils [13].

Drawbacks related to continuous cropping can be resolved by implementing agricultural practices. Crop rotation is one of the most simplest and efficient methods to prevent such problems [14-17]. As a worldwide trend, growing crops in monoculture has been developed, and this method is affected by several factors, such as plant habits, mechanization, economic market trends, government policies, and chemical inputs [18]. Phytoremediation by crop rotation during summer fallow to remediate retrogressed soils has been performed as an effective measure to solve problems associated with the greenhouse industry [19-21]. One of the pivotal steps in phytoremediation involves the selection of suitable catch crops for summer fallow. Scallion (*Allium fistulosum var.*gigantum*)*, sweet corn (*Zea mays* L.), common bean (*Phaseolus vulgaris* L.), *Garland chrysanthemum*, and edible amaranth (*Amaranthus mangostanus* L.) have been selected as ideal summer catch crops in greenhouse soil remediation. Summer catch crops can improve microflora, increase the number of bacteria and actinomycetes, decrease the number of fungi, enhance bacterium-to-fungus ratio (number of bacteria/number of fungi; B/F), and reduce the number of *Fusarium*, as well as prevent nutrient loss in soils [22]. Wu *et al.* [23] also reported that summer catch crops, such as wheat (*Triticum aestivum* L.), rye (*Lolium perenne*), sudangrass (*Sorghum sudanense (Piper) Stapf*), sage (*Salvia officinalis*), basil (*Ocimum basilicum*), and rape (*Brassica campestris* L.), can reduce electrical conductivity (EC) and available N and K contents in soils but effectively improve available P content and soil enzyme activities in soils. Rotation with *Garland chrysanthemum* during summer fallow also increases the yield of cucumber fruits by affecting the biomass and population of rhizosphere soil microbes in cucumbers [21]. Thus, planting crops in summer fallow positively affects soil rhizosphere microflora compared with growing crops in bare fallow; in this process, retrogressed soils can also be remediated by changing soil environment [24-27]. However, information regarding the long-term effects of planting summer crops on soil nutrients, microbial communities, and soil enzyme activities is limited; the relationship of these parameters with the yield and quality of red pepper in greenhouses remains unclear. As such, we aim to determine the mechanism by which soil environments are affected by planting summer catch aquatic crops in the fallow period under greenhouse conditions. Gu *et al.* [28] found that flooding with added wheat straw for 20 d not only increases ammonium nitrogen, available phosphorus, and available potassium contents but also effectively reduces the incidence of chili pepper blight; as a result, pepper growth is promoted and chili pepper yield is enhanced. These results suggest that paddy-upland rotation may be an effective method to alleviate problems related to continuous cropping.

In this study, rice (*Oryza.sativa* L.), water spinach (*Ipomoea aquatica*), and cress (*Oenanthe clecumbens*) were used as summer catch crops to investigate their effects on red pepper yield and quality, soil nutrients, microbial communities, and enzyme activities in greenhouses with problems related to continuous cropping. This study aimed to screen optimum summer catch crops for crop rotation in red pepper planted in greenhouses; this study also aimed to provide a theoretical basis to cultivate red pepper and establish soil micro-ecological environments.

2. Materials and methods

2.1. Materials

The pepper cultivar 'Huai pepper 1108' was obtained from Vegetable Research Center of Huaiyin Institute of Agricultural Sciences in Xuhuai, Jiangsu Province. The seeds were sterilized in 10% hydrogen peroxide for 15 min and washed thoroughly with distilled water, then bud-forced in a germinating box at 28 °C in a culture dish with single filter paper. Uniform sprouted seeds were chosen and sown in a 70-hole seedling tray with vegetable seedling substrates on September 10, 2012 and transplanted on October 24, 2012 with a planting space of 30 cm × 50 cm. The transplanted samples were irrigated regularly. After the plants were transplanted, the greenhouse was ventilated 9 h daily and shaded with straws.

2.2. Experimental design

This study was conducted in greenhouses at the Horticultural Experimental Station, Huai'an, Jiangsu, China, from October 2012 to June 2013. Pepper was planted in the greenhouses for four consecutive years, and the experiment was performed on the fifth year of pepper cultivation. The soil texture in the greenhouse was sandy loam with pH 7.6, and the soil layer at depths between 0 and 200 mm contained 26.61 g•kg^{-1} total organic matter, 147 mg•kg^{-1} available N, 45.4 mg•kg^{-1} available phosphate, and 302 mg•kg^{-1} available potassium. Four treatments were prepared: planting rice, water spinach, or cress during summer fallow, and fallow as the control.

2.3. Plant sampling and measurements

Plant height and stem diameter of red pepper were measured in different growth stages, including early fruiting stage (December 15, 2012), middle fruiting stage (February 18, 2013), and late fruiting stage (April 15, 2013). The fruits were harvested when red pepper reached maturity; fruit yields per plot were recorded. The total yield of each plot was calculated and converted to kg per ha.

Red pepper fruits were collected in the middle fruiting stage (February 18, 2013) to determine quality. Vitamin C, soluble sugar, soluble protein, and nitrate contents were determined through 2,6-dichlorophenolindophenol titration, anthrone colorimetry, Coomassie brilliant blue G-250 staining, and salicylic acid colorimetry, respectively. These procedures were described in details in a previous study of Li [29]. Each treatments were replicated three times.

2.4. *Soil sampling*

Soil samples were collected by Liu *et al.* [30]. Soil samples (at depths between 2 and 25 cm) were collected before aquatic crops were planted (June 10, 2012), after aquatic crops were harvested, and before pepper was transplanted (October 15, 2012). Pepper rhizosphere soils in each test plot was collected in different growth stages, including early fruiting stage (December 15, 2012), middle fruiting stage (February 18, 2013), and late fruiting stage (April 15, 2013). At each sampling site, soil samples (at depths between 2 and 25 cm) at pepper rhizosphere were collected from five spots. The collected samples were combined into a sterilized polyethylene bag and transported to the laboratory at 4 °C . One portion of each sample was stored at -40 °C for microbial analysis, and the remainder was air-dried at room temperature (25 °C to 30 °C) and sifted using a 1 mm sieve to analyze physicochemical properties and enzyme activity.

2.5. *Analysis of soil enzyme activity*

Soil urease (EC 3.5.1.5) activity was determined by incubating 10 g of soil with 10 mL of 10% urea solution for 24 h at 37 °C. Ammonium formation was spectrophotometrically determined at 578 nm, and ammonium activity was expressed as NH_4-N mg•g^{-1} soil 24 h^{-1} [31]. Soil invertase (EC 3.2.1.26) activity was determined by incubating 5 g of soil with 15 mL of 8% sucrose solution for 24 h at 37 °C. The suspension was allowed to react with 3,5-dinitrosalicylic acid, and absorbance was detected at 508 nm. Activity was expressed as glucose mg•g^{-1} soil 24 h^{-1} [31]. Soil catalase (EC 1.11.1.6) activity was detected by incubating 5 g of soil with 5 mL of 0.3% H_2O_2 for 30 min at 30 °C. The suspension was titrated with 0.1 mol•L^{-1} $KMnO_4$ solution. Activity was expressed as 0.1 mol•L^{-1} $KMnO_4$ mL•g^{-1} soil 30 min^{-1} [31]. Soil natural phosphatases were determined using phenylphosphate as a substrate, which was incubated in citrate buffer (pH 7.0) at 37 °C for 3 h [31]. Each sample was subjected to analysis in triplicate; data were the average of the three determined values expressed on the basis of oven-dried soils.

2.6. *Evaluation of soil microorganism population*

Bacterial, fungal, and actinomycete populations in the soil samples were determined via the standard dilution plate method in terms of colony forming units (cfu) [31]. A portion (5 g) of the soil samples was weighed and placed in a 200 mL shaking bottle containing 45 mL of sterile distilled water and agitated for 15 min. The homogeneous solution was then serially diluted to 10^{-2} to 10^{-5} to inoculate bacterial, fungal, and actinomycete cultures. The bacteria extracted from the soil sample were incubated in beef broth peptone substrate. The fungi extracted from the soil sample were incubated in potato dextrose agar (PDA) substrate (3 mL of lactic acid was added per 1000 mL of substrate to control bacterial growth). The actinomycetes extracted from the soil sample were incubated in a Gause No. 1 substrate (1 mL of 3% potassium bichromate was added per 1000 mL of substrate to suppress bacterial growth). Bacterial, fungal, and actinomycete populations were counted at 2, 3, and 7 d after these microorganisms were incubated at 28 °C, respectively. Three plates in each treatment were counted in each replicate.

2.7. Statistical analysis

The experiments were conducted in a completely randomized design. Analysis of variance (ANOVA) was performed with DPS v14.10 statistical software; means were compared by the least significant difference, tested at a significance level of $P < 0.05$.

3. Results

3.1. Effect of planting aquatic crops during summer fallow on the growth of red pepper in greenhouses

Table 1. Effect of planting aquatic crops in summer fallow period and fallow on the growth of the following red pepper in greenhouse.

Treatments	Early fruit stage		Middle fruit stage		Late fruit stage	
	Plant height (cm)	Stem diameter (mm)	Plant height (cm)	Stem diameter (mm)	Plant height (cm)	Stem diameter (mm)
RP	60.18±2.42a	10.37±0.22b	81.75±3.26a	12.67±0.76a	83.17±2.95a	13.96±0.47a
WSRP	59.76±1.78a	9.48±0.62c	80.19±3.16a	11.98±0.91b	81.96±1.88a	13.15±0.58b
CRP	60.23±1.26a	10.84±0.31a	81.66±2.43a	12.05±0.86b	82.55±1.77a	14.18±0.83a
FRP	52.87±2.56b	8.24±0.41d	70.15±2.18b	9.18±0.72c	71.69±2.86b	10.92±0.64c

RRP: rice-red pepper; WSRP: water spinach-red pepper; CRP: cress-red pepper; FRP: fallow-red pepper. Values (mean± standard error) with different small letters in the column are not significantly different at 0.05 levels according to least significant difference (LSD) test ($n = 3$).

Plant height and stem diameter are two main indexes of plant vegetative growth. Table 1 shows that different aquatic crops planted during summer fallow could significantly improve the growth of red pepper. In the whole fruiting stage of red pepper, plant height and stem diameter in each treatment were larger than those in the control group. The largest plant height and stem diameter of CRP (the treatment of planting cress during summer fallow) were observed in early fruiting stage; these parameters increased by 7.36 cm and 2.6 mm, respectively, compared with plants grown via fallow treatment. By contrast, the largest plant height and stem diameter of RRP (the treatment of planting rice during summer fallow) were in the middle fruiting stage; these parameters increased by 11.6 cm and 3.49 mm, respectively. The largest plant height of RRP was observed in the late fruiting stage, but the largest stem diameter of red pepper was observed in CRP. The results indicated that planting aquatic crops during summer fallow could effectively promote the growth of red pepper.

3.2. Effect of planting aquatic crops during summer fallow on the quality of red pepper in greenhouses

Compared with the control group, the three aquatic crops planted in the fallow period significantly improved the quality of red pepper by increasing vitamin C, soluble sugar, and soluble protein contents and by reducing nitrate content (Table 2). The four treatments significantly increased vitamin C content and significantly reduced nitrate content; by contrast, soluble sugar and soluble protein contents did not significantly increase. Rice was subjected to crop rotation with red pepper and most significantly affected red pepper quality; vitamin C, soluble sugar, and soluble protein contents increased by 15.4%, 3.0%, and 4.7%, respectively. As a negative index of pepper quality, fruit nitrate content decreased by 37.0% in rice-red pepper rotation.

Table 2. Effect of planting aquatic crops in summer fallow period and fallow on nutritive quality of the following red pepper in greenhouse.

Treatments	Vitamin C content ($mg \cdot kg^{-1}$ FW)	Soluble sugar content ($mg \cdot g^{-1}$ FW)	Soluble protein content ($mg \cdot kg^{-1}$ FW)	Nitrate content ($mg \cdot kg^{-1}$ FW)
RRP	347.75±10.78a	11.15±0.58a	2.89±0.24a	141.78±10.43d
WSRP	323.58±11.49c	10.58±0.67b	2.78±0.21a	178.62±12.37b
CRP	334.41±12.65b	10.96±0.88a	2.91±0.41a	159.71±10.98c
FRP	301.29±10.23d	10.83±0.49ab	2.76±0.36a	225.16±11.25a

RRP: rice-red pepper; WSRP: water spinach-red pepper; CRP: cress-red pepper; FRP: fallow-red pepper; FW: fresh weight. Values (mean± standard error) with different small letters in the column are not significantly different at 0.05 levels according to least significant difference (LSD) test ($n = 3$).

3.3. Effect of planting aquatic crops during summer fallow on soil nutrients of red pepper in greenhouses

Table 3 shows that the nutrients in the soil sample were fairly enriched. Soil density, pH, total N, and available N, P, and K were also high because an expired compound fertilizer was used in previous growing seasons in the greenhouse; by contrast, planting aquatic crops and fallow could improve soil microenvironment by reducing soil density, pH, total N, and available N, P, and K. The three aquatic crops significantly decreased soil density, pH, total N, and available N, P, and K compared with the control crop because no fertilizer was applied during rotations. Planting rice during summer fallow was the most effective among the four treatments. After the fallow period was completed, treatment by planting rice significantly improved soil microenvironment compared with the base soil of the control group; as a result, soil density, pH, total N, and available N, P, and K decreased by 7.43%, 3.96%, 12.12%, 50.48%, 21.47%, and 15.17%, respectively.

Table 3. Effect of planting aquatic crops in summer fallow period and fallow on soil nutrients of the following red pepper in greenhouse.

Sampling	Density	pH	Total N	Available	Available	Available K

time	Treatments	($g{\bullet}cm^{-3}$)		($g{\bullet}kg^{-1}$)	N ($mg{\bullet}kg^{-1}$)	P ($mg{\bullet}kg^{-1}$)	($mg{\bullet}kg^{-1}$)
Before fallow	Before treatment	1.21±0.02	7.82±0.45	1.65±0.02	186.6±5.5	65.3±2.1	425.8±10.3
After fallow	RRP	1.12±0.01c	7.51±0.38b	1.45±0.03d	92.4±2.1c	51.2±1.2c	361.2±11.6d
	WSRP	1.13±0.02bc	7.56±0.41b	1.52±0.01b	89.3±1.6d	53.6±1.5b	378.5±9.8b
	CRP	1.14±0.01b	7.69±0.52a	1.48±0.02c	96.8±1.5b	52.3±1.6c	368.4±10.2c
	FRP	1.20±0.00a	7.72±0.46a	1.58±0.03a	124.6±3.9a	58.4±1.5a	411.6±14.6a
Early fruit stage	RRP	1.18±0.02d	7.49±0.32a	1.83±0.03b	218.6±3.3c	70.5±1.9b	498.5±13.5c
	WSRP	1.20±0.01b	7.52±0.28a	1.79±0.01d	235.6±4.1b	72.5±1.8b	510.3±11.4b
	CRP	1.19±0.01c	7.58±0.35a	1.81±0.02c	220.9±5.3c	70.2±2.1b	491.9±12.1d
	FRP	1.23±0.01a	7.61±0.44a	1.85±0.01a	245.2±6.6a	78.6±2.5a	551.4±13.8a
Middle fruit stage	RRP	1.16±0.01c	7.62±0.51a	1.61±0.02c	168.5±2.9c	62.3±1.4c	441.6±10.5c
	WSRP	1.17±0.01bc	7.63±0.48a	1.68±0.03b	194.2±3.5b	65.4±1.3b	461.2±9.6b
	CRP	1.16±0.00c	7.63±0.23a	1.62±0.01c	170.6±5.2c	62.6±1.5c	441.9±10.3c
	FRP	1.18±0.01a	7.73±0.45a	1.72±0.00a	201.4±5.5a	68.9±2.1a	491.6±11.5a
Late fruit stage	RRP	1.13±0.01b	7.68±0.22a	1.45±0.00c	145.5±4.5c	53.4±1.1d	372.5±9.8c
	WSRP	1.14±0.00b	7.71±0.41a	1.48±0.01b	165.8±3.8b	55.7±1.3b	384.2±10.8b
	CRP	1.14±0.01b	7.72±0.26a	1.42±0.01d	146.3±5.4c	54.6±1.1c	378.7±9.7bc
	FRP	1.16±0.01a	7.83±0.43a	1.52±0.02a	192.5±6.8a	59.5±1.2a	422.1±11.5a

RRP: rice-red pepper; WSRP: water spinach-red pepper; CRP: cress-red pepper; FRP: fallow-red pepper. Values (mean± standard error) with different small letters in the column are not significantly different at 0.05 levels according to least significant difference (LSD) test (n = 3).

In the whole growing season of red pepper, the same patterns were observed among the dynamic changes in soil density, pH, total N, and available N, P, and K in the four treatments (Table 3). Soil density, pH, total N, and available N, P, and K increased in the early fruiting stage; afterward, these parameters decreased in the middle fruiting stage until the end of the growing season. By contrast, soil pH initially decreased in the early fruiting stage, increased in the middle fruiting stage, and remained constant until the end of the growing season. In the late fruiting stage of red pepper, the lowest soil density, pH, and available N, P, and K in RRP were observed in the four treatments. Compared with those of the control base soils, these parameters decreased by 6.61%, 1.8%, 22.03%, 18.22%, and 12.52%, respectively. However, total N in CRP was lower than that in the base soils, and the decrease was 13.94%. These results indicated that paddy-upland rotations could improve soil environment, and this trend could also be observed in crops planted in the next growing season.

3.4. Effect of planting aquatic crops during summer fallow on soil microbial community of red pepper in greenhouses

Quantity was closely related to composition changes in soil microbial flora, absorption, and soil nutrient transformation. Compared with those grown in the base soils, crops grown in three paddy-upland rotations increased bacterial population and reduced fungal and actinomycete populations. However, fallow reduced all of the tested soil microbial communities (Table 4). The planted aquatic crops were plowed and irrigated to increase soil permeability and microbial populations. In the late fruiting stage, RRP yielded the highest bacterial and actinomycete populations and the lowest fungal population. Fallow treatment resulted in the highest fungal population. Therefore, different aquatic crops rotated with pepper could significantly affect soil microflora, effectively improve community structure, and enhance soil remediation (Table 4).

After the summer fallow period, bacterial, fungal, and actinomycete populations initially increased and then decreased in the whole growing season of red pepper (Table 4); dynamic changes in single microorganism did not differ among the four treatments. Bacterial population significantly increased and significantly differed among the four treatments. The highest quantity of RRP was detected in the early fruiting stage of red pepper. The quantity of RRP also increased by 10.71% compared with that of the control group. Soil bacterial populations in WSRP (the treatment of planting water spinach during summer fallow) and CRP were higher than those in the control group; in particular, soil bacterial populations in WSRP and CRP increased by 5.31% and 6.76% in the early fruiting stage, respectively.

Table 4. Effect of planting aquatic crops in summer fallow period and fallow on soil microbial community of the following red pepper in greenhouse.

Sampling time	Treatments	Bacteria (10^7cfu•g^{-1})	Fungi (10^4cfu•g^{-1})	Actinomyces (10^4cfu•g^{-1})
Before fallow	Before treatment	3.35±0.13	2.86±0.09	2.65±0.11
After fallow	RRP	3.96±0.16a	2.25±0.05c	2.01±0.07d
	WSRP	3.88±0.12b	2.36±0.04b	2.16±0.05c
	CRP	4.01±0.15a	2.19±0.10d	2.31±0.06b
	FRP	3.15±0.09c	2.49±0.06a	2.49±0.05a
Early fruit stage	RRP	4.55±0.14a	2.65±0.08d	3.78±0.07a
	WSRP	4.36±0.10b	2.86±0.07b	3.56±0.05c
	CRP	4.42±0.12b	2.77±0.03c	3.68±0.06b
	FRP	4.14±0.18c	3.16±0.10a	3.25±0.04d
Middle fruit stage	RRP	4.23±0.16a	2.45±0.03c	4.15±0.14a
	WSRP	4.12±0.11b	2.64±0.05b	3.99±0.10b
	CRP	4.22±0.14a	2.41±0.12c	4.11±0.14a
	FRP	3.65±0.08c	3.46±0.10a	3.78±0.11c

Late fruit stage	RRP	4.1±0.11a	2.12±0.05d	3.46±0.15a
	WSRP	3.98±0.09c	2.23±0.06b	3.28±0.12c
	CRP	4.02±0.08b	2.19±0.04c	3.34±0.11b
	FRP	3.41±0.08d	3.11±0.08a	2.92±0.14d

RRP: rice-red pepper; WSRP: water spinach-red pepper; CRP: cress-red pepper; FRP: fallow-red pepper. Values (mean± standard error) with different small letters in the column are not significantly different at 0.05 levels according to least significant difference (LSD) test (n = 3).

The fungal population in the soils of three aquatic crop rotations significantly decreased in the whole growing season of red pepper; fungal populations also significantly differed among the three treatments. The number of fungi of RRP was lower than that of the control group; in particular, the number of fungi decreased by 16.14%, 29.19%, and 31.83% in early, middle, and late fruiting stages of red pepper, respectively. The number of fungi in CRP was also lower than that of WSRP and the control group.

Soil actinomycete populations in the four treatments increased in the early fruiting stage and reached the maximum number; soil actinomycete populations decreased in the late fruiting stage. Compared with the control group, the treatments of RRP, WSRP, and CRP significantly increased the number of soil actinomycetes. The number of soil actinomycetes in RRP was higher than that of the control group; in particular, the number of soil actinomycetes in RRP increased by 16.3%, 9.79%, and 18.49% in early, middle, and late fruiting stages of red pepper, respectively.

3.5. Effect of planting aquatic crops during summer fallow on soil enzyme activity of red pepper in greenhouses

Soil enzymes are important factors of soil ecosystem metabolism, and these enzymes can affect soil biological and chemical processes [32]. Soil enzyme activities varied among the treatments (Table 5). The highest activity in the soil of each treatment varied in terms of enzymes and soil samples from a specific growing season. Compared with those grown in the control soil, the three aquatic crops planted in the fallow period could significantly increase soil urease, phosphatase, invertase, and catalase activities. However, the effects of the fallow treatment were not significant.The results showed that the dynamic changes in different soil enzyme activities did not differ in individual treatments; by contrast, these changes significantly differed among the four treatments. Soil urease, phosphatase, and catalase activities in individual treatments increased in the early fruiting stage and decreased in the middle fruiting stage. Soil invertase activity in each treatment increased until the middle fruiting stage and decreased in the late fruiting stage. Compared with the control group, the three aquatic crops planted during fallow could significantly increase soil urease, phosphatase, invertase, and catalase activities in the growth stage of red pepper. Soil urease and catalase activities of RRP and phosphatase activity of CRP were the highest in the early fruiting stage; soil invertase activity of RRP peaked in the middle fruiting stage. Therefore, aquatic crop rotations could help improve soil activity and fertility.

Table 5. Effect of planting aquatic crops in summer fallow period and fallow on soil enzyme activities of the following red pepper in greenhouse.

Sampling time	Treatments	Urease (NH_3-N $mg•g^{-1}•d^{-1}$ DS)	Neutral phosphatase (Phenol $mg•g^{-1}•d^{-1}$ DS	Invertase (Glucose $mg•g^{-1}•d^{-1}$ DS)	Catalase ($0.1mol·L^{-1}KMnO_4$ $mL·g^{-1}•h^{-1}$ DS)
Before fallow	Before treatment	1.75±0.09	3.26±0.13	1.78±0.22	1.45±0.04
After fallow	RRP	2.98±0.08a	3.82±0.12b	2.89±0.13a	1.78±0.09a
	WSRP	2.76±0.10c	3.66±0.14c	2.57±0.11b	1.65±0.03b
	CRP	2.87±0.08b	3.96±0.09a	3.11±0.09a	1.81±0.06a
	FRP	1.98±0.11d	3.41±0.08d	2.21±0.06c	1.51±0.01c
Early fruit stage	RRP	3.97±0.12a	4.78±0.11a	4.42±0.12a	2.89±0.11a
	WSRP	3.66±0.13c	4.48±0.09b	3.54±0.09b	2.35±0.08b
	CRP	3.86±0.15b	4.82±0.06a	4.37±0.13a	2.85±0.05a
	FRP	2.98±0.12d	4.16±0.05c	3.12±0.21c	1.95±0.02c
Middle fruit stage	RRP	3.69±0.17a	4.33±0.11b	4.98±0.16a	1.76±0.03b
	WSRP	3.57±0.15b	4.06±0.04c	4.27±0.22c	1.55±0.02c
	CRP	3.71±0.16a	4.47±0.12a	4.79±0.21b	1.85±0.05a
	FRP	2.45±0.11c	3.56±0.09d	3.46±0.13d	1.41±0.10d
Late fruit stage	RRP	3.45±0.12b	3.96±0.10a	4.76±0.21a	1.82±0.04a
	WSRP	3.42±0.11b	3.82±0.11c	3.98±0.17c	1.58±0.05c
	CRP	3.52±0.15a	3.88±0.09b	4.55±0.15b	1.78±0.07b
	FRP	2.28±0.08c	3.33±0.11d	3.35±0.12d	1.44±0.08d

RRP: rice-red pepper; WSRP: water spinach-red pepper; CRP: cress-red pepper; FRP: fallow-red pepper; DS: dry soils. Values (mean± standard error) with different small letters in the column are not significantly different at 0.05 levels according to least significant difference (LSD) test (n = 3).

3.6. Correlation between soil properties and enzyme activities in microbial communities

Soil properties were largely determined by enzymes and microbial communities in soils (Table 6). Soil bulk density was positively correlated with soil N and available N, P, and K; soil bulk density did not also significantly affect soil enzymes and abundance of soil-borne bacteria and actinomycetes. By contrast, soil bulk density is positively correlated with the abundance of fungi. This result indicated that a decrease in soil bulk density might be related to enhance nutrient amounts in soils. Soil pH negatively correlated with soil urease, neutral phosphatase, and catalase activities. Therefore, an increase in pH likely inhibits soil enzyme activities. pH was also negatively correlated with soil

IOP Conf. Series: Earth and Environmental Science **41** (2016) 012022 doi:10.1088/1755-1315/41/1/012022

bacterial community; thus, an increase in soil pH could reduce soil-borne bacterial populations. Total N and available N, P, and K were positively correlated with soil-resident enzymes and soil microflora. Available N, P, and K were significantly correlated with fungal and actinomycete populations. Furthermore, soil nutrients could directly affect soil microflora and composition and soil-resident enzyme activities. Bacterial population in the soil of red pepper planted in the greenhouse was positively correlated with urease, phosphatase, invertase, and catalase activities. Fungi did not significantly affect soil enzymes; however, actinomycete population was positively correlated with urease, phosphatase, and invertase activities. These results showed that an increase in soil bacteria and actinomycetes could promote soil-resident enzyme activities. Therefore, appropriate soil pH and available N, P, and K contents are key factors to promote soil-resident enzyme activities and to create microbial communities.

Table 6. Correlation analysis among soil nutrients, enzyme activity and microbial community.

Index	DS	pH	TN	AN	AP	AK	Uase	NPase	Inv	Cat	BC	FC
DS	1											
pH	0.066	1										
TN	0.821**	-0.374	1									
AN	0.744**	-0.058	0.829**	1								
AP	0.867**	-0.222	0.963**	0.918**	1							
AK	0.818**	-0.284	0.962**	0.899**	0.988**	1						
Uase	-0.212	-0.650**	0.215	0.269	0.170	0.224	1					
NPase	0.121	-0.734**	0.536*	0.427	0.444	0.489*	0.890**	1				
Inv	-0.281	-0.261	0.041	0.263	0.067	0.127	0.860**	0.672**	1			
Cat	0.219	-0.685**	0.572*	0.423	0.447	0.468	0.668**	0.867**	0.387	1		
BC	-0.097	-0.754**	0.370	0.324	0.304	0.348	0.942**	0.932**	0.711**	0.762**	1	
FC	0.712**	0.187	0.690**	0.715**	0.766**	0.764**	-0.326	-0.100	-0.247	-0.005	-0.203	1
AC	0.25	-0.12	0.48	0.67	0.54	0.57	0.64	0.59	0.80	0.31	0.54	0.24

4	4	1	3**	2*	0*	2**	9*	0**	4	0*	5

DS: density; TN: total N; AN: available N; AP: available P; AK: available K; Uase: urease; NPase: neutral phosphatase; Inv: invertase; Cat: catalase; BC: bacteria community; FC: fungi community; AC: actinomycetes community.

3.7. *Effect of planting aquatic crops during summer fallow on red pepper yield and economic benefit analysis*

The three aquatic crops planted during fallow significantly improved the yield of red pepper in greenhouses (Table 7). RRP, WSRP, and CRP increased pepper yield by 15.4%, 10.2%, and 14.0%, respectively, compared with the fallow treatment (FRP). However, the final total income of CRP was the highest among all of the treatments; the final total income, which increased by 57%, significantly differed compared with that of the control group. The final total incomes of RRP and WSRP were significantly higher than those of the control group; the final total incomes also increased by 24.7% and 46.5%, respectively.

Table 7. Analysis of yield and income.

Treatments	RRP	WSRP	CRP	FRP
Summer yield (kg·ha^{-1})	9552	53514	35677.5	0
Summer incomes (¥·ha^{-1})	13585.2	87040.8	101605.5	0
Yield of red pepper (kg·ha^{-1})	62683.5	59827.5	61884	54297
Incomes of red pepper (¥·ha^{-1})	308601	291465	303804	258282
Total incomes (¥·ha^{-1})	322186.2	378505.8	405409.5	258282

RRP: rice-red pepper; WSRP: water spinach-red pepper; CRP: cress-red pepper; FRP: fallow-red pepper. Values with different small letters in the same line are not significantly different at 0.05 levels according to least significant difference (LSD) test (n = 3).

4. Discussion

Continuous cropping with specific crops has been considered as the main factor that contributes to cropping limitations in greenhouses [33], resulting in soil nutrient loss and imbalance. These phenomena increase the prevalence of multiple soil-borne diseases and seriously affect the normal growth of protected vegetables by impairing yield and quality [34-35]. Our result revealed that paddy-upland rotation is an effective way to prevent continuous cropping problems in the red pepper industry. He *et al.* [36] used two aquatic vegetables with long life spans, namely, water chestnuts (*Eleocharis dulcis*) and arrowhead (*Sagittaria sagittfolia*), to replace rice for paddy-upland rotation in vegetable facilities; this treatment can overcome cropping obstacles and significantly improve the yield of subsequently planted crops. Qian *et al.* [37] applied a similar strategy to investigate strawberry (*Fragaria ananassa*) and *Ipomoea* rotation and their effects on water spinach yield; strawberry and *Ipomoea* rotation can also effectively overcome continuous cropping problems, increase cropping land index, and improve land utilization. Thus, the economic value of water spinach increases.

In this study, several aquatic summer catch crops could effectively improve the quality and growth indicators of red pepper, which is a housekeeper crop in local greenhouses. Among these crops, rice and pepper rotation (RRP) most significantly improved the growth and quality of pepper in the following growing season, followed by cress and water spinach (Tables 1 and 2). This result suggested that rice, water spinach, and cress were rotated with pepper and could effectively alleviate the adverse effects of continuous cropping soils on the growth of pepper in the following growing season; crop rotation also increased plant height, stem thickness, leaf area, and chlorophyll content. Each treatment improved the quality of peppers by reducing nitrate content and by significantly increasing soluble sugar and protein contents of red pepper. Rice, water spinach, and cress planted during fallow significantly improved pepper yield in the following growing season . RRP showed the highest pepper yield in the following growing season. Cress provided the highest economic benefit when this crop planted during fallow, followed by water spinach and rice. The most economical combination is the rice-pepper rotation, followed by cress and water spinach (Table 7). Therefore, different aquatic crops planted during fallow could increase the total income of greenhouses.

Soil secondary salinization, nutrient imbalance, increased auto-toxicity, and soil-borne diseases are the main causes of continuous cropping problems, which seriously hamper crop growth and reduce yield and quality [38]. Paddy-upland crop rotation can alleviate continuous cropping problems. The results from xerophyte and aquatic crop rotation revealed that aquatic plants not only uptake excess nutrients in soils; however, agricultural measures help distribute topsoil nutrients through irrigation, which is necessary to maintain balance of soil nutrients and reduce soil salinization [39-40]. In this study, soil bulk density and pH of rice, water spinach, and cress were significantly lower than those of fallow treatment; thus, rice, water spinach, and cress planted during fallow not only actively reduced soil bulk density and pH but also increased soil permeability. In another treatment, rice was rotated with pepper and yielded optimum results. Rice, water spinach, and cress planted during fallow could significantly reduce total nitrogen and available N, P, and K of soils, particularly in rice-pepper rotation. The lowest reduction was caused by the fallow treatment (Table 3). Therefore, planting aquatic crops help maintain soil nutrient composition and nutrient balance through excess soil nutrition uptake caused by continuous cropping.

The type and population of soil microorganisms are mainly determined by plant root exudates [41]. A specific soil microflora of a particular ecosystem was determined by nutrient composition in soils and by plant root exudates and plant biomass degradation. In agricultural practice, different crops prefer various soil nutrients; paddy-upland rotation was considered as the most efficient technique in terms of improving soil properties, maintaining nutrient balance, and increasing diversity and stability of soil microbial populations. Moreover, crop rotations and fallow treatment are also effective ways to limit pathogen proliferation because of the lack of hosts and the changes in their living environment [42]. Our results showed that the selected aquatic crops planted during fallow could significantly affect soil microbial population and composition (Table 4). Rice, water spinach, and cress planted during summer fallow significantly increased bacterial populations; by contrast, fallow treatment slightly increased bacterial populations. Among the treatments, cress-pepper rotation is the most effective method to decrease soil fungal populations; rice-pepper rotation is the most effective mechanism to decrease soil actinomycete populations. Agricultural measures, such as plowing, fertilization, and irrigation, are implemented before or after crop planting is performed; these measures positively affect soil permeability and fertilizer availability. Fallow treatment did not significantly affect soils. Rice, water spinach, and cress planted during fallow significantly affected soil microflora structure. These crops were conducive to microbial growth; thus, the proliferation of pathogenic fungi was inhibited and the spread of soil-borne diseases was prevented. Therefore, paddy-upland rotation can improve soil microbial composition, maintain diversity and stability of soil microbial populations, and reduce the occurrence of cropping problems in greenhouses.

Soil enzymes mainly come from soil microbial metabolic processes [43], plant root exudates, and animal and plant decomposition. Crop rotation can increase microbial diversity and enzyme activity in soils [44]. Our data demonstrated that aquatic crops planted during fallow enhanced soil urease, phosphatase, invertase, and catalase activities compared with those planted on the control sample soil. The increase in urease and invertase activities indicated that soil biological activity, hydrolysis activity of soil enzymes, and available organic nitrogen increased after aquatic crops were rotated. Inorganic nitrogen content was also enhanced possibly because organic nitrogen was released from soils. Increased soil phosphatase activity could promote decomposition and transformation of soil organic phosphorus, as well as conversion and utilization of phosphate fertilizer. Increased soil catalase activities indicated that soil oxidation was enhanced; thus, this enzyme may protect plant roots against oxidation [45]. Our data showed that soil phosphatase and catalase activities were enhanced after rice-pepper rotation was completed; this result may explain the process by which available phosphorus accumulates in soils. Sun *et al.* [45] found that the proliferation of fungal population in peanut planted in continuous cropping soils can significantly inhibit soil alkaline phosphatase, invertase, and urease activities; by contrast, soil bacterial and actinomycete populations increase and promote soil alkaline phosphatase, invertase, and urease activities. Yang *et al.* [46] reported that soil invertase and urease activities are positively correlated with the number of bacteria and actinomycetes in the rotation of green Chinese onion and cucumber, along with waxy corn and cucumber. In our study, the number of soil bacteria and actinomycetes was positively correlated with urease, phosphatase, invertase, and catalase activities in the rotation of rice, water spinach, and cress with pepper; by contrast, the number of fungi was not correlated with soil enzyme activities. This result is consistent with the conclusions of Sun *et al.* [45] and Yang *et al.* [46]. Soil bulk density was also positively correlated with total nitrogen and available N, P, and K of soils. By contrast, soil bulk density was not significantly related to enzyme activities and bacterial and actinomycete titer in soils; nevertheless, soil bulk density was significantly positively correlated with fungal titer. However, soil pH was negatively correlated with soil urease, phosphatase, invertase, and catalase activities; soil pH was also negatively correlated with soil bacterial and actinomycete titer. Therefore, soil enzyme activities were closely related to soil nutrition. Furthermore, total N and available N, P, and K were positively correlated with soil enzyme activities and microbial communities; this result suggested that available N, P, and K contents in soils should be controlled to promote soil enzyme activities and to establish beneficial microbial communities.

Therefore, paddy-upland crop rotation can improve physical and chemical properties of soil and microbial community structures in greenhouses that experience continuous cropping problems; these properties are enhanced by increasing microbial populations, soil urease, phosphatase, invertase, and catalase activities, and soil fertility. Thus, the growth and development of red pepper in the following growing season could be facilitated.

5. Conclusions

This study showed that planting aquatic crops during summer fallow could balance soil nutrient, reduce salt accumulation, increase soil microbial populations, and promote soil ecological condition. As such, growth, yield, and quality of red pepper are improved. Therefore, appropriate paddy-upland crop rotation, such as rice-pepper rotation, could be a useful method to alleviate continuous cropping obstacles experienced when red peppers are grown in greenhouses. However, whether effects can be maintained after rotation is performed should be further analyzed.

Acknowledgments

This study was financially supported by Jiangsu Agriculture Science and Technology Innovation Fund [CX(11)4056] and Strategic Key Project of Jiangsu Provincial Science and Technology (BE 2012417).

References

[1] Tian Y Q, Zhang X Y, Wang J G and Gao L H 2013 Soil microbial communities associated with the rhizosphere of cucumber under different summer cover crops and residue management: A 4-year field experiment *Sci. Hortic.* **150** 100-9

[2] Gao L H, Qu M, Ren H Z, Sui X L, Chen Q Y and Zhang Z X. 2010 Structure, function, application, and ecological benefit of a single-slope, energy-efficient solar Greenhouse in China *HortTechnology* **20** 626-31

[3] Yu J Q 2011 Progress in protected vegetable production and research during "The Eleventh Five-year Plan" in China *China Vegetable* **2** 11-23 (in Chinese)

[4] Rodoni L M, Zaro M J, Hasperué J H, Concellón A and Vicente A R 2015 UV-C treatments extend the shelf life of fresh-cut peppers by delaying pectin solubilization and inducing local accumulation of phenolics *LWT - Food Science and Technology* **1** 408-14

[5] Singh R, Giri S K and Kotwaliwale N 2014 Shelf-life enhancement of green bell pepper (*Capsicum annuum* L.) under active modified atmosphere storage. *Food Packaging and Shelf Life* **2** 101-12

[6] Ahamad I, Cheng Z H, Meng H W, Liu T J, Wu C N, Ali K M, Wasila H and Rehman K A 2013 Effect of intercropped garlic (*Allium sativum*) on chlorophyll contents, photosynthesis and antioxidant enzymes in pepper *Pak. J. Bot.* **6** 1889-96

[7] Yang Y H 2009 Studies on the prevention effect of PGPR in the greenhouse pepper obstacles *J. Hebei Univ. Eng. (Nat. Sci. Ed.)* **2** 103-105 (in Chinese)

[8] Wu C W,Du X F,Gu D L,Wen T G, Yang W F,Guo X S,Wang G L, Xie Z Y and Wang W Z 2015 Effects of mulching modes on soil environment and growth of red pepper in greenhouse *Jiangsu J. Agric. Sci.* **2** 407-14 (in Chinese)

[9] Liu L,Huang B J,Sun J,Guo S R,Li L Q and Guo H W 2013 Relationship between soil

microbial quantity,enzyme activity and soil fertility in hot pepper greenhouse soils of different continuous cropping years *Soil and Fertilizer Sciences in China* **2** 5-10 (in Chinese)

[10] Guo H W,Guo S R, Liu L,Sun J and Huang B J 2012 Effects of continuous cropping on physical and chemical properties of soil, physiological resistance and ion absorption of pepper *Soils* **6** 1041-47 (in Chinese)

[11] Guo H W,Guo S R and Huang B.2011. Study on soil physical and chemical properties in pepper greenhouse of different continuous cropping years. *Jiangsu Agric. Sci.*, 5,452-455 (in Chinese)

[12] Yao H Y, Jiao X D and Wu F Z 2006 Effects of continuous cucumber cropping and alternative rotations under protected cultivation on soil microbial community diversity *Plant Soil* **284** 195-203

[13] Zhou X G, Gao D M, Liu J, Qiao P L, Zhou X L, Lu H B, Wu X, Liu D, Jin X and Wu F Z 2014 Changes in rhizosphere soil microbial communities in a continuously monocropped cucumber (*Cucumis sativus* L.) system *Eur. J. Soil Biol.* **60** 1-8

[14] Congreves K A, Hayes A, Verhallen E A and Eerd V L L 2015 Long-term impact of tillage and crop rotation on soil health at four temperate agroecosystems. *Soil Till. Res.* **1** 17-28

[15] Spencer J L, Hughson S A and Levine E 2014 Chapter 7-Insect Resistance to Crop Rotation. *Insect Resistance Management (Second Edition)* (Amsterdam: Elsevier) p 233-78

[16] Melero S, López-Bellido R J, López-Bellido L, Muñoz-Romero V, Moreno F and Murillo J M 2011 Long-term effect of tillage, rotation and nitrogen fertiliser on soil quality in a Mediterranean Vertisol. *Soil Till. Res.* **2** 97-107

[17] Lehuger S, Gabrielle B, Cellier P, Loubet B, Roche R, Béziat P, Ceschia E and Wattenbach M.2010. Predicting the net carbon exchanges of crop rotations in Europe with an agro-ecosystem model. *Agric. Ecosys.t Environ.* **3**, 384-395.

[18] Bennett A J, Bending G D, Chandler D, Hilton S and Mills P 2012 Meeting the demand for crop production: the challenge of yield decline in crops grown in short rotations *Biol. Rev. Camb. Philos. Soc.* **87** 52-71

[19] Li W, Cheng Z H, Meng H W, Zhou J, Liang J and Liu X J 2012 Effect of rotating different vegetables on microbiomass and enzyme in tomato continuous cropped substrate and after culture tomato under plastic tunnel cultivation. *Acta Hortic.Sin.* **1** 73-80 (in Chinese)

[20] Guo R Y, Li X L, Christie P, Chen Q, Jiang R F and Zhang F S 2008 Influence of root zone nitrogen management and a summer catch crop on cucumber yield and soil mineral nitrogen dynamics in intensive production systems *Plant Soil* **313** 55–70

[21] Wu Y, Zhang X, Gao L, Yuan L and Wei W 2007 Effect of different cultivation patterns on continuous cropping soil environment and cucumber growth *Chin. J. Eco-agric. (ISHS)* **761** 547-54

[22] Li Y, Gao L H, Wu Y F, Guo R Y and Zhang X Y 2006 Effect of Summer Catch Crops on Soil Environment in Solar Greenhouse *J. Shenyang Agric. Univ.* **3** 531-4 (in Chinese)

[23] Wu F Z, Guo X, Liu S W, Zhou X G and Yi Z H 2014 Effect of summer catch mode on root zone soil enzyme activities and soil chemical properties of continuously monocropped cucumber *J. Northeast Agric. Univ. (English Edition)* **10** 29-34

[24] Marinari S, Mancinelli R, Brunetti P and Campiglia 2015 Soil quality, microbial functions

and tomato yield under cover crop mulching in the Mediterranean environment *Soil Till. Res.* **145** 20-8

[25] Campiglia E, Radicetti E, Brunetti P and Mancinelli R 2014 Do cover crop species and residue management play a leading role in pepper productivity *Sci. Hortic.* **166** 97-104

[26] Gabriel J L, Garrido A and Quemada M 2013 Cover crops effect on farm benefits and nitrate leaching: Linking economic and environmental analysis *Agric. Systems* **121** 23-32

[27] Nair A and Ngouajio M 2012 Soil microbial biomass, functional microbial diversity, and nematode community structure as affected by cover crops and compost in an organic vegetable production system *Appl. Soil Ecol.* **58** 45-55

[28] Gu Z G, Ma Y, An X, Wang G F, Sun D and Wang Q J 2014 Effects of wheat straw with flooding on soil properties and phytophthora blight control in continuous chili pepper cropping field *J. Agro-Environ. Sci.* **9** 1762-9(in Chinese)

[29] Li H S 2000 *Experimental Principles and Techniques of Plant Physiology and Biochemistry* (Beijing:Higher Education Press) p 123-125,184,185,195-197, 246-9 (in Chinese)

[30] Liu S W,Liu S Q,Pan K,Wang L L and Wu F Z 2011 Effects of different preceding crops on soil micro-ecological environment and yield of cucumber *J. Northeast Agric. Univ. (English Edition)* **4** 6-14

[31] Li Z G, Luo Y M and Teng Y 2008 *Research Method of Soil and Environmental Microbiology* (Beijing: Science Press) p 90-99,395-413 (in Chinese)

[32] Cao H, Sun H, Yang H, Sun B and Zhao Q G 2003 A review: soil enzyme activity and its indication for soil quality *Chin. J. Appl. Environ. Biol.* **1** 105-9(in Chinese)

[33] Wu X, Liang Y L, Hao W L, Luo A R, Peng Q and Chen C 2011 Effect of mulching mode on growth and water use of greenhouse pepper during fruiting stage *Chin. J. Eco-agric.* **1** 54-8 (in Chinese)

[34] Wu F Z, Zhao F Y and Gu S Y 2002 Effect of the continuous cultivating cucumber on the bio-chemical properties of soil in the plastic greenhouse *System Sci. Comprehensive Studies in Agric.* **18** 20-2

[35] Liang Y L, Chen Z J, Xu F L, Yan Y G, Du S N, Zhang C E 2003The effect of continuous cropping year on physiological characteristics of cucumber in sunlight greenhouse *Acta Botanica Boreali-occidentalia Sin.* **23** 1398-1401 (in Chinese)

[36] He S M, Yang Y J, Li B Y and Xu M F 2005 Application of an efficient planting mode of greenhouse vegetables and water chestnut rotation *Jiangsu Agric. Sci.* **1** 74-5 (in Chinese)

[37] Qian Y M, Zhao M Z, Wu W M, Wang Z W, Wang J, Yu H M,Cai W J and Zhang Z 2012 Preliminary study on water spinach cultivation under the mode of greenhouse strawberry and water spinach rotation *Jiangsu Agric. Sci.* **12** 167-8 (in Chinese)

[38] Yu J Q and Du X S 2000 Soil-sickness problem in the sustainable development for the protected production of vegetables *J. Shenyang Agric. Univ.* **2** 124-6 (in Chinese)

[39] Xiong H Y, Li T X and Yu H Y. 2008. Amount of soil microorganism and influencing factor of different no-tillage years in ''paddy-upland'' rotation systems *J. Wuhan Univ. (Nat.sci.ed.)* **2** 244-8 (in Chinese)

[40] Xiong H Y, Li T X, Yu H Y and Zhang X Z 2008 Microbial characters in no-tillage soil with paddy-upland rotation *Plant Nutr. fertilizer Sci.* **1** 145-150 (in Chinese)

[41] Hu Y S, Wu K, Liu N, Chen H G and Jia X C 2004 Studies in microbial population dynamics

in the cucumber rhizospheres at different developmental stages *Sci. Agric. Sin.* **10** 1521-6 (in Chinese)

[42] Wang Y, Shen Q R and Shi R H 1996 Soil microbial biomass and its ecological effects *J. Nanjing Agric. Univ.* **4** 45-51 (in Chinese)

[43] Zimmermann S and Frey B 2002 Soil respiration and microbial properties in an acid forest soil: effects of wood ash *Soil boil. Biochem.* **34** 1727-37

[44] Miller M and Dick R P 1995 The stability and activity of soil enzymes as influenced by crop rotations *Soil boil. Biochem.* **27** 1161-6

[45] Sun X S, Feng H S, Wan S B and Zuo X Q 2001 Changes of main microbial strains and enzymes activities in peanut continuous cropping soil and their interactions *Acta Agro. Sin.* **5** 617-20 (in Chinese)

[46] Yang F J, Wu H T, Wei M and Shi Q H 2009 Effects of rotation and fallowing on the microbial communities and enzymes activities in a solar greenhouse soil under continuous cucumber cropping *Chin. J. Appl. Ecol.* **12** 2983-8 (in Chinese)

IOP Conf. Series: Earth and Environmental Science **41** (2016) 012023 doi:10.1088/1755-1315/41/1/012023

Stock characteristics of soil organic carbon pools under three subtropical forests in South China

X Y Zhang[1,3], D S Guan[2] and M Z Xiao[1]

[1]South China Institute of Environmental Sciences, Ministry of Environmental Protection, Guangzhou 510655, China

[2]School of Environmental Science and Engineering, Sun Yat-sen University, Guangzhou 510275, China

E-mail: zhangxiuyu@scies.org

Abstract. Vegetation biomass and soil organic carbon (SOC) pools for the three representative forest types, i.e. conifer forest (CF), mixed conifer and broad-leaf forest (CBF), evergreen broad-leaf forest (EBF) in South China were investigated. We found that SOC stock of the three chief forest ranged from 55.54 to 151.16 MgC·ha^{-1}, and it increased with increasing vegetation biomass under the same type forest within 100cm depth. The organic carbon contents at an equivalent level of forest maturity tended to be in the following decreasing order: EBF > CBF > CF, various active organic carbon (AOC) fractions in the 0-20cm topsoil layer tended to be in the following decreasing order: light fraction carbon (LFC) ≈ particulate organic carbon (POC) > easily oxidisable carbon (EOC) > microbial biomass carbon (MBC) > water-soluble carbon (WSC). At an equivalent level of forest maturity, there was a trend that each of these five AOC fractions increased from CF to CBF to the EBF.

1. Introduction

It is generally believed that estimates of quantity and distribution of forest soil carbon pool are essential to better understand the role of the forest ecosystem in global carbon cycling [3]. A major proportion of atmospheric CO_2-C sequestrated by forest vegetation eventually enters into soils and is stored as soil organic carbon (SOC), and a research reveals that site forest vegetation biomass alone provides no indication of carbon-sequestering capacity of a given forest ecosystem because about two-thirds atmospheric CO_2-C sequestrated by forest vegetation is stored as SOC within the top 100cm of the soil layer [9]. However, less studies emphasis on active organic carbon (AOC) fraction in the surface layer, which has an important role to play in regulating global carbon balance, especially in tropical and subtropical forest ecosystem [16], therefore, gaining content characteristics of AOC fraction in the surface layer under site forest vegetation would be helpful to define the role of SOC in regulating atmospheric CO_2 concentrations. Differing degrees of intervention on forest ecosystems adds uncertainty to estimate the forest soil carbon pool [19]. Taking vegetation carbon stock (tree, undergrowth vegetation, coarse woody debris and litter) into account, the objective of our study was to stress on characterizing the SOC stock within 100cm and its AOC content within 20cm under three representative secondary subtropical forest types in South China.

2. Materials and methods

2.1. *Site descriptions*
The study site is located in Guangzhou of Guangdong Province (112°57′～114°3′E , 22°26′～23°56′N), southern China, where the regional climate is subtropical monsoon climate. The original forests vanished and currently there are three dominant types of secondary forests in the area: conifer forest (CF), mainly *Pinus massonianai*, evergreen broad-leaf forest (EBF), mainly *Engelhardtia roxburghiana, Schima superba* and *Castanopsis chinensis*, and mixed conifer and broad-leaf forest (CBF). Red soil is approximately 100 cm thick in the study site, which was formed from arenaceous shale. According to Soil Taxonomy of China, soil at the site is classified as a fine loamy, hyperthermic, acidic and Udic Cambisol.

2.2. *Experimental design, sampling and measurements*
Nine plots were selected in this study. They represented young forest stands (plot 1, 4, 7) with less than 10 years old, middle forest stands (plot 2, 5, 8) with 40-50 years old and mature forest stands (plot 3, 6, 9) with more than 80 years old for each of the three forest types. Each plot was 50 × 50 m^2. Within each plot, six sampling locations were established along the "S-shaped" pattern with each sampling location was 10m apart. Soil samples were collected at each sampling location from the following four soil layers: 0-20, 20-40, 40-60 and 60-100 cm. Soil samples were sealed in plastic bags and transported to the laboratory for determining bulk density and carbon contents. Soil samples were air dried and sieved to 2mm, in order to remove litter, large pieces of roots and stones (>2 mm).

The method described by Zheng et al. (2008) has been used to estimate the forest vegetation biomass of each plot. SOC content from 0 to 100-cm was measured while the active organic carbon (AOC) was measured at the 0-20cm layer. SOC was determined by wet combustion with $K_2Cr_2O_7$ [5, 8]. Various AOC fractions were determined the follows methods: (1) water-soluble carbon (WSC) was determined at 25℃ using the distilled water-extraction method [8]; (2) microbial biomass carbon (MBC) was determined by using the fumigation-extraction method [8, 15]; (3) easily oxidable carbon (EOC) was determined after oxidation of the sample with 333 mmol L^{-1} $KMnO_4$ at 25 ℃ [18]; (4) light fraction carbon (LFC) was determined by using the method showed by Shi et al. (2007); and (5) particulate organic carbon (POC) was determined by using the method described by Cambadella and Elliott (1992).

2.3. *Statistical analysis*
Average density of soil organic carbon (SOCD) of the top 100cm soil profile was estimated by using the following formula:

$$\mathrm{SOCD} = \sum_{i=1}^{n} T_i \times \theta_i \times C_i \times (1-\delta_i)/100 \qquad (1)$$

Where n denotes the number of soil layers involved, δ_i% is the gravel content (>2mm) in Soil layer i; θ_i signifies the soil bulk density in Soil layer i ($g \cdot cm^{-3}$); C_i represents the organic carbon content in Soil layer i ($g \cdot kg^{-1}$); and Ti stands for the thickness of Soil layer i (cm).

All data statistical analysis were performed with the SPSS16.0 software package. Significant difference of a soil organic carbon parameter among the nine plots were tested using a Duncan's multiple range test method; Pearson linear correlations has been used to assess the relationships between various AOC fractions and SOC in the soil surface layer(0-20cm).

3. Results

3.1. Vegetation biomass and carbon stock in the nine investigated plots

The living tree biomass dominated the above-ground plant biomass for all the nine investigated plots (Table 1). Results show that living tree biomass, litter biomass and coarse woody debris (CWD) increased with increasing forest maturity. At an equivalent forest age stage, tree biomass tended to be higher in EBF plot than in CF plot or in CBF plot. Litter biomass in CF and CBF were higher than EBF did at an equivalent level of forest age stage while the coarse woody debris biomass tended to be in the following decreasing order at an equivalent level of forest age stage: EBF >CBF >CF. Undergrowth vegetation biomass tended to be lower in CF than in CBF or in EBF. For all the three forest types, undergrowth vegetation biomass was in the following decreasing order: middle-age stands >mature-age stands >young-age stands.

Vegetation carbon stock ranged from 25.39 $Mg\cdot ha^{-1}$ to 194.74 $Mg\cdot ha^{-1}$ in the nine investigated plots, and EBF tended to be higher than the other two forest type when they at equivalent forest age stage, which was consistent with the allocation of their total vegetation biomass.

Table 1. Vegetation biomass and carbon density of the three investigated forest types.

Forest type		Biomass ($Mg\cdot ha^{-1}$)					Carbon density
		Tree	UV[a]	CWD[b]	Litter	Total	($MgC\cdot ha^{-1}$)
	Plot 1	36.87	8.93	0.25	6.25	52.30	25.39
CF	Plot 2	95.38	13.21	0.36	8.58	117.53	57.68
	Plot 3	190.62	11.65	1.07	11.79	215.13	106.39
	Plot 4	43.59	12.47	2.41	5.87	64.34	31.25
CBF	Plot 5	97.85	16.65	6.48	7.46	128.44	63.01
	Plot 6	179.46	15.13	10.55	10.54	215.68	106.56
	Plot 7	45.74	12.06	3.84	4.74	66.38	32.35
EBF	Plot 8	160.35	15.24	11.03	7.03	193.65	95.71
	Plot 9	350.23	14.39	17.55	9.72	391.89	194.74

[a] Undergrowth vegetation.
[b] Coarse woody debris.

3.2. Soil organic carbon stock in the nine investigated plots

For CF (Table 2), there was no statistical difference in SOC contents among the three plots (Plots 1-3) representing the different stage of forest maturity for all the four soil layers. For the CBF, except for the surface soil layer (0-20cm) where SOC contents in Plot 6 was significantly higher than those in Plots 4-5, all the soil layers showed no statistical difference in SOC contents among the three plots (Plots 4-6) representing the different stage of forest maturity. For EBF, except for the bottom soil layer (60-100cm) where no statistically significant difference in SOC contents existed among the three plots (Plots 7-9) representing the different stage of forest maturity, all the soil layers showed certain statistical difference in SOC contents among these three plots; there is a trend that the SOC contents increased with increasing forest maturity.

Table 2. Comparison of SOC contents and SOC density among the nine investigated plots.

Forest type		SOC contents ($g\cdot kg^{-1}$)				SOC density
		0～20cm	20～40cm	40～60cm	60～100cm	($MgC\cdot ha^{-1}$)
	Plot 1	12.59±2.85a	4.47±0.79a	2.11±0.34a	1.75±0.24a	55.54±4.81a
CF	Plot 2	13.57±2.06a	4.26±0.96a	2.14±0.28a	1.96±0.31a	58.53±7.46a
	Plot 3	14.79±3.57a	5.63±1.27a	2.70±0.41a	2.05±0.17a	66.69±6.39a
	Plot 4	12.83±1.78a	5.27±1.35a	2.31±0.25a	1.99±0.26a	59.76±8.55a
CBF	Plot 5	15.09±3.32a	6.44±1.49a	2.98±0.50a	2.01±0.32a	69.90±10.34a
	Plot 6	20.85±4.54b	8.52±2.29a	3.17±0.62a	2.05±0.43a	89.77±9.51b

	Plot 7	19.28±5.25b	8.13±1.61a	3.09±0.47a	2.08±0.33a	84.91±7.87b
EBF	Plot 8	25.19±5.43bc	12.51±2.29b	5.98±0.69a	3.46±0.56a	123.97±20.42c
	Plot 9	31.06±4.58c	14.37±2.74b	9.03±1.15b	4.12±0.68a	151.16±34.95c

Values within a column followed with different letters are significantly different at $P<0.05$.

When the nine plots are considered together, there is no significant difference in SOC among each of the nine plots for the bottom soil layer (60-100cm). Except for Plot 9, no significant difference in SOC existed among each of the nine plots for the soil layer 40-60cm. Further upward to the soil layer 20-40cm, SOC contents was significantly higher in Plots 8-9 than in any other plots. SOC contents was significantly higher in Plots 8-9 than in Plots 6-7 which, in turn, had significantly higher SOC than the other five plots did; there is no significant difference in SOC among Plots 1-5.

3.3. Various active organic carbon fractions in the nine investigated plots

For CF (Table 3), there was no statistically significant difference in each of the five AOC fractions among the three different plots (Plots 1-3) representing different forest development stages. For CBF, there was no statistically significant difference in each of the five AOC fractions between Plot 4 and Plot 5 except for EOC which was significantly higher in Plot 5 than in Plot 4. Each of the five AOC fractions was significantly higher in Plot 6 than in either Plot 4 or Plot 5 except for EOC which showed no significant difference between Plot 5 and Plot 6. For EBF, mixed situations were observed: there was no significant difference in LFC among the three plots; WSC, EOC and POC were significantly lower in Plot 7 than in Plots 8 and Plot 9 but no significant difference in WSC, EOC and POC existed between Plot 8 and Plot 9; MBC was significantly higher in Plot 9 than in Plots 7 and Plot 8 but no significant difference in MBC existed between Plot 7 and Plot 8.

Table 3. Contents ($g \cdot kg^{-1}$) of various AOC fractions in 0-20cm soil layer of three forest types.

Forest type		WSC	MBC	EOC	LFC	POC
	Plot 1	0.10±0.03a	0.30±0.05a	0.36±0.05a	2.16±0.15a	2.34±0.14a
CF	Plot 2	0.13±0.05a	0.37±0.03a	0.43±0.07a	2.48±0.20a	2.56±0.22a
	Plot 3	0.14±0.02a	0.46±0.09a	0.55±0.09a	2.81±0.25a	2.79±0.18a
	Plot 4	0.14±0.04a	0.32±0.06a	0.71±0.11a	2.54±0.38a	2.88±0.16a
CBF	Plot 5	0.17±0.05a	0.39±0.05a	1.24±0.10ab	2.82±0.41a	3.22±0.27a
	Plot 6	0.24±0.06b	0.59±0.11b	2.22±0.13b	4.05±0.57ab	4.20±0.41ab
	Plot 7	0.21±0.04b	0.57±0.08ab	2.00±0.14b	3.81±0.49a	4.28±0.39ab
EBF	Plot 8	0.29±0.06bc	0.81±0.17b	3.37±0.23c	5.49±0.66ab	6.04±1.04b
	Plot 9	0.39±0.08c	1.10±0.22c	4.35±0.37c	7.24±1.24b	7.50±0.95b

Values within a column followed with different letters are significantly different at $P<0.05$.

4. Discussion

In this study, SOC density of the nine plots ranged from 55.54 $MgC \cdot ha^{-1}$ to 151.16 $MgC \cdot ha^{-1}$ within 100 cm of soil profiles. It was reported that, average SOC density of Chinese forests was 107.8 $MgC \cdot ha^{-1}$ [10], by comparison, only SOC density of plot 8 and plot 9 under EBF is higher than the results mentioned above, which indicates SOC density of subtropical forest in South China was relatively low. The reason is that in China, soil had a lower capacity for carbon sequestration under subtropical forest than that under temperature forests [20], which was consistent with the distribution pattern in the northern hemisphere [11].

It is interesting to note that the proportion of SOC stock in the total forest ecosystem carbon pool decreased with increasing maturity of forest development within a forest type (Figure. 1). At the young forest stands, the vegetation carbon stock was smaller than the SOC stock, suggesting that it is unlikely that a major proportion of SOC was contributed by the existing forest vegetation. Therefore, the SOC in these plots mainly consisted of residual soil organic carbon originated from previous forest vegetation. The reason is that ecological components such as coarse root biomass, large amounts of CWD, and a thick forest floor layer are important contributors to long-term C storage within these ecosystems [1].

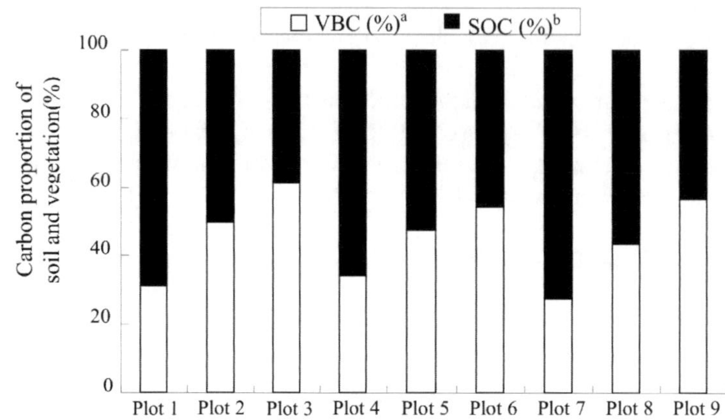

Figure 1. Proportion of soil carbon pool and vegetation carbon pool in the total forest ecosystem carbon pool .

[a] Vegetation biomass carbon.
[b] Soil organic carbon.

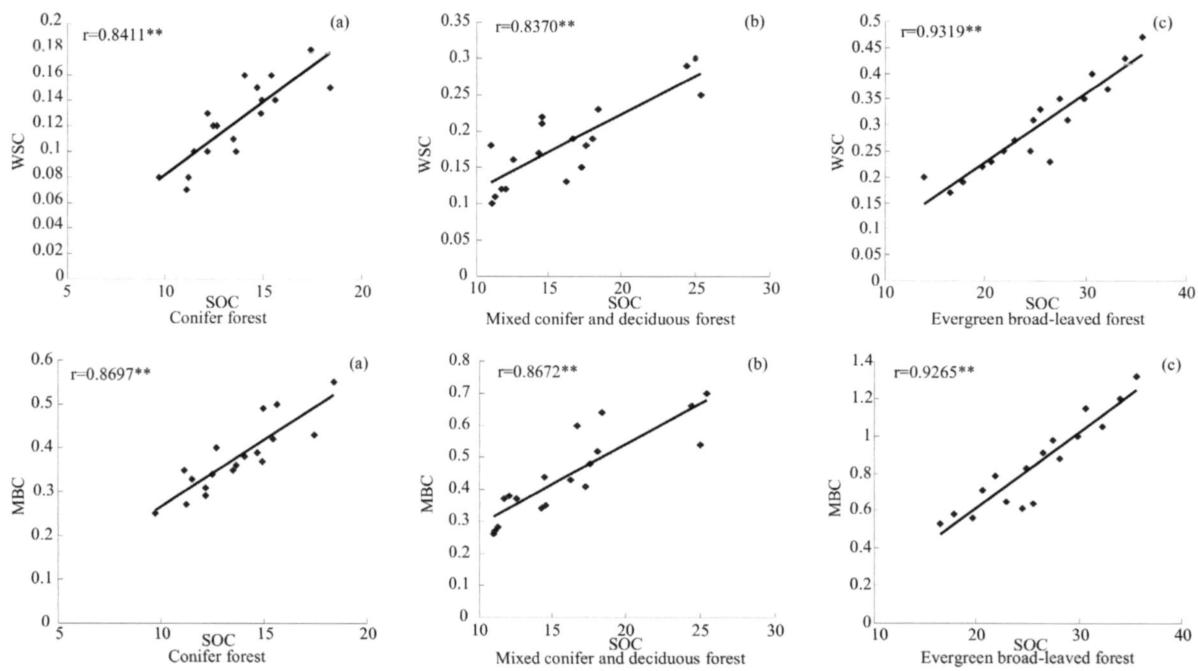

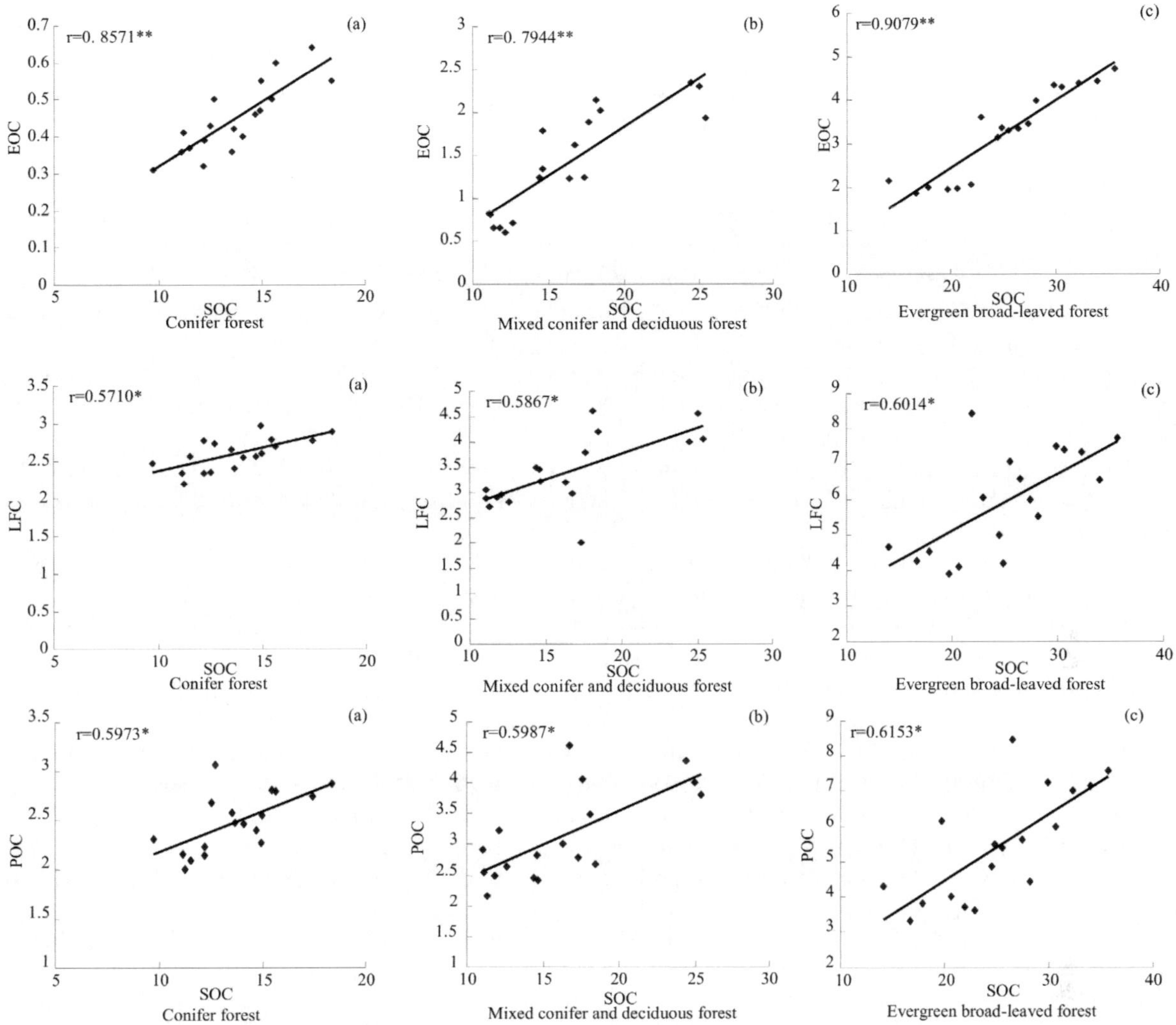

Figure 2. Relationship between AOC fractions and SOC in the topsoil layer (0-20cm).

Forest net primary productivity (NPP) can affect the contents of various AOC fractions by regulating SOC allocation so the correlations between AOCs and SOC can mirror forest NPP to certain degree [4]. In this study, correlations between WSC and SOC, EOC and SOC, MBC and SOC were at significant level (p<0.01) (Figure 2), and the correlation coefficient followed the order: EBF > CF > CBF while correlations between LFC and SOC, POC and SOC, were at significant level (p<0.05), and the correlation coefficient followed the order EBF >CBF >CF.

The correlation coefficients of LFC vs SOC and POC vs SOC were lower than those of WSC vs SOC, EOC vs SOC and MBC vs SOC. In the field, LFC and POC were found to be frequently associated with gravels in the soils. This observation suggests that in addition to microbial activities, soil physical properties might also affect the contents of LFC and POC [12]. Ingo and Ingrid (2006) suggested that the long-term stabilization of soil organic matter (SOM) was mainly controlled by various protection mechanisms provided by soil matrix and soil minerals rather than by the chemical structure of SOM itself.

Acknowledgments
We would like to thank the funding support provided by the National Natural Science Foundation of China (No. 40971054) and the commonweal Project of South China Institute of Environmental Sciences, Ministry of Environmental Protection (No. zx_200912_04).

References

[1] Bisbing S M, Alaback P B and DeLuca T H 2010 Carbon storage in old-growth and second growth fire-dependent western larch (Larix occidentalis Nutt.) forests of the Inland Northwest, USA *Forest Ecol. Manage.* **259**(5) 1041-9
[2] Cambadella C A and Elliott E T 1992 Particulate soil organic matter changes across grassland cultivation sequence *Soil Science Society of America Journal* **56** 777-83
[3] Clough B J 2014 *Bayesian approaches for mapping forest soil organic carbon: addressing spatial and model uncertainty* (Doctoral dissertation, Rutgers University-Graduate School-New Brunswick) pp 1-13
[4] Franzluebbers A J, Haney R L, Honeycutt C W, Arshad M A, Schomberg H H and Hons F M 2001 Climatic influences on active fractions of soil organic matter *Soil Biol. Biochem.* **33** 1103-11
[5] Heanes D L 1984 Determination of total organic-C in soils by an improved chromic acid digestion and spectrophotometric procedure *Commun. Soil Sci. Plant. Anal.* **15** 1191-213
[6] Hedde M, Aubert M, Decaens T and Bureau F 2008 Dynamics of soil carbon in a beechwood chronosequence forest *Forest Ecol. Manage.* **255** 193-202
[7] Ingo S and Ingrid K K 2006 Chemical composition of young and old carbon pools throughout Cambisol and Luvisol profiles under forests *Soil Biol. Biochem.* **38** 2411-24
[8] Jiang P K and Xu Q F 2006 Abundance and Dynamics of Soil Labile Carbon Pools Under Different Types of Forest Vegetation *Pedosphere* **16**(4) 505-11
[9] Lal R 2005 Forest soils and carbon sequestration *Forest Ecol. Manage.* **220** 242-58
[10] Liu S, Wang H, Luan J 2011 A review of research progress and future prospective of forest soil carbon stock and soil carbon process in China *Shengtai Xuebao/Acta Ecol. Sin.* **31**(19) 5437-48
[11] Morisada K, Ono K and Kanomata H 2004 Organic carbon stock in forest soils in Japan *Geoderma* **119** 21-32
[12] Richards A E, Dalal R C and Schmidt S 2007 Soil carbon turnover and sequestration in native subtropical tree plantations *Soil Biol. Biochem.* **39** 2078-90
[13] Schlesinger W H and Lichter J 2001 Limited carbon storage in soil and litter of experimental forest plots under increased atmospheric CO_2 *Nat.* **411**(6836) 466-9
[14] Shi Y, Chen X and Shen S M 2007 Light Fraction Carbon and Water-Stable Aggregates in Black Soils *Pedosphere* **17**(1) 97-100
[15] Vance E D, Brooks P C and Jenkinson D S 1987 An extraction method for measuring soil microbial biomass C *Soil Biol. Biochem.* **19** 703-7
[16] Wang Q K and Wang S L 2007 Soil organic matter under different forest types in Southern China *Geoderma* **142** 349-56
[17] Li X and Han S 2008 Preservation of broadleaf species in Korean pine (Pinus koraiensis) plantations affects soil properties, carbon storage, biomass allocation, and available nitrogen storage *Canadian J. Forest Res.* **38**(8) 2227-35
[18] Xu Q F and Xu J M 2003 Changes in Soil Carbon Pools Induced by Substitution of Plantation for Native Forest *Pedosphere* **13**(3) 271-8
[19] Yadava P S and Thokchom A 2014 Carbon Sequestration Potential of the Forests of North-eastern India *Carbon Capt., Storage and, Util.: Poss. Clim. change Sol. Energ. Ind.* 149
[20] Yang L, Pan J, ShaoY, Chen J M, Ju W M, Shi X and Yuan S 2007 Soil organic carbon decomposition and carbon pools in temperate and sub-tropical forests in China *J. Environ. Manage.* **85** 690-5

[21] Zheng H, Ouyang Z Y, Xu W H, Wang X K, Miao H, Li X Q and Tian Y X 2008 Variation of carbon storage by different reforestation types in the hilly red soil region of southern China *Forest Ecol. Manage.* **255** 1113-21
[22] Zhou G Y, Liu S G, Li Z A, Zhang D Q, Tang X L, Zhou C Y, Yan J H and Mo J M 2006 Old-growth Forests can Accumulate Carbon in Soil *Sci.* **314**(5804) 1417

Sod-podzolic soils with a complex organic profile of the southern Vyatka River basin

A M Prokashev[1, 2], E S Soboleva[1], R R Chepurnov[1], A S Matushkin[1], N D Ohorzin[1], I L Borodaty[1], I A Zhuikova[1], I Y Alalykina[1], G A Russkikh[1], S A Pupysheva[1], S L Mokrushin[1] and I A Vartan[1]

[1]Geography Department, Institute of Chemistry and Ecology, Vyatka State University, 36 Moskovsky Str., Kirov, Russia

E-mail: amprokashev@gmail.com

Abstract. The article is devoted to the morphology and substantive properties of the sod-podzolic soils with a complex organic profile having the second humus horizons and found on the southern right bank of the lower Vyatka River. The research results of the mineral and organic phases are the evidence of the relict origin of the second humus horizon as well as the evidence of profile polygenicity of the given soils. They went through two fundamentally different pedogenesis stages during the postglacial period: 1) the developmental accumulative evolution stage in the first half of the Holocene and 2) the accumulative-eluvial stage of erasing evolution including the elements of inheriting evolution in the second half of the Holocene. Keywords: granulometric composition, fractional and group composition, humus age, genesis, soil evolution

1. Introduction

Soils with a problematic profile structure are widespread in the soil mantle of the lower Vyatka River basin, which is situated between the southern taiga and mixed forests. This soil type was originally attributed to the gray forest type. The reason for this was a transitional accumulative-eluvial horizon (Ahh) in the profile. However, the verification of these soils was called in question due to the following special studies [1-4]. The information below confirms this.

2. Object, subject and methods of research

The object of the study is the sod-podzolic soils with a complex organic profile on the covering silt found between the rivers Vyatka and Yaran under the virgin fir and spruce forests, secondary birch forests and agricultural landscapes. The subject of the research is morphology, the mineral and organic phases of soils. The authors of the article used the following research methods: comparative soil-geographical method, morphological and genetic profile method, balance method, method of granulometry and bulk chemical analysis of the mineral part, method of fractional-and-group and radiocarbon humus analysis.

3. Results of the study

3.1. Soil morphology

IOP Conf. Series: Earth and Environmental Science **41** (2016) 012024 doi:10.1088/1755-1315/41/1/012024

The objects under research vary considerably in their appearance. The Ya-1 cut of the sod-podzolic silty clay loam residual calcareous soil with the second humus horizon on the carbonate covering silt is situated in the upper flattened part of the undulating Yaran-Izhsk water divide. It lies under the virgin spruce and fir forest, wood sorrel and aise-weed forest. Atmospheric moistening is sufficient; the type of soil moisture regime is periodically flushing.

Profile structure:

The O horizon (0-2 cm) contains wet pine needles, twigs, cones, leaves, stalks of grass and moss. It is brown, slightly decomposed and loose. The transition is clear, straight and distinguishable by color and degree of organic decomposition.

The OAY horizon *(2-5 cm)* is wet, black-and-brown, silty clay loam, loose, grainy, friable, enriched with humified organic residues. The roots are abundant; the transition is clear and distinguishable by color.

The AY horizon (5-14 cm) is wet, dark gray, variegated in color. Intensely colored lumps are found when the soil is dry. Some aggregates are light gray outside and dark gray inside. The horizon is silty clay loam, granular and lumpy, loose, grainy; it contains many small rounded gray-brown hardpans sized 2-3 mm and more. There are many roots; the transition is clear and distinguishable by color and structure.

The Elh horizon (14-24 cm) is wet, flaky laminated, silty clay loam, loose, smoky-gray, but becoming light gray after drying. The inner part of the aggregates is darker. There is intense whitish silica powder on the structural units. There are many small grayish brown hardpans; the roots are sparse; the transition is gradual, wavy and having a large local pocket.

The Bh horizon (24-52 cm) has a humified local pocket of 40 cm wide in the upper part and 25 cm at the bottom. The horizon is wet, carbon-black having whitish powder at the top which gradually disappears at the depth of 48 cm. The aggregates are evenly impregnated with humus. The impregnation is less intense along the pocket periphery. There are whitish spots of silica up to 3-4 cm in diameter at the depth of 32-50 cm. The horizon is clay loam, and clay in a deeper layer having a heterogeneous structure. Downstream the platy aggregates gradually change to fine- and coarse nutty ones. The roots are sparse; the transition is clear, distinguishable by color.

The El horizon (24-32 cm) is wet, whitish pale, clay loam, laminal-platy, loose, friable; bleached silica powder is abundant. There are many small grayish brown hardpans. The roots are sparse; the transition is gradual and tongue-like.

The ElB horizon (32-41 cm) is wet, whitish brown, clay, platy, nutty, compact, friable with lots of whitish bleached silica weakening in a deeper layer. There are few hardpans; the roots are sparse; the transition is gradual, tongue-like.

The Bt1 horizon (41-58 cm) is wet, brown, clay, nutty, dense. It has dark humus varnishing and easily falls into structural units. On the surface of aggregates there is whitish silica weakening in a deeper layer; in the lower parts of the aggregates there are small clay papillary cutans. The roots are sparse; the transition is gradual.

The Bt2 horizon (58-94 cm) is wet, yellowish brown, clay, prism-shaped and nutty, dense, slightly viscous. It has dark humus varnishing which is less intense compared to the Bt1 horizon. Whitish silica powder is distinguishable only on the large cracks between aggregates. There are few cutans; the roots are single, the transition is gradual.

The BCca horizon (94-135 cm) is wet, yellowish brown, clay loam, prism- shaped, dense; it is more viscous than the Bt2 horizon. There is slight humus varnishing on the surface of aggregates and along the walls of root pores. There are few clay cutans; at the depth of 120 cm there are small calcareous pipes as well as weak effervescence caused by HCl solution. The roots are single, the transition is gradual.

The Cca horizon (135-160 cm) is wet, yellow-brown, variegated in color, clay loam, viscous, non-structured and has slight reticulated humus varnishing in root pores. At the depth of 150 cm there are mottlings of reddish and grayish green units consisting of Perm clay eluvium. The horizon contains soft pink-white calcareous grit. In the lower part the soil has signs of mixing and dislocation. The roots are single; the transition is gradual.

The CDca horizon (160–220 cm) is wet, brownish-red, heterogeneous, clay, nutty, dense. It contains pinkish and grayish-greenish mottlings, but there are less calcareous units in comparison with the Cca horizon. The mineral matter has signs of mixing.

Morphological analysis indicates that the profile clearly differentiates between accumulative-eluvial and illuvial parts. The upper and lower humus horizons vary in their appearance considerably. The AY horizon of the virgin soil has features of a typical accumulative horizon. The plowing layers of the arable soil have lost their natural structure.

The AElh horizon is characterized by a contrasting combination of accumulative and eluvial horizon features, i.e. dark gray or carbon-black coloring, brittle slab-shaped or slab-shaped granular structure, silica powder on the surface of aggregates impregnated with humus, and sparse roots.

The morphological appearance of the ElB horizon is typical of transient sub-eluvial parts of soil profiles with heavy granulometric composition which are distinguished by whitish-brown coloring, slab-shaped or slab-shaped and fine-nutty structure, and whitish silica. Humus varnishing and clay cutans are specific features of the texture-illuvial horizons.

Depth of the sod-podzolic soil horizons with a complex organic profile can be summarized as follows: O (2 cm) + OAY (3 cm) + AY (10 cm) + AElh (12 cm) + ElB (8 cm) + Bt_1 (24 cm) + Bt_2 (33 cm) + BC(ca) (28 cm) + C(ca) + (D).

3.2. Granulometric composition

The granulometric composition shows a clear differentiation of the profile (Table 1). The upper 30-35 cm stratum is considerably depleted of physical clay and has a silty clay loam composition. There is almost no difference between the humus horizons regarding the content of particles less than 0.01 mm. Their quantity increases substantially in the ElB horizon. The middle part of the profile (Bt horizons) is characterized by the highest number of physical clay particles and always has a silty clay granulometric composition. In the C horizons the level of physical clay slightly decreases.

The reason for eluvial and illuvial differentiation is the similar profile distribution of clay with the minimum level in the A horizons. Both humus horizons are close in their characteristics, although the lower one tends to have more clay. The Bt horizons are defined by the maximum concentration of the clay fraction. The amount of clay in the soil-forming material is reduced. Negative clay balance was found in the profile of the soils under research by means of balance calculations of the absolute clay fraction content (Table 2).

Bulk chemical analysis confirms sharp textural differentiation which is especially noticeable in the virgin soils (Table 3). Their organogenic horizons O and OAY are clearly isolated due to the considerable amount of biophile elements (Ca, Mn, K, P, S, etc.) and due to the relatively low amount of silica. The underlying humus horizons A and AElh are characterized by the maximum concentration of SiO_2; as a result they are depleted of iron oxide, aluminum, alkaline earth cations. The number of the last mentioned substantially increases in the illuvial stratum and, on the contrary, the content of silica decreases.

A characteristic feature of the clay fraction is the fact that its composition is stable. And it is proved by the profile distribution of bulk forms of silicon and aluminum oxide as well as SiO_2:Al_2O_3 ratio (Table 3). The latter shows that there is no significant chemical decay in the aluminosilicate core of the secondary clay minerals. Substantial iron subtraction from eluvial horizon clay is an indicator of surface gleyification [5-6].

This fact is confirmed by a high content of hardpans sized 0.25-5 mm – in the lower humus horizon the amount is up to 18%, and in the upper humus horizon it is up to 8%. It is proved also by a high amount of polyvalent cations – Mn, P, and Fe [1]. Both the negative balance of clay and the presence of large amounts of displaceable aluminum can possibly indicate acid hydrolysis of fine-dispersed mass of eluvial horizons.

It is most probably that leaching, illimerization, and exogleying processes as well as acid hydrolysis influenced the activation of the vertical differentiation of the soil profile in the Late Holocene.

Table 1. Granulometric composition of the sod-podzolic soils with a complex organic profile in the lower Vyatka River basin.

Horizon, depth of the sample, cm	Bulk weight, g/cm^3	Hygroscopic moisture, %	Fraction quantity, mm, % of dry soil							Clay accumulation, %
			1-0.25	0.25-0.05	0.05-0.01	0.01-0.005	0.005-0.001	<0.001	< 0.01	
Ya-1 cut										
OA 2-5	-	8.3	2.0	14.4	47.1	13.0	13.7	9.8	36.5	–72
AY 5-14	0.96	4.6	0.6	13.5	47.3	16.2	14.6	7.8	38.6	–77
Elh 14-24	1.23	2.0	0.3	8.0	52.2	13.6	20.3	5.6	39.5	–84
El 24-32	1.42	1.7	0.2	20.3	38.9	13.1	15.7	11.8	40.6	–66
Bh 28-32	-	1.4	0	14.1	43.4	13.5	16.3	12.7	42.5	–63
Bh 32-45	-	2.3	0	18.0	41.2	10.2	17.1	23.5	50.8	–32
ElB 32-41	1.47	3.7	0	12.1	34.6	9.1	13.9	30.3	53.3	–12
Bt1 41-50	1.48	4.7	0	8.3	32.0	9.1	12.7	37.9	59.7	+10
Bt1 50-58	1.48	4.9	0	11.6	30.4	9.0	11.4	37.6	58.0	+9
Bt2 70-80	1.49	4.6	0	12.7	31.2	8.4	13.3	34.4	56.1	0
BCca 110-120	-	4.1	0.5	19.1	37.0	9.4	5.5	28.5	43.4	-
Cca 135-145	-	4.4	3.0	18.6	29.7	6.3	18.7	23.7	48.7	-
CDca 180-190	-	5.2	6.5	16.0	28.0	14.7	25.9	8.9	49.5	-
CDca 210-220	-	6.2	1.2	13.0	25.9	14.5	31.6	13.8	59.9	-
Ya-2 cut										
PY 0-26	1.21	2.0	0.6	15.6	46.7	13.6	16.0	7.5	37.1	–75
AElh 26-32	-	1.8	0	14.1	44.3	15.9	16.8	8.9	41.7	–70
ElB 26-32	1.48	2.8	0	10.4	42.6	9.2	14.2	23.6	47.0	-22
Bt1 32-42	1.46	4.0	0	9.0	34.9	9.9	13.5	32.7	56.1	+9
Bt1 43-53	1.46	4.8	0	8.7	30.7	9.0	13.9	38.1	61.0	+27
Bt2 70-80	-	4.5	0	11.5	34.1	8.1	12.4	33.9	54.4	+13
BC 95-105	-	4.2	0.6	12.1	33.9	8.7	14.6	30.1	53.4	0
Cca 115-125	-	4.0	6.1	22.7	22.9	6.3	16.1	25.9	48.3	-
CDca 140-150	-	3.9	11.5	28.2	15.0	5.3	16.1	23.9	45.3	-

3.3. Humus content and composition

The second humus horizon in the sod-podzolic soils with a complex organic profile predetermines a significant total depth of their humus accumulative stratum of about 27 cm (Table 4). The lower humus horizon retains quite a high amount of humus (1.7% on average), but it is less than that of the AY (PY) horizon.

A lot of observations have proven a strong fluctuation in the humus content in the AY horizon, and a slight fluctuation in the AElh horizon.

This fact denotes fundamental differences in the conditions of humus formation and humus accumulation processes in the comparable horizons: the upper ones are closely related to the modern pedogenesis ecology, while the lower ones are autonomous. The latter indirectly indicates the formation of organic matter in the AElh horizon which took place under conditions different from the present-day ones.

The analysis of the humus group composition in the soils under research reveals a number of fundamental differences between some parts of the profile. The upper humus accumulative stratum is characterized by fulvate-humate and humate composition of organic matter while the illuvial part of the profile has a fulvate nature. The average Cha:Cfa ratio (carbon of humic acid to carbon of fulvic acid) for the AElh horizon is 2.5. In the upper accumulative horizon this ratio is close to 1. In the AElh horizon the ratio is much higher: from 2.5 to 5. There is a correlation between the degree of bleaching and humus group composition in the AElh horizon. The humate composition results in a strong dark color of the lower humus horizon while the organic matter quantity is relatively small. The humus group composition in the upper horizon is similar to the organic matter in sod-podzolic soils in the southern taiga. The humate nature of the AElh horizon shows its discrepancy with the modern conditions.

Table 2. Balance of the clay fraction in the sod-podzolic soils with a complex organic profile in the lower Vyatka River basin [7].

Horizon, depth, cm		Reserves, kg/sq. m (and %)					
		Excluding tracking substance[a]			Including tracking substance[a]		
		Initial	Real	Balance	Initial	Real	Balance
				Ya-1 cut			
AY	5-14	38.2	6.7	-31.5	29.7	6.7	-23.0
Elh	14-24	42.5	6.9	-35.6	41.1	6.9	-34.2
El	24-32	34.0	13.4	-20.6	36.4	13.4	-23.0
ElB	32-41	38.2	40.1	+1.9	41.6	40.1	-1.5
Bt1	41-50	38.2	50.5	+12.3	41.4	50.5	+9.1
Bt1	50-58	34.0	44.5	+10.5	36.8	44.5	+7.7
Bt2	58-94	152.9	184.5	+31.6	152.9	184.5	+31.6
BCca	94-135	174.1	174.1	0	-	-	-
	5-94	377.9	346.5	-31.4	380.0	346.5	-33.5
	5-135	552.0	520.7	-31.3 (-5.7%)	-	-	- (-9.7%)
				Ya-2 cut			
PY	0-26	114.3	23.6	-90.7	92.6	23.6	-69.0
AElh	26-32	26.4	7.9	-18.5	28.2	7.9	-20.3
Bt1	32-58	114.3	134.4	+20.1	109.2	134.4	+25.2
Bt2	58-87	127.4	143.5	+16.1	124.6	143.5	+18.9
BC	87-112	94.5	109.9	+15.4			
CDca	112-125	49.1	49.1	0			
	0-87	382.4	309.4	-73.0	345.7	309.4	-43.5
	0-125	526.2	468.4	-57.8 *(-12.3%)*	-	-	-

[a] Total content of TiO_2 is assumed as tracking substance

Table 3. Bulk chemical composition of the sod-podzolic soils with a complex organic profile in the lower Vyatka River basin.

Horizon, depth of the sample, cm		Percentage of calcined material weight										
		SiO_2	CaO	MgO	Fe_2O_3	Al_2O_3	TiO_2	P_2O_5	MnO	Na_2O	K_2O	SO_3
						Ya-1 cut (soil)						
O	0-2	69.21	6.86	1.29	3.40	12.39	0.82	0.88	0.69	1.66	2.52	0.16
OA	2-5	72.02	3.72	1.23	3.32	12.28	0.83	0.56	0.58	1.45	2.26	0.15
AY	5-14	75.04	1.18	1.18	4.37	12.49	0.93	0.14	0.49	1.15	2.04	0.14

Elh	14-24	78.52	0.74	0.91	3.80	9.90	0.91	0.06	0.16	1.36	2.20	0.05
El	24-32	78.41	0.73	1.22	4.19	10.83	0.88	0.05	0.07	1.32	2.16	0.05
ElB	32-41	73.38	0.89	1.78	5.80	13.25	0.85	0.07	0.08	1.21	2.14	0.05
Bh	35-45	75.01	0.92	1.35	4.91	12.66	0.79	0.09	0.09	1.35	2.37	0.05
Bt1	41-50	70.56	0.76	2.21	6.56	14.62	0.84	0.05	0.09	1.11	2.09	0.05
Bt2	70-80	71.03	1.36	1.92	6.62	14.06	0.77	0.06	0.09	1.22	2.11	0.05
BCca	110-120	69.38	3.19	2.25	6.48	13.73	0.77	0.03	0.09	1.21	2.02	0.08
Cca	135-145	68.73	3.90	2.56	6.88	13.55	0.72	0.04	0.09	1.24	1.92	0.05
CDca	210-220	61.98	1.57	3.74	10.57	17.10	0.80	0.15	0.11	1.39	1.97	0.05
Ya-2 cut (soil)												
PY	0-26	78.63	0.96	0.91	3.95	10.10	0.89	0.04	0.19	1.31	2.13	0.06
AElh	26-32	78.50	0.88	0.88	3.69	9.99	0.93	0.05	0.11	1.35	2.09	0.13
ElB	26-32	74.14	0.88	1.66	5.15	12.37	0.99	0.03	0.05	1.24	2.09	0.05
Bt1	32-42	72.38	0.98	1.97	6.42	13.85	0.98	0.03	0.06	1.16	2.04	0.08
Bt1	43-53	70.19	1.14	2.19	7.13	15.00	0.96	0.04	0.08	1.08	2.14	0.05
Bt2	70-80	71.68	1.35	2.04	6.42	13.66	0.89	0.06	0.08	1.25	1.96	0.05
BC	95-105	71.95	1.50	1.46	6.50	13.52	0.91	0.04	0.09	1.29	2.10	0.05
CDca	115-125	72.32	2.42	2.22	5.88	12.50	0.79	0.05	0.09	1.19	1.75	0.06
CDca	140-150	73.90	1.35	2.20	5.84	12.79	0.74	0.05	0.10	1.20	1.71	0.05
Ya-1 cut (clay)												
AY	5-14	54.21	2.19	2.95	9.92	24.86	1.00	0.12	0.43	0.40	2.33	0.24
AY	5-14	54.21	2.19	2.95	9.92	24.86	1.00	0.12	0.43	0.40	2.33	0.24
Elh	14-24	53.92	0.71	3.29	11.73	24.60	0.99	0.16	0.44	0.40	2.43	0.10
El	24-32	53.88	1.00	3.40	12.89	24.13	1.09	0.08	0.12	0.32	2.35	0.09
ElB	32-41	54.87	1.16	3.66	12.21	23.95	0.89	0.06	0.09	0.30	2.45	0.12
Bt1	41-50	55.29	1.17	3.67	12.23	23.88	0.82	0.06	0.09	0.28	2.13	0.07
Bt2	70-80	54.97	1.67	4.24	12.16	22.98	0.81	0.08	0.11	0.33	2.17	0.05
BCca	110-120	53.64	2.67	4.12	12.95	22.33	0.86	0.08	0.11	0.37	2.28	0.06
Cca	135-145	54.43	2.50	4.12	12.45	22.07	0.81	0.04	0.10	0.36	2.12	0.05
Ya-2 cut (clay)												
PY	0-26	54.90	1.59	3.36	10.73	24.16	1.07	0.11	0.09	0.45	2.39	0.15
AElh	26-32	54.23	1.41	3.34	11.58	23.98	1.11	0.12	0.11	0.41	2.56	0.12
ElB	26-32	54.11	1.17	3.44	12.24	23.67	0.94	0.06	0.09	0.37	2.28	0.07
Bt1	32-42	54.84	1.36	3.92	12.27	23.88	0.87	0.08	0.07	0.26	2.22	0.13
Bt1	43-53	54.88	1.35	3.56	12.18	23.82	0.79	0.09	0.09	0.33	2.28	0.06
Bt2	70-80	55.23	1.68	3.79	12.11	23.53	0.78	0.07	0.12	0.34	2.27	0.08
BC	95-105	54.91	2.00	4.01	11.98	22.57	0.78	0.10	0.12	0.37	2.29	0.11
CDca	115-125	54.27	2.51	3.89	11.90	22.55	0.81	0.07	0.14	0.35	2.20	0.07
CDca	140-150	54.88	2.52	4.15	12.04	22.84	0.78	0.08	0.13	0.37	2.09	0.07

The fractional composition of HA indicates fundamental differences between the eluvial and illuvial parts of the profile. In the upper part humic acids are represented primarily by a fulvous HA1 fraction which is associated with iron and aluminum. And in the lower part they are represented only by the second black fraction (calcium humate). The latter illustrates that they do not correspond to the modern bioclimatic conditions.

IOP Conf. Series: Earth and Environmental Science **41** (2016) 012024 doi:10.1088/1755-1315/41/1/012024

Table 4. Humus composition of the sod-podzolic soils with a complex organic profile in the lower Vyatka River basin, % of carbon total quantity [2], [8].

Horizon, depth of the sample, cm	C total, %	HA1[a]	HA2[a]	FA 1a[b]	FA1[b]	FA2[b]	IR[c]	Cha: Cfa
				Ya-1 cut				
O 0-2	25.39	15.3	3.4	1.8	13.4	-	66.1	1.2
OA 2-5	17.77	18.6	3.7	1.4	16.8	-	59.5	1.2
AY 5-14	5.50	20.0	2.0	4.0	17.5	3.8	52.7	0.9
Elh 14-24	1.09	28.4	-	11.9	12.8	1.9	45.0	1.1
El 24-32	0.46	13.7	4.1	13.0	2.2	1.8	65.2	1.0
ElB 32-41	0.36	12.2	2.5	13.9	3.3	12.5	55.6	0.5
Bh 28-32	0.83	25.3	19.3	8.4	3.6	6.0	37.4	2.5
Bh 32-45	0.83	8.4	30.1	7.2	7.2	3.6	43.5	2.1
Bt1 41-50	0.41	-	12.4	9.8	18.5	0.8	58.5	0.4
Bt1 50-58	0.38	-	11.8	10.5	14.7	5.1	57.9	0.4
Bt2 70-80	0.31	-	11.9	9.7	7.1	10.0	61.3	0.4
BCca 110-120	0.21	-	9.5	9.5	-	-	81.0	1.0
Cca 135-145	0.19	-	7.9	10.5	-	-	81.6	0.8
				Ya-2 cut				
PY 0-26	1.40	23.6	-	6.4	11.6	10.5	47.9	0.8
AElh 26-32	1.0	43.0	-	8.0	-	11.0	37.0	2.3
ElB 26-32	0.35	-	20.0	20.0	8.6	-	51.4	0.7
Bt1 32-42	0.34	-	10.0	19.1	15.0	3.0	52.9	0.3
Bt1 43-53	0.37	-	8.1	13.5	15.7	5.9	56.8	0.2
Bt2 70-80	0.28	-	9.3	7.1	10.7	1.5	71.4	0.5
BC 95-105	0.23	-	12.6	13.0	3.2	5.7	66.5	0.6
CDca 115-125	0.22	-	11.4	9.1	-	4.0	75.5	0.8
CDca 140-150	0.19	-	13.2	10.5	-	4.7	71.6	0.8

[a] HA – humic acid
[b] FA – fulvic acids
[c] IR – insoluble residue

According to the mentioned above we can claim that there are at least two distinct phases in the history of organic matter formation in the soils with the AElh horizon. The humate nature of the lower humus horizon as well as large quantities of the black HA2 fraction in its composition point out the initial formation of the studied soils under conditions of a neutral or even weakly alkaline reaction of the medium. Apparently, intense organic-accumulative processes took place at the same time that resulted in a relatively thick and dark-colored horizon. Its thickness approximately conformed to the total depth of the modern AY and AElh horizons. The similar conditions took place in the Boreal-Atlantic Holocene stage about 9-5 thousand years ago [9].

On the basis of the ratio difference between the first and second HA fractions which correlate with the relictness degree of the AElh horizon we can make an assumption that the fraction HA1 transforms into HA2 in case of leaching of the bases with a high eluvial character of the soil formation processes. The precondition for this is the increased mobility of iron and aluminum after calcium and magnesium subtraction.

The profile FA distribution is characterized by the highest absolute level (% of the soil weight) of the most active (FA1) and aggressive (FA1a) fractions in the accumulative-eluvial stratum. They are especially abundant in the forest litter and in the upper humus horizon. The absolute level of active FA forms is considerably lower in the AElh horizon. In the textural illuvial stratum the absolute level of active and aggressive FA fractions is substantially reduced.

According to the content of humus carbon in various horizons their distribution is different. In this case the relative increase in the content of FA1 and FA1a fractions at the top of the illuvial stratum becomes an important issue. It is considered as an area of intense mineral phase destruction of the profile in the modern period.

3.4. Humus age

Table 5 shows the results of radiocarbon dating of humic acids in the AY, AElh and Bh horizons of the sod-podzolic soils with a complex organic profile.

The analysis points out sharp differences of the HA age of the upper and lower humus horizons. According to the most dated HA fractions (second and third) the formation of the AElh horizons took place in the Atlantic Holocene stage (7.5-5.5 thousand years ago).

A younger age of the fraction HA1 (humates associated with iron and aluminum) can possibly be a result of its formation in the upper profile leached bases. The radiocarbon dates of the first HA fraction of the AElh horizon indicate the time when the accumulative Boreal-Atlantic phase of soil formation changed to accumulative-eluvial between the Atlantic and Subboreal stages (about 5000 years ago).

The fact that there are several horizons in one and the same profile which are qualitatively different in their age (and the humus composition) indicates a quick direction change of soil formation processes which took place about 5000 years ago between the Atlantic and Subboreal stages.

Table 5. The age of humic acid of the sod-podzolic soils with a complex organic profile in the lower Vyatka River basin.

Cut, №	Horizon	Depth of the sample, cm	HA fraction	Age acc. to ^{14}C, years	Laboratory sample, №
Ya-1	AY	5 – 14	HA1	600±90	LU-1502
			HA2	1580±180	LU-1501
			HA3	2590±140	LU-988
	Elh	15 – 25	HA1	5580±140	LU-994
			HA2	6600±430	LU-993
			HA3	6740±300	LU-984
	Bh	35 – 50	HA1	5530±160	LU-995
			HA3	7630±390	LU-985

Combining heterochronous humus compounds in one and the same profile explains a shift of landscape zones on the territory of the Vyatka-Kama region in the last 10 thousand years. In the early Holocene the boundary of soils with a complex organic profile ran to the Cheptsa River valley, and now it is moving to the south.

4. Conclusion

The sod-podzolic soils under research went through two major stages of pedogenesis: 1) accumulative-eutrophic – it is a developmental evolution stage in the Atlantic phase of the Middle Holocene, and 2) accumulative-eluvial stage of erasing evolution with the elements of inheriting evolution in the Subboreal-Subatlantic phase of the Late Holocene.

In the last 300-400 years, the processes of spontaneous evolution have been superimposed by agro-anthropogenic degradation resulted in mechanical destruction of relict humus horizons caused by plowing and soil erosion.

Reference

[1] Prokashev A M 1999 *The soils with a complex organic profile in the south of the Kirov region* (Kirov: Vyatka State Pedagogical University Press) p 176
[2] Belchikova N P 1951 Some patterns of content, humus composition and properties of humic acids in the main groups of the USSR soils *Proc. of the Soil Institute named after V V*

Dokuchaev vol 38 pp 33–58
[3] Tyulin V V and Okhorzin N D 1980 Structural features of the soil cover of the south-western part of the Kirov region *Soil structure and its importance for soil mapping, recording and use of soil resources* (Chisinau) 197–9
[4] Tyulin V V, Rossokhina M V and Prokashev A M 1981 Some peculiarities of soil formation and physical and chemical properties of soils with the second humus horizon of the Yaran-Nemda interfluve of the Kirov region *Problems of improving soil fertility and productivity of agricultural crops in the Kirov region* (Perm) **70** 15–25
[5] Zaidelman F R 1970 Diagnostics, the differences and similarities of podzolic and lessivaged soils, gleyed podzols, pseudo-podzolic soils, pseudo-gleyed soils *Soil Science* **12** 169–183
[6] Zaidelman F R 1974 Podzol- and gley-formation (Moscow: Nauka) p 208
[7] Rozanov B G 1983 Soil morphology (Moscow Publishing House of Moscow University) p 320
[8] Kononova M M 1963 *Soil organic matter* (Moscow Publishing House of the USSR Academy of Sciences) p 314
[9] Prokashev A M, Zhuikova I A and Pakhomov M M 2003 *The history of the living soil cover of the Vyatka-Kama region in the post-glacial period* (Kirov: Vyatka State Humanities University) p 143

IOP Conf. Series: Earth and Environmental Science **41** (2016) 012025 doi:10.1088/1755-1315/41/1/012025

Bioactive compounds of fourth generation gamma-irradiated *Typhoniumflagelliforme* Lodd. mutants based on gas chromatography-mass spectrometry

N F Sianipar[1,4], R Purnamaningsih[2] and Rosaria[3]

[1] Research Interest Group Food Biotechnology, Bina Nusantara University, 15325 Tangerang, Indonesia

[2] Indonesian Center for Agricultural Biotechnology and Genetic Resources Research and Development (BB-Biogen), 16111 Bogor, Indonesia

[3] Alumni of Biology Department, Faculty of Science and Technology, Pelita Harapan University, 15811 Tangerang, Indonesia

E-mail: nestipro@yahoo.com

Abstract. Rodent tuber (*Typhonium flagelliforme*Lodd.) is an Indonesian anticancer medicinal plant. The natural genetic diversity of rodent tuber is low due to vegetative propagation. Plant's genetic diversity has to be increased for obtaining clones which contain a high amount of anticancer compounds. *In vitro* calli were irradiated with 6 Gy of gamma ray to produce *in vitro* mutant plantlets. Mutant plantlets were acclimated and propagated in a greenhouse. This research was aimed to identify the chemical compounds in the leaves and tubers ofthe fourth generation of rodent tuber's vegetative mutant clones (MV4)and control plantsby using GC-MS method. Leaves and tubers of MV4 each contained 2 and 5 anticancer compounds which quantities were higher compared to control plants. MV4 leaves contained 5 new anticancer compounds while its tubers contained 3 new anticancer compoundswhich were not found in control. The new anticancer compounds in leaves were *hexadecanoic acid, stigmast-5-en-3-ol, ergost-5-en-3-ol, farnesol isomer a*, and *oleic acid* while the new anticancer compounds in tubers were *alpha tocopherol, ergost-5-en-3-ol*, and *beta-elemene*. Rodent tuber mutant clones are very potential to be developed into anticancer drugs.

1. Introduction

Rodent tuber (*Typhonium flagelliforme* Lodd.) is a native Indonesian medicinal plant from Araceae family [1]. Rodent tuber contains antineoplastic/anticancer and antivirus chemical agents[2]. Bioactive compounds in this plant arealkaloids, saponins, steroids, and glycosides [3]. Rodent tuber's extract was cytotoxic against the cancer of lung, breast [4], liver [5], blood(leukemia)[6], colon, prostate gland, and cervix [7]. Its hexane extract was cytotoxic against *Artemia salina* larvae[8]. It also had antibacterial and antioxidant properties [9].

Rodent tuber from Indonesia has a low genetic diversity because it is usually propagated by conventional vegetative methods[1]. Genetic diversity must be increased in order to obtain plant clones which contain a high amount of anticancer compounds. Mutation induction is one of the ways to increase plant's genetic diversity. Mutation could be induced by irradiating plants with physical mutagens, such as gamma ray.

In vitro embryogenic somatic cell population/calli of rodent tuber plant from Bogor have been induced, proliferated, and regenerated by single node culture method[10]. The mutation induction of those embryogenic calli were done by combining the effects of 6 Gy gamma irradiation and *in vitro* somaclonal variation. *In vitro* plantlets, which were regenerated from those induced calli, showed various morphological characters [11] and had genetic differences from control plants according to Randomly Amplified Polymorphic DNA (RAPD) molecular marker analysis [12]. *In vitro* plantlets were acclimated in a greenhouse. There were 37 clones of the first generation ofrodent tuber vegetative mutant (MV1). MV1 clones had various morphological characters [13]. Out of those 37 MV1 clones, there were 17 clones which were genetically different from control plants according to RAPD [14].

Genetic mutation may result in a change of the composition and quantities of phytochemical compounds in a plant, which could be detected by Gas Chromatography-Mass Spectrometry (GC-MS). GC uses gas as the mobile phase to separate chemical compounds. MS will identify those separated compounds by referring to a database[15]. GC-MS has been used to identify the chemical contents of herbal plants, such as *Meliaorientalis*[16], *Maranta arudinacea* L. [17], and the nonpolar fraction of the Malaysian rodent tuber [18].

MV1 clones of rodent tuber have been propagated and regenerated into the fourth generation mutant clones (MV4). The purpose of this research was to identify the chemical composition andanalyzing the anticancer bioactive components of the polar fraction of leaves and tubers of rodent tuber MV4 clones by using GC-MS method.

2. Materials and methods

2.1. Plant materials and extract preparation

Rodent tuber plant from Bogor, Indonesia was isolated, propagated *in vitro*[10], and irradiated with gamma ray to induce genetic mutation [11]. Rodent tuber plantlets were acclimatedin a greenhouse [13]. Mutant and control plants were harvested for extraction.One control plant and eight MV4 mutant clones, i.e. 6-9-1, 6-3-3-6, 6-1-1-6, 6-3-2-5, 6-6-3-6, 6-2-8-2, 6-1-1-2, 6-1-2 were analyzed in this research. Shoots and tubers were dried and homogenised. Homogenised samples were macerated in 100ml of ethanol 96% and incubated overnight for two days to obtain chemical components in its ethanol fractions. Ethanol fractions were filtered with Whatman filter paper.

2.2. GC-MS analysis

Sample's ethanol fractions were injected into the GC column. Injection volume was 5μl with 5:1 split ratio and the injection temperature was 250°C. Helium was used as carrier gas with velocity 0.8 μl per minute. Column temperature was set at 70°C with 5°C per minute increment. When the temperature reached 200°C, it remained constant for 1 minute and then will be increased at the rate of 20°C per minute until the temperature reached 280°C. The temperature remained constant for another 28 minutes. Mass spectrometer was operated in electron impact ionisation mode with 70 eV voltage.

2.3. Identification of phytochemical compounds

Identification of the mass spectrum of GC-MS was performed by referring to the National Institute Standard Technique (NIST) database with ≥90% fit factor. The relative abundance percentages of each compound were calculated by comparing its average peak area to the total area.

3. Results and discussion

GC-MS has successfully identified phytochemicals in the leaves and tubers of rodent tuber's control and MV4 clones (Figure 1). There were 32 different chemical compounds in theethanolic fraction of a control plant. The 5 most abundant compounds were neophytadiene (10.6%), phytol isomer (13.07%), alpha tocopherol (8.85%), campesterin (7.23%), and stigmasterol (8.90%). Ethanolic fractions of the tubers contained at least 33 different chemical compounds, the most abundant were 9,12-octadecadienoic acid (7.67%), octacosane (9.35%), nonacosane (11.03%), octacosane (8.66%), and stigmasterol (7.88%).

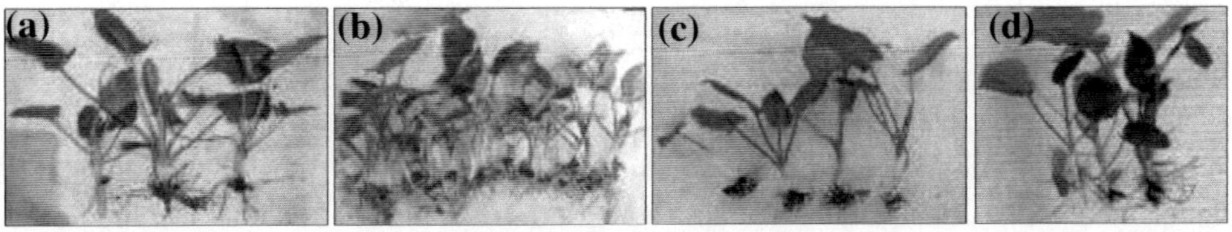

Figure 1. A: control; B: MV4 clone 6-1-2; C: MV4 clone 6-3-2-5; D: MV4 clone 6-1-1-6.Rodent tuber control and mutant clones after 20 weeks of growing in a green house.

GC-MS analysis revealed that leaves and tubers of control plants contained a range of different phytochemicals. The pytochemicals which were contained in higher amounts in leaves than in tubers were phytol, squalene, and stigmasterol. Whilst, tubers contained higher amounts of hexadecanoic acid methyl ester, hexadecanoic acid ethyl ester, hexadecanoic acid, 9,12-octadecadienoic acid methyl ester, 9,12,15-octadecatrienoic acid methyl ester, 9,12-octadecadienoic acid ethyl ester, hexadecanamide, 4,22-stigmastadiene-3-one, and stigmast-4-en-3-one compared to leaves.

Based on GC-MS analysis, there were chemical composition differences between control and mutant clones and between each of the different clones of mutant plants. Some anticancer bioactive compounds were present in higher amounts in the mutant clones compared to in control plants. Mutant clones also contained several new anticancer compounds which were not found in control (figure 2, table 1 and 2). Leaves of mutant clones contained higher amounts of two anticancer compounds compared to control, i.e. *hexadecanoic acid ethyl ester* and *hexadecanoic acid methyl ester*. They also contained five new anticancer compounds, i.e. *hexadecanoic acid*, *stigmast-5-en-3-ol*, *ergost-5-en-3-ol*, *farnesol isomer a*, and*oleic acid*. Tubers of mutant clones contained higher amounts of five anticancer compounds, i.e. hexadecanoic acid, hexadecanoic acid methyl ester, hexadecanoic acid ethyl ester, squalene, and 2-methoxy-4-vinylphenol. Among those five compounds, the relative abundance of hexadecanoic acid in 6-6-3-6 clone was the most different compared to control, i.e. increasing 30.63% from control plants. In addition, tubers of mutant clones contained three new anticancer compounds, i.e. *alpha tocopherol*, *ergost-5-en-3-ol*, and *beta-elemene*.

The higher amounts of some anticancer compounds in the mutant compared to control plants has also been observed in GC-MS profile of rodent tuber first generation mutant clones (MV1) [19]. Leaves and tubers of MV1 contained higher amounts of three and four anticancer compounds compared to control, respectively. In addition, leaves and tubers of MV1 each contained four new anticancer compounds which were not found in control plants.

Hexadecanoic acid (palmitic acid) was selectively cytotoxic against leukemia cancer cells MOLT-4 due to its interaction with DNA topoisomerase I and its ability to induce apoptosis. Hexadecanoic acid had*in vivo*antitumor activity[20]. Biological activities of hexadecanoic acid ethyl ester were antioxidant, antimicrobial[21], and could reduce the risk of coronary heart disease. Hexadecanoic acid methyl ester was able to inhibit the growth and induce apoptosis of human gastric cancer cells[22]. Stigmast-5-en-3-ol (3.beta.,24s) (beta-sitosterol) is a phytosterol with various biological activities. It could reduce the cell's cholesterol level, modify membrane lipid profile [23], and it was antidiabetic [24], and able to inhibit the growth of cancer cells. Ergost-5-en-3-ol (3 beta) (campesterol) is a phytosterol which has been proven to be able to inhibit various cancer cells, such as lung [25], gastric [26], and ovary [27] cancers. Farnesol is a non-sterol isoprenoid which is commonly found in various fruits and aromatic plants, such as citrus, sage, spearmint, nutmeg, basil, lemon grass, and chamomile. Farnesol could selectively inhibit the proliferation and induce apoptosis of leukemia and cervical cancer cells[28]; [29]. Farnesol had *in vivo*antitumor and anticarcinogenic activities [30]; [31].

Table 1. Comparison of the composition of anticancer compounds in leaves of rodent tuber control and MV4 mutant clones based on GC-MS. NA is not available, which means that the quantity of a compound wasnot high enough to be detected by GC-MS. The quantities of anticancer bioactive compounds in MV4 mutant clones which were higher than control were indicated by the yellow highlights.

Name of Compound	Relative abundance (%) in control	Relative abundance (%) in mutant							
		6-9-1	6-3-3-6	6-1-1-6	6-3-2-5	6-6-3-6	6-2-8-2	6-1-1-2	6-1-2
Hexadecanoic acid, methyl ester	0.39	1.33	0.82	0.53	0.88	0.71	0.32	0.41	0.48
Hexadecanoic acid, ethyl ester	0.50	0.51	0.86	0.64	1.07	1.00	0.86	0.86	1.01
Hexadecanoic acid	NA	16.7	20.89	19.46	29.09	19.33	24.53	19.36	32.85
Stigmast-5-en-3-ol	NA	3.64	NA	NA	NA	NA	NA	NA	2.14
Ergost-5-en-3-ol	NA	NA	3.17	3.40	NA	NA	2.29	NA	2.47
Farnesol isomer a	NA	NA	NA	NA	0.33	NA	NA	NA	NA
Oleic acid	NA	NA	NA	NA	NA	2.03	NA	NA	NA

Table 2. Comparison of the composition of anticancer compounds in tubers of rodent tuber control and MV4 mutant clones based on GC-MS. NA is not available, which means that the quantity of a compound wasnot high enough to be detected by GC-MS. The quantities of anticancer bioactive compounds in MV4 mutant clones which were higher than control were indicated by the yellow highlights.

Name of Compound	Relative abundance (%) in control	Relative abundance (%) in mutant							
		6-9-1	6-3-3-6	6-1-1-6	6-3-2-5	6-6-3-6	6-2-8-2	6-1-1-2	6-1-2
Hexadecanoic acid	7.23	19.52	26.62	10.72	27.16	37.86	32.36	36.76	25.50
Hexadecanoic acid, methyl ester	1.46	0.83	0.68	2.51	0.69	0.54	0.34	0.66	NA
Hexadecanoic acid, ethyl ester	0.87	1.13	1.18	2.02	1.39	1.94	1.31	1.23	3.69
Squalene	0.86	NA	NA	1.24	NA	0.49	0.55	1.32	NA
Alpha tocopherol	NA	0.24	NA	NA	0.59	NA	NA	0.61	NA
Ergost-5-en-3-ol	NA	2.95	2.85	3.31	2.43	2.56	2.20	0.36	1.35
2-methoxy-4-vinylphenol	0.90	NA	NA	0.96	0.45	NA	NA	0.46	NA
Beta-elemene	NA	4.33	NA	NA	0.13	NA	NA	NA	NA

Oleic acid is a n-9 monounsaturated fatty acid which was cytotoxic against several types of cancer cells. Oleic acid could inhibit the growth [32] and induce apoptosis [33] of breast cancer cells and inhibit colon adenocarcinoma[34]. Squalene was able to inhibit carcinogenesis of various cancer cells, such as colon cancer [35]. Vitamin E (α dan γ-*tocopherol*) has been proven to be able to reduce the risk of carcinogenesis [36]. 4-vinyl-2-methoxy-phenol was able to inhibit carcinogenesis which is induced by Polycyclic Aromatic Hydrocarbons (PAHs) benzo(a)pyrene (BaP) by regulating the cell cycle protein in order to prevent hyper-phosphorilation of retinoblastome's *tumor suppressor protein* [37].

Beta-elemene could induce apoptosis of non-small lung carcinoma by activating caspase-3, -7 and -9, reducing the expression of Bcl-2, inducing the release of cytochrome c, and increasing the level of caspase-9 and poly(ADP-ribose) polimerase[38]. Bete-elemenealso had antiproliferative activity agains prostate carcinoma cell DU145 and PC-3. In addition, it was able to inhibit the growth of lung, colon, cervix, breast, and brain carcinoma cells [39]. Based on GC-MS analysis, mutant clone 6-1-2 had the highest total amount of anticancer compounds in leaves (38.95%) compared to control (0,89%) and the other mutant clones (table 1). GC-MS identified 20 chemical compounds in the ethanolic fraction of 6-1-2's leaves (Tabel 5). The five most abundant compounds in leaves of 6-1-2 were hexadecanoic acid (32.85%), (2E)-3,7,11,15-tetramethyl-2-hexadecen-1-ol (8,71%), ethyl (9z,12z)-9,12-octadecadienoate (5,67%), hexadecanoic acid (5,74%), and 9,12-octadecadienoic acid (Z,Z) (16,07%).

A

B

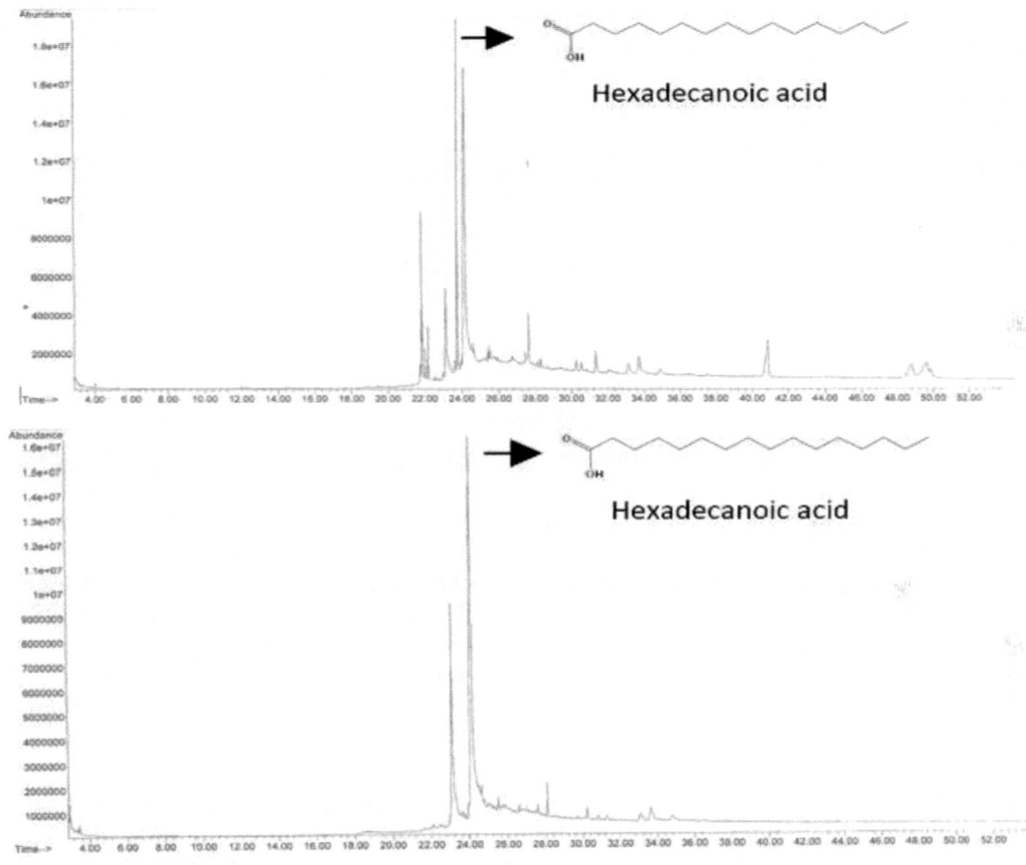

Figure 2. GC-MS chromatogram of the leaves and tubers of MV4 mutant clones. a) leaves of 6-1-2; b) tubers of 6-6-3-6. X-axis represents retention time while Y-axis represents relative abundance. The chemical structures of compounds with the highest relative abundances were shown (chemical structure was obtained from NIST database).

Table 3. Chemical compounds in leaves of rodent tuber MV4 mutant clone 6-1-2 based on GC-MS. [a]Compounds were identified by comparing retention time data with authentic standard database of NIST/EPA/NIH fit factor ≥90%. [b]Relative abundance was determined based on area percentage of each compound.

Retention Time	Name of Compound[a]	Relative Abundance[b]
31.106	Neophytadiene	1.39
31.161	(2E)-3,7,11,15-tetramethyl-2-hexadecene	0.81
31.513	Neophytadiene	0.58

31.906	Hexadecanoic acid, methyl ester	0.48
32.402	Hexadecanoic acid, ethyl ester	1.01
32.657	Hexadecanoic acid	7.03
32.850	Hexadecanoic acid	4.29
32.940	Hexadecanoic acid	2.86
33.057	9,12-octadecadienoic acid (Z,Z)-,methyl ester	4.37
33.099	Hexadecanoic acid	3.92
33.209	(2E)-3,7,11,15-tetramethyl-2-hexadecen-1-ol	8.71
33.430	Ethyl (9z,12z)-9,12-octadecadienoate	5.67
33.471	Hexadecanoic acid	3.49
33.581	Hexadecanoic acid	5.74
33.699	9,12-octadecadienoic acid (Z,Z)	16.07
33.823	Hexadecanoic acid	5.52
34.119	14-methyl-8-hexadecyn-1-ol	4.75
35.181	Methyl 20-methyl-heneicosanoate	0.57
36.312	Tetracosanoic acid, methyl ester	0.33
37.022	Squalene	0.70
39.415	7-bromo-5-(2-bromo-phenyl)-1,3-dihydro-benzo(e)(1,4)diazepin-2-one	3.05
40.125	Stigmastan-3,5-diene	2.68
40.483	Alpha-tocopherol	0.86
42.200	Ergost-5-en-3-ol	2.47
42.759	Stigmasterol	2.96
43.883	Stigmast-5-en-3-ol	2.14
45.455	4,22-stigmastadiene-3-one	1.56

43.39% of compounds identified in the tuber of mutant clone 6-6-3-6 were known to have anticancer activity, which was the highest amount of anticancer compounds compared to control, which anticancer compounds only comprised 11.32% of the identified compounds, and the other mutant clones (table 1). GC-MS identified 18 chemical compounds in theethanolic fraction of the tuber of MV4 6-6-3-6 clone (table 4). The five most abundant compounds in leaves of MV4 6-6-3-6 clone were hexadecanoic acid (37.86%), 9,12-octadecadienoic acid (Z,Z)-methyl ester (3.39%), ethyl (9z,12z)-9,12-octadecadienoate (8.19%), (9E,12E)-9,12-octadecadienoic acid (25%), and stigmasterol (3.85%).

Table 4. Chemical compounds in tubers of rodent tuber MV4 mutant clone 6-6-3-6 based on GC-MS. [a]Compounds were identified by comparing retention time data with authentic standard database of NIST/EPA/NIH fit factor ≥90%. [b]Relative abundance was determined based on area percentage of each compound.

Retention Time	Name of Compound[a]	Relative Abundance[b]
31.899	Hexadecanoic acid, methyl ester	0.54
32.409	Hexadecanoic acid, ethyl ester	1.94
32.747	Hexadecanoic acid	14.58
32.837	Hexadecanoic acid	3.87
32.940	Hexadecanoic acid	2.50
33.057	9,12-octadecadienoic acid (Z,Z)-methyl ester	3.39
33.099	Hexadecanoic acid	3.38
33.202	Hexadecanoic acid	3.56
33.264	Hexadecanoic acid	1.33
33.443	Ethyl (9z,12z)-9,12-octadecadienoate	8.19

33.485	Hexadecanoic acid	3.39
33.588	Hexadecanoic acid	5.25
33.781	(9E,12E)-9,12-octadecadienoic acid	22.51
33.885	9,12-octadecadienoic acid (Z,Z)	2.49
34.560	(R)-(-)-14-methyl-8-hexadecyn-1-ol	2.03
35.188	Methyl 20-methyl-heneicosanoate	0.53
36.312	Tetracosanoic acid, methyl ester	0.35
36.739	Methyl 17-methyl-octadecanoate	0.31
37.022	Squalene	0.49
39.401	7-bromo-5-(2-bromo-phenyl)-1,3-dihydro-benzo(e)(1,4)diazepin-2-one	1.91
40.097	Stigmastan-3,5-diene	1.46
42.207	Ergost-5-en-3-ol	2.56
42.821	Stigmasterol	3.85
43.890	(23s)-ethylcholest-5-en-3beta-ol	1.95
44.752	Ergost-4-en-3-one,(24R)-	0.86
45.469	4,22-stigmastadiene-3-one	1.77

Some chemical compounds, i.e. ethyl (9z,12z)-9,12-octadecadienoate, 14-methyl-8-hexadecyn-1-ol, methyl 20-methyl-heneicosanoate, tetracosanoic acid methyl ester, stigmastan-3,5-diene, and ergost-5-en-3-ol, were detected in leaves of MV4 clone 6-1-2 but not in leaves of control. Tubers of MV4 clone 6-6-3-6 contained seven chemical compounds, i.e. ethyl (9z,12z)-9,12-octadecadienoate, tetracosanoic acid methyl ester, methyl 17-methyl-octadecanoate, 7-bromo-5-(2-bromo-phenyl)-1,3-dihydro-benzo(e)(1,4)diazepin-2-one, ergost-5-en-3-ol, (23s)-ethylcholest-5-en-3beta-ol, and ergost-4-en-3-one,(24R)-, which were not detected in control plants. This finding indicated the difference in metabolomic profilebetween mutant and control plants due to gamma irradiation-induced genetic mutation.

The difference in chemical contents between control and mutant plants was due to gamma irradiation and somaclonal variation during plant tissue culture process. The application of auxin-type plant growth regulators, such as 2,4-D, in plant culture medium could increase heritable genetic diversity [40]. In addition, the combination of stress condition in *in vitro* culture and physical mutagen irradiation was able to induce retrotransposon activity, i.e. the movement of DNA from one chromosome to the other. This phenomenon could alter the genetic makeup of a cell, either by increasing or silencing the expression rate of nearby genes or genes which are located next to the retrotransposon [41]. In this respect, the expression of nearby genes which might be responsible for expressing certain organic compounds couldalso be increased. This research has shown that the usage of plant growth regulators in *in vitro* culture and the irradiation of calli with gamma ray were able to produce mutant clones which contained a high amount of anticancer compounds. Beside that, gamma irradiation and somaclonal variation were also able to alter the biochemical process [40]. In this research, some MV4 mutant clones showed an increase in the amount of primary and secondary metabolites compared to control plants.

4. Conclusion

The chemical compounds in control and fourth generation vegetative mutant clones (MV4) of rodent tuber have been successfully detected by GC-MS. There were metabolomic profile differences between leaves and tubers, and between mutant and control plants. Leaves and tubers of MV4 each contained 2 and 5 anticancer compounds whose quantities were higher compared to control plants. MV4 leaves contained 5 new anticancer compounds while its tubers contained 3 new anticancer compounds which were not found in control. The new anticancer compounds in leaves were *hexadecanoic acid*, *stigmast-5-en-3-ol*, *ergost-5-en-3-ol*, *farnesol isomer a*, and *oleic acid* while the new anticancer compounds in tubers were *alpha tocopherol*, *ergost-5-en-3-ol*, and *beta-elemene*. The difference in chemical compounds between mutant and control plants was due to the 6-Gy gamma ray irradiation and somaclonal variation of *in vitro* calli.

Acknowledgement
The authors would like to thank the Indonesian Directorate General of Higher Education (DIKTI) who funded our research through competitive grant (hibah bersaing) project. We would also like to thank Prof. Ika Mariska for reviewing this manuscript.

References

[1] Essai1986 *Medicinal Herbs Index in Indonesia*(Jakarta: PT Essai indonesia)p357
[2] Teo CKH and Ch'ng BI1996 *Cancer Yet They Live*(Penagn: CA Care Publication)
[3] Syahid SF 2007 Perbanyakan keladi tikus (*Typhonium flagelliforme* Lodd) secara *in vitroWarta Puslitbangbun***13**(3) 19-20
[4] Lai CS, Mas RH, Nair NK, Mansor SM and Navaratnam V 2010Chemical constituents and *in vitro* anticancer activity of *Typhoniumflagelliforme*(Araceae) *J. Ethnopharmocol.***127**486-94
[5] Lai CS, Mas RHMH, Nair NK, Majid MIA, Mansor SM and Navaratnam V 2008 *Typhoniumflagelliforme*inhibits cancer cell growth *in vitro* and induces apoptosis: an evaluation by the bioactivity guided approach *J. Ethnopharmacol.***118** 14-20
[6] Mohan S, Abdul AB, Abdelwahab SI, Al-Zubairi AS, Aspollah SM, Abdullah R, Taha MM, Beng NK and Isa NM 2010 *Typhoniumflagelliforme*inhibits the proliferation of murine leukemia WEHI-3 cells *in vitro* and induces apoptosis *in vivoLeuk. Res.***34** 1483-92
[7] Hoesen DSH 2007 Pertumbuhan dan perkembangan tunas *Typhonium* secara in vitro.*Berita Biologi***8**(5) 413-22
[8] Sianipar NF, Maarisit W and Valencia A 2013a Toxic activities of hexane extract and column chromatography fractions of rodent tuber (*Typhonium flagelliforme* Lodd.) on *Artemia salinaIndones. J. Agric. Sci.***14**(1) 1-7
[9] Mohan S, Bustamam A, Ibrahim S, Al-Zubairi AS and Aspollah M 2008 Anticancerous Effect of *Typhonium flagelliforme* on human T4-lymphoblastoid cell line CEM-ss. *JPT***3**(6) 449-56
[10] Sianipar NF, Rustikawati, Maarisit W, Wantho A and Sidabutar DNR 2011 *Proc. Int. Conf. on Biological Science BIO-UGM* (Jogjakarta: Universitas Gajah Mada)
[11] Sianipar NF, Wantho A, Rustikawati and Maarisit W 2013b The effect of gamma irradiation on growth response of rodent tuber (*Typhonium flagelliforme* Lodd.) mutant in in vitro culture *HAYATI J. Biosci.***20**(2) 51-6
[12] Sianipar NF, Ariandana and Maarisit W 2015a Detection of gamma-irradiated mutant of rodent tuber (*Typhonium flagelliforme* Lodd.) in vitro culture by RAPD molecular marker *Procedia Chem.***14** 285-94
[13] Sianipar NF, Laurent D, Purnamaningsih R and Darwati I 2013c T*Proc. Int. Conf. on Biological Science ICBS UGM* (Jogjakarta: Universitas Gajah Mada)
[14] Sianipar NF, Laurent D, Purnamaningsih R and Darwati I 2015b Genetic variation of the first generation of rodent tuber (*Typhonium flagelliforme* Lodd.) mutants based on RAPD molecular markers *HAYATI J. Biosci.***22**(2) 98-104
[15] Kayser O and Quax W 2007 *Medicinal Plant Biotechnology* (Germany: Wiley VCH)
[16] Marimuthu S,Padmaja Band Nair S 2013 Phytochemical screening studies on *Melia orientalis* by GC-MS analysis *Pharmacognosy Res.* **5**(3) 216
[17] Nishaa S, Vishnupriya M, Sasikumar JM andGopalakrishnan VK 2013 Phytochemical screening and GC-MS analysis of ethanolic extract of rhizomes of *Maranta arudinacea* L. *RJPBCS.***4**(2) 52-9
[18] Mohan S, Bustamam A, Ibrahim S, Al-Zubairi A, Aspollah M, Abdullah R and Elhassan MM 2011 In vitro ultramorphological assessment of apoptosis on CEMss induced by linoleic acid-rich fraction from *Typhonium flagelliforme* tuber *Evid. Based Complement. Alternat. Med.***2011** 1-12

[19] Sianipar NF, Purnamaningsih R, Darwati I and Laurent D 2016*Appl. Mathematics in Science and Engineering Int. Conf.* (Malaysia: Malaysia Technical Scientist Association)

[20] Harada H, Yamashita U, Kurihara H, Fukushi E, Kawabata J and Kamei Y 2002 Antitumor activity of palmitic acid found as a selective cytotoxic substance in a marine red alga *Anticancer Res.***22** 2587-90

[21] Bodoprost J and Rosemeyer H 2007 Analysis of Phenacylester derivatives fatty acids from human skin surface sebum by reversed-phase HPTLC: chromatographic mobility as a function of physicochemical properties *Int. J. Mol. Sci.***8** 1111-24

[22] Yu F, Lian X, Guo H, McGuire P, Li R, Wang R and Yu F 2005 Isolation and characterzation of methyl esters and derivatives from *Euphorbia kansui* (Euphorbiaceae) and their inhibitory effects on the human SGC-7901 cells *J. Pharm. Pharmaceut Sci.***8**(3) 528-35

[23] Awad AB, Chen YC, Fink CSand Hennessey T 1996 Beta-sitosterol inhibits HT-29 human colon cancer cell growth and alters membrane lipids *Anticancer Res.***16** 2797-804

[24] Sujatha S, Anand S, Sangeetha KN, Shilpa K, Lakshmi J, Balakrishnan A and Lakshmi BS 2010 Biological evaluation of (3β)-stigmast-5-en-3-ol as potent anti-diabetec agent in regulating glucose transport using *in vitro* model*Int. J. Diabetes Mellit.***2** 101-9

[25] Mendilaharsu M, Stefani ED, Deneo-Pellegrini H, Carzoglio J and Ronco A 1998 Phytosterols and risk of lung cancer: A case-control study in Uruguay*Lung cancer***21** 37-45

[26] De Stefani E, Boffetta P, Ronco AL, Brennan P, Deneo-Pellegrini H, Carzoglio JC and Mendilaharsu M 2000 Plant sterols and risk of stomach cancer: a case-control study in Uruguay *Nutr. Cancer***37** 140-4

[27] McCann SE, Freudenheim JL, Marshall JR and Graham S 2003 Risk of human ovarian cancer is related to dietary intake of selected nutrients, phytochemicals and food groups *J. Nutr.***133** 1937-42

[28] Rioja A, Pizzey AR, Marson CM and Thomas NS 2000 Preferential induction of apoptosis of leukaemic cells by farnesol*FEBS Letter* **467**(2-3)291-5

[29] Yazlovitskaya EMand Melnykovych G 1995 Selective farnesol toxicity and translocation of protein kinase C in neoplastic HeLa-S3K and non-neoplastic CF-3 cells *Cancer Lett.***88**(2) 179-83

[30] Belanger JT 1998 Perillyl alcohol: applications in oncology *Altern. Med. Rev.***3**(6) 448-57

[31] Crowell PL 1999 Prevention and therapy of cancer by dietary monoterpenes *J. Nutr.* **129** 775S-8S

[32] Escrich E, Solanas M, Moral R, Grau L, Costa I andVela EER 2008 Dietary LIPIDS and breast cancer: Scientific clinical, anatomopathological and molecular evidences*Revista espanola de obesidad***6** 129-38

[33] Menendez JA, Vellon L, Colomer R and Lupu R 2005 Oleic acid, the main monounsaturated fatty acid of olive oil, suppresses Her-2/neu (erbB-2) expression and synergistically enhances the growth inhibitory effects of trastuzumab (Herceptin) in breast cancer cells with Her-2/neu oncogene amplification *Ann. Oncol.***16** 359-71

[34] Carrillo C, Cavia MM and Alonso-torre SR 2011 Oleic acid inhibits store-operated calcium entry in human colorectal adenocarcinoma cells*Eur. J. Nutr.***51** 677-84

[35] Rao CV, Newmark HL and Reddy BS 1998 Chemopreventive effect of squalene on colon cancer*Carcinogenesis***19** 287-90

[36] Jiang Q, Chisten S, Shigenaga MK and Ames BN 2001 γ-tocopherol, the major form of vitamin E in the US diet, deserves more attention*Am. J. Clin. Nutr.***74** 714-22

[37] Jeong JB and Jeong HJ 2010 2-methoxy-4-vinylphenol can induce cell cycle arrest by blocking the hyper-phosphorylation of retinoblastoma protein in benzo(a)pyrene-treated NIH3T3 cells *Biochem. Biophys. Res. Commun.***400** 752-7

[38] WangG, Li X, Huang F, Zhao J, Ding H, Cunningham C, Coad JE, Flynn DC, Reed E and Li Q Q 2005 Antitumor effect of beta-elemene in non-small-cell lung cancer cells is mediated via induction of cell cycle arrest and apoptotic cell death*Cell Mol. Life Sci.***62** 881-93

[39] Li QQ, Wang G, Huang F, Banda M and Reed E 2010 Antineoplastic effect of beta-elemene on prostate cancer cells and other types of solid tumour cells *J. Pharm. Pharmacol.* **62** 1018-27

[40] Jayasankar S 2005 *Variation in Tissue Culture*, ed R N Trigiano and DJ Gray (New York: CRC Press LLC) pp 301-9

[41] Kaeppler SM, Kaeppler HF and Rhee Y 2000 Epigenetic aspects of somaclonal variation in plants *Plant. Mol. Biol*. **43** 179-88

IOP Conf. Series: Earth and Environmental Science **41** (2016) 012026 doi:10.1088/1755-1315/41/1/012026

Ecological and Geochemical Assessment of Woody Vegetation in Tungsten-Molybdenum Mining Area (Buryat Republic, Russia)

N S Kasimov[1, 4], N E Kosheleva[2] and I V Timofeev[3]

[1] Head of the Dept., Faculty of Geography, Lomonosov Moscow State University, Moscow, Russia

[2] Leading researcher, Faculty of Geography, Lomonosov Moscow State University, Moscow, Russia

[3] Junior researcher, Faculty of Geography, Lomonosov Moscow State University, Moscow, Russia

E-mail: natalk@mail.ru

Abstract. Biogeochemical studies performed in the impact zone of the Dzhida tungsten–molybdenum mining and processing enterprise in Zakamensk (Buryat Republic, Russia) in 2013 showed that the needles and bark of *Lárix sibírica Ledeb.* and the leaves and bark of *Bétula platyphýlla Sukacs* are characterized by certain changes in their trace element (TE) composition. The total index of the biogeochemical transformation Z_v which sums all the positive and negative deviations of TE contents from the background values for larch and birch in the city averaged 95 and 46 for their assimilating organs and 30 and 25 for their perennial organs, respectively. This was caused by the increased uptake of W, Mo, V, Pb, Bi, Cd, and Co in the city. The close correlation between TE concentrations in soils and plants was observed for the elements of the strong and moderate biological capturing, including cationic Cu, Zn, Sr, Cd, and Ba and anions of the ore elements W and Mo and the associated Bi. The most intensive TE accumulation was found in the larch needles due to the wax layer which firmly fixes the dust enriched with TEs. Indication of the ecological state of urban woody plants revealed that their organs contain the increased concentrations of Pb and Fe and suffer from the Mn deficiency, which attests to disturbance of photosynthetic processes, most pronounced in the residential area. Birch leaves are characterized by a very low Cu/Zn ratio which detects the imbalance of these elements participating in the synthesis of enzymes.

1. Introduction

The development of nonferrous metal deposits exerts a significant impact on the environment and contributes to the emergence of cities and towns, mines, and factories that are exposed to high concentrations of pollutants. Being involved in the biogeochemical cycle, these pollutants enter through the soil, hydrosphere, and atmosphere into the plants. Getting trace elements (TEs) from the soil and the atmosphere, the plants are indicative of the degree of the environmental pollution. Therefore, the biogeochemical monitoring has become widely used in recent decades [1–4]. Woody assimilating organs (leaves and needles) are most rich in ash elements and are extremely sensitive to changes in the environmental conditions [5-7]. TE concentrations in perennial organs (bark, root system) are insufficiently studied; there are few works devoted to the concentration of heavy metals and metalloids (HMMs) in these organs [6, 9, 10].

Our work was aimed at the study of the TE composition of assimilating (needles, leaves) and perennial (bark) organs of Asian white larch (*Lárix sibírica Ledeb.*) and Siberian birch (*Bétula platyphýlla Sukacs*) in the impact zone of the Dzhida tungsten–molybdenum factory (DTMF) in the Zakamensk city (the Buryat Republic, Russia), where the mining of nonferrous metals continued for more than 60 years. In 2001, the DTMF was closed. However, necessary environmental protection measures—elimination of mines, rehabilitation of disturbed lands, and purification of contaminated mine waters—were not realized. As a result, the zone of ecological disaster appeared in this area [10]. In our work, we attempted to (a) determine the degree of technogenic disturbances of the micro-elemental composition of tree organs, (b) specify the intensity of the uptake of TEs by the trees from soils, (c) evaluate the ecological state of larch and birch in mining landscapes, and (d) determine indicative potential of the considered tree organs for the geochemical monitoring of the environment.

2. Study objects

Environmental conditions. The city of Zakamensk occupies the area of 45 km^2 in 460 km to the southwest of the Ulan-Ude city in the Trans-Baikal region. It is allocated to the boundary between two regional geological structures represented by the Early Paleozoic calcareous–terrigenous deposits of the Dzhida synclinorium and granitoid intrusions of the Modonkul massive with a considerable (300–400 m) ruggedness of the local topography. The climate is sharply continental, with cold and low-snow winters and relatively short and warm summer seasons. Mean annual precipitation is about 250–300 mm; predominant winds blow from the west and southwest. Mountainous soddy taiga and soddy calcareous soils are developed in autonomous geomorphic positions and on steep slopes covered by forests. *Lárix sibírica* and *Bétula platyphýlla* are the major cenosis-forming tree species. In the intermontane depressions, on the footslopes, and in the valleys of the Modonkul and Dzhida rivers, soddy soils are developed under the anthropogenically disturbed meadow and meadow-bog vegetation, and meadow alluvial soils are formed under meadow vegetation with willow groves [11].

Technogenic impacts. From 1934 to 2001, the DTMF produced up to 80% of tungsten concentrate obtained in the USSR. The tungsten ore contains HMMs of the 1st–3rd hazard classes, such as Pb, Zn, F, Mo, W, Be, Bi, As [12]. Various reagents, including kerosene and sulfuric acid were applied for ore dressing. Waste products are stored in the Dzhida tailing (filled), the Barun-Naryn tailing (sludge pond), and the emergency tailing. In 2000, the first two tailings got the status of technogenic deposits because of high concentrations of W and Mo in them. In 2009, rehabilitation works begun, and some part of the stored wastes was removed toward the Barun-Naryn tailing, where in 2010, the Zakamensk JSC started to extract W concentrate. The new Zyn-Naryn tailing was formed for storage of wastes.

Land-use structure of the city. Three land-use zones—residential, industrial, and natural recreation —can be distinguished in Zakamensk. The industrial zone includes the DTMF, the "Foundry" Smelter, the thermal power plant to the south of the residential zone and the Barun-Naryn, Zun-Naryn, and Dzhida tailings to the west of the residential zone on the right bank of the Modonkul River. On the left bank, in place where the river channel turns from the meridional direction to the sublatitudinal one, the Modonkul deposit of technogenic sands was shaped. The tops and gentle slopes of the hills in the Modonkul and Zimka river valleys were considered as background natural territories.

3. Materials and methods

Assimilating (needles and leaves) and perennial (bark) organs of Asian white birch and Siberian larch were sampled in the middle of the summer of 2013 in different land-use zones of the city. The mixed samples were taken from three or more trees of the same age after the flowering stage and after the five-day long rainless period. Needle and leave samples were dried for 6 hours at 70–80°C. In contrast to larch needles, birch leaves were initially washed by distilled water in order to remove elements precipitated on the leave surface and not participating in plant metabolism [12, 13]. Overall, we collected 32 samples for birch leaves and bark and 21 samples for larch needles and bark (Figure 1).

Bulk contents of 54 HMMs in the dry samples of plant material were determined by mass and atomic emission spectrometry with inductively coupled plasma methods in the All-Russian Institute of Mineral Ores on Elan-6100 and Optima-4300 devices (Perkin Elmer, USA). For a detailed analysis, 16 priority pollutants typical of Mo–W deposits [14] and characterized by the high toxicity for living organisms were selected. They belong to the first (Zn, As, Pb, Cd), second (Cr, Co, Ni, Cu, Mo, Sb), and third (V, Sr, Ba, W) hazard classes; in addition, data on Sn and Bi were analyzed.

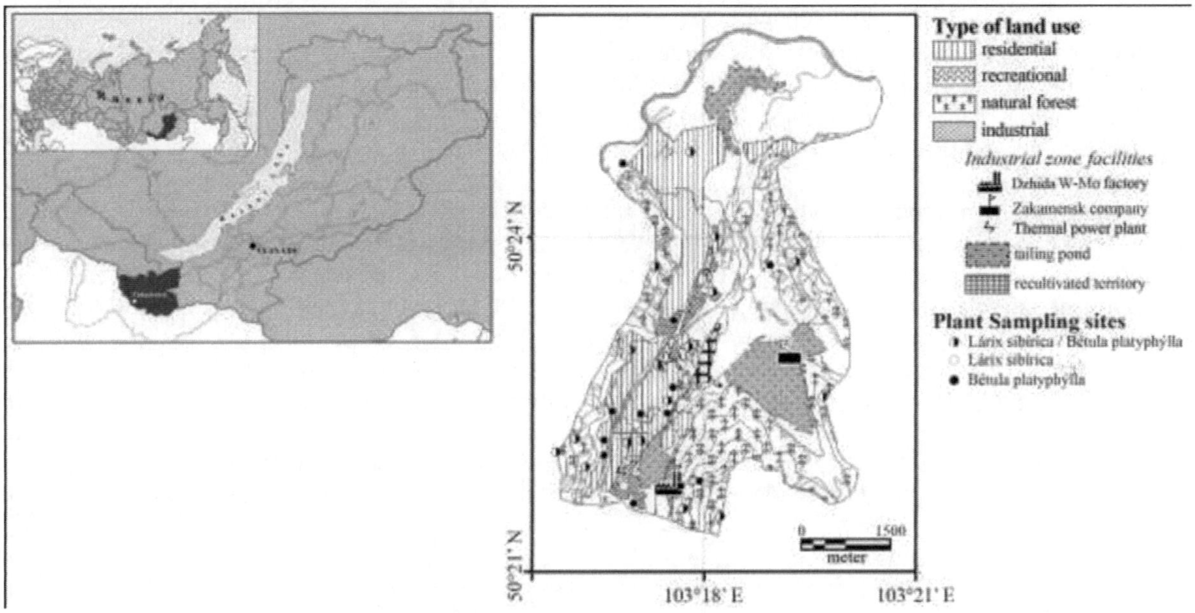

Figure 1. Map of sampling points and land-use zoning of Zakamensk.

The intensity of biological uptake of elements was judged from the coefficient of biological accumulation CBA= l_x/C_s, where l_x is an element concentration in plant ash; C_s is its concentration in the soil [15]. The changes in the TE composition of plant materials were evaluated with the help of the total index of biogeochemical transformation: $Z_v = \sum_{l=1}^{n_1} EF_l + \sum_{l=1}^{n_2} DF_l - (n_1 + n_2 - 1)$, where $EF_l = C_i/C_b$ and $DF_l = C_b/C_i$ are local enrichment and dispersion factors, respectively; C_i and C_b are the concentrations of TEs in the urban and background samples, respectively; and n_1 and n_2 are the numbers of TEs with $EF_l > 1$ and $DF_l > 1$, respectively [16]. The Z_v index reflects the disturbance of normal proportions of TEs in plant organs typical of their phylogenetic and ontogenetic specialization. It is a quantitative measure of the imbalance of elements in plants which appears under the impact of anthropogenic loads. The Z_v index has five grades corresponding to minimal (10–20), moderate (20–30), strong (30–40), very strong (40–60), and extremely strong (>60) disturbance. The ecological state of the plants was specified from the Fe/Mn, Pb/Mn, and Cu/Zn ratios [17–19].

4. Results and discussion

4.1.1. Assessment of the technogenic geochemical transformation of plants

The response of the tree species to technogenic pollution was judged from the total Z_v index. As the woody vegetation and, hence, sampling points in the city had an uneven distribution pattern, the compiled maps (Figure 2) can only present the most general spatial trends of the multielemental contamination of the trees. Coniferous and small-leaved deciduous species differ greatly in geochemical transformation. Larch needles are characterized by higher values of the Z_v index: its average value in the city reaches 95, i.e., it corresponds to the extremely strong technogenic transformation of the TE composition of larch needles. For birch leaves, the average Z_v index is 46. The major contribution to the Z_v value for larch needles is paid by the following element association: $Cr_{19}W_{16}V_{14}Pb_{12}Bi_9Mo_8Sb_7Ni_{6.5}Cd_{4.3}Co_{3.3}Sn_{2.5}$; for birch leaves, this association is somewhat different: $W_8V_7Mo_7Bi_{5.9}Pb_{5.4}Sb_{4.5}Cd_{2.1}Co_{2.0}$ (subscripts denote EF_l values averaged for the city).

There are two stable biogeochemical anomalies in the center and north of the city (Figs. 2A and 2C). The first anomaly is formed in the industrial and residential zones, where Z_v values reach 320 and 205 for larch needles and 100 and 71 for birch leaves, respectively. This anomaly accumulates pollutants from several sources: the Dzhida tailing, the material of which is subjected to erosion and deflation; the thermal power plant which utilizing black oil, and the "Foundry" Smelter producing foundry iron, steel, and bronze and reworking scrap metal and old cars. The pollutants emitted into the atmosphere and discharged with waste water are taken up by the trees from the soils and absorbed from the air.

The northern anomaly with Z_v values reaching 260 in larch needles and 72 in birch leaves is allocated to the residential zone, to the southwest from the Modonkul deposit of technogenic sands. Its origin is mainly related to the aerial pollution with TEs. Local weather conditions are characterized by frequent and relatively strong (up to 20 m/s) winds, and the narrow valley of the Modonkul River with high banks, on which the city is found, creates the wind corridor, along which fine dust particles are transferred from the places of waste storage. Thus, the major factor of the biogenic accumulation of the pollutants is the precipitation of dust particles enriched in W, Mo, Pb, Bi, Cd, Sb, and V on the surface of assimilating plant organs; the higher accumulation of the pollutants by larch needles is explained by the presence of the wax film favoring firm fixation of the dust on their surface [20–26].

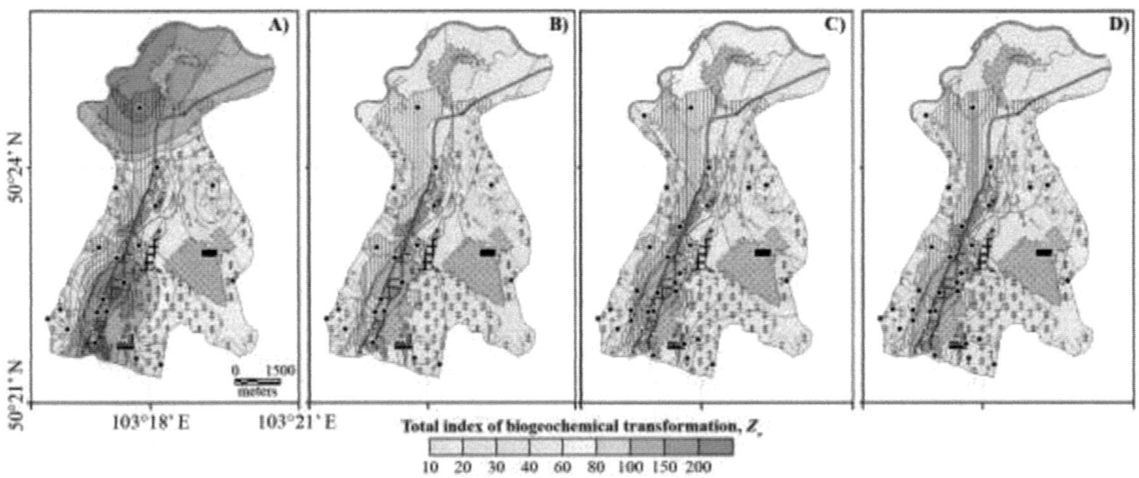

Figure 2. The degree of geochemical transformation of the (A) needles and (B) bark of *Larix sibirica* and (C) leaves and (D) bark of *Bétula platyphýlla* in Zakamensk.

The accumulation of the pollutants in perennial organs of the trees is much weaker; the corresponding Z_v values are smaller than those of the assimilating organs (Figure 2). Thus, Z_v averages 30 for the bark of larch and 25 for the bark of birch, which corresponds to the strong and moderate biogeochemical transformation, respectively. The major contribution is due to the W–Pb–Bi (EF_l=3–7.5) and Cd–Mo–Zn (1.7–3) element associations, respectively. Maximum Z_v values—85 for larch bark and 55 for birch bark—were determined in the biogeochemical anomaly in the south of the city, near the "Foundry" Smelter and the thermal power plant.

4.2. Relationships between TE composition of soils and plants

The trees can selectively absorb some elements from the environment and accumulate them in their organs. The geochemical specialization of the particular tree species is reflected in the coefficients of the biogenic accumulation of elements (CBA). To calculate the CBA, soil profiles were examined in the sites of tree sampling, and soil samples were taken for the determination of TEs with the ICP-MS method. For each soil profile, the mean weighted (according to the thickness of various soil horizons) concentrations of elements were calculated, because the tree roots take up mineral and organic substances from various depths. Then, the averaging of these mean weighted concentrations was performed for each land-use zone of the city.

In all the land-use zones, larch needles actively accumulate mainly cation-forming elements (Cu, Zn, Sr, Cd, Ba), which is typical of the Trans-Baikal region with acid soils [27]. This group also includes the anion-forming elements: major ore element W and his associate Bi (Figure 3A). Their high concentrations in soils and plants are explained by the natural lithogeochemical anomaly. The group of elements with weak biological uptake in the background and natural recreation zones of the city includes anion-forming V and Cr. The maximum number of elements (11) is accumulated in the larch needles in the residential zone, where TEs enter by lateral migration from the crumbling nearby tailings.

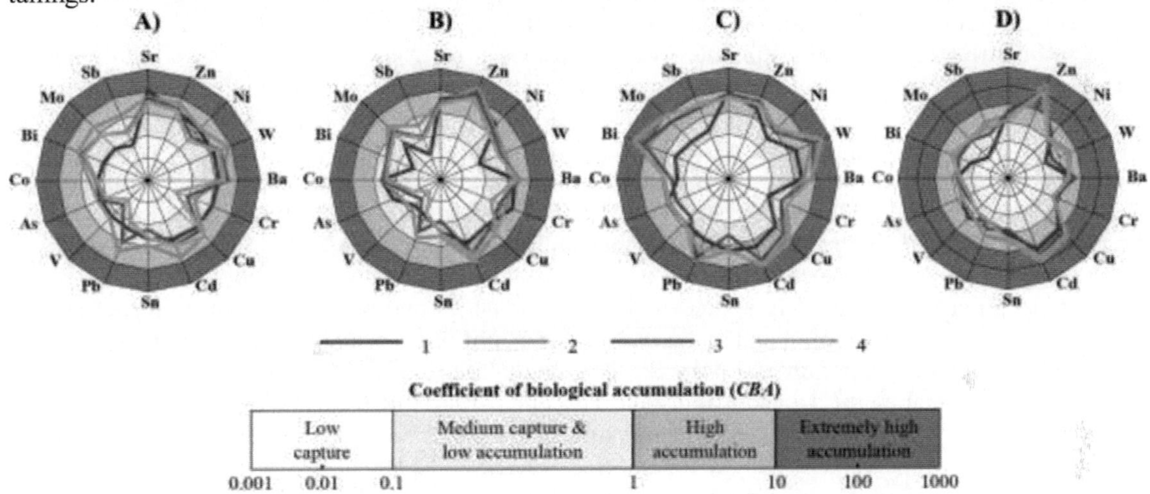

Figure 3. Diagrams of the accumulation of TEs in the (A) needles and (B) bark of *Larix sibirica* and (C) leaves and (D) bark of *Bétula platyphýlla* in Zakamensk. Land-use zones: (1) industrial, (2) residential, (3) natural recreation, and (4) background territories.

The geochemical specialization of larch bark is analogous to that of larch needles (Figure 3B) and is characterized by the accumulation of cation-forming (Cd, Ni, Cu, Zn, Sr, Sn, Ba, Pb), ore (W, Mo), and accompanying (Bi) elements. However, if we analyze CBA values calculated with respect to the soils of the same land-use zones, where plant samples were taken, we shall not see the intense biological accumulation of TEs in the industrial zone, where the soils have very high bulk element concentrations, so that element abundance in the plants relative to the soils (CBA value) becomes relatively low. For example, the concentrations of Mo, W, and Bi in the soils reach 127, 199, and 21 mg/kg, i.e., they are 30 to 120 times higher than their clarkes in the lithosphere. Under such high soil content of TEs, no significant bioconcentration of these elements takes place.

The accumulation of elements by assimilating and perennial organs of Asian white birch displays some common features with that of larch; at the same time, the differences between these two species are significant (Figure 3C, 3D). Both tree species absorb mainly cation-forming (Cu, Zn, Sr, Cd, Ba), ore (W, Mo), and accompanying (Bi) elements. In the industrial zone, ore elements belong to the groups of the moderate and strong element "capturing" by the plants.

A specific feature of birch leaves and bark (as compared with larch needles and bark growing in the same urban ecosystem) is a more intense accumulation of Cd and Zn which are typical pollutants of urban environments. Our results are in agreement with the earlier obtained data [28, 29].

4.3. Ecological state of the plants

The *Fe/Mn ratio* is the most informative indicator of photosynthesis with the optimum range of 1.5–2.5 necessary for the normal development of plants [30]. The Fe/Mn ratio in assimilating organs of the trees in Zakamensk depends on the level of anthropogenic load (Table 1). In larch needles, this ratio changes from 0.11 on the background plots to 4.7 in the residential area. Maximum values (12.4) in the latter zone attest to a sharp deficit of Mn and biological accumulation of active ferrous oxide inducing plant chlorosis [31] and the stress state of the plants [32]. The optimum ratio (1.75) was found only in the industrial zone. A reverse situation—the high Mn concentration—leads to the decreased concentrations of ferrous oxides that are mobilized in plant cells in the form of iron organophosphates, which also induces chlorosis related to the deficit of iron [31]. This situation is typical of the background and natural recreation areas, where the Fe/Mn in larch needles is about 0.11–0.22. Birch leaves in all the land-use zones suffer from the deficit of iron and excess of Mn (Fe/Mn = 0.11–0.63), only the maximum values of this ratio (2.4) are within the optimum range.

The distribution of Fe/Mn ratios in the perennial organs of the trees has much in common with that in the assimilating organs. Thus, minimum values are observed in the bark of larch trees growing on the natural recreation and background territories (0.97 and 1.99 respectively); with an increase in the anthropogenic load, the Fe/Mn ratio sharply increases up to 4.5 in the residential zone and 5.7 in the industrial zone, i.e., it becomes 1.8 and 2.3 times higher than the upper limit of the optimum range, respectively. The bark of birch trees is characterized by the even more pronounced deficit of iron and excess of manganese than that in the birch leaves. The Fe/Mn ratio in the birch bark is 0.02 in the natural recreation zone. In the other land-use zones, it varies from 0.04 to 0.21.

The maximum permissible concentration of Fe in plants is 240 mg/kg of dry phytomass [33]. In the assimilating organs of the trees it is exceeded in the industrial and residential zones; in the bark of larch, virtually in all the samples; and in the bark of birch, only in the samples from the industrial zone. The critically high concentration of 750 mg/kg dry phytomass [33] is exceeded in the organs of larch in some local anomalies: in the industrial zone Fe concentration in the bark of larch is up to 1800 mg/kg of dry phytomass, in the residential zone, the critical concentration is exceeded in the needles.

The phytotoxic concentration of Mn for the trees reaches 500 mg/kg of dry phytomass [34]. It is exceeded in the bark and leaves of the birch in all the land-use zones. For larch, it is only exceeded in the needles in the background and natural recreation zones. It is known that Siberian larch has a "ferrallitic" composition of ash and is a concentrator of Mn [35]. The concentration of Mn by birch even under minimal anthropogenic loads is probably related to the fact that in the acid medium (at pH <5.7), this element is transformed into the mobile and available form Mn^{2+} [36]. In the leaves of birch growing in the city, the concentration of Fe decreases simultaneously with an increase in the concentration of Mn. Change of the natural ratios of these elements–antagonists in assimilating organs of the trees subjected to the contamination was also observed by other authors [30, 37–40].

Table 1. Indicators of the ecological state of assimilating and perennial organs of the trees in various land-use zones of the Zakamensk city

Land-use zone	Indicator					
	Fe/Mn		Pb/Mn		Cu/Zn	
	average	min-max	average	min-max	average	min-max
			Needles of *Larix sibirica*			
B	0.11	0.06-0.17	0.0005	0.0002-0.0008	0.29	0.28-0.31
R	0.22	0.06-0.56	0.002	0.0004-0.005	0.31	0.28-0.34
S	4.68	0.41-12.4	0.042	0.003-0.125	0.29	0.2-0.36
I	1.75	0.68-2.82	0.011	0.006-0.34	0.28	0.23-0.34

			Leaves of *Bétula platyphýlla*			
B	0.11	0.08-0.16	0.0004	0.0003-0.0004	0.07	0.05-0.1
R	0.14	0.07-0.4	0.001	0.0002-0.002	0.07	0.03-0.11
S	0.56	0.09-1.62	0.006	0.0006-0.032	0.05	0.03-0.1
I	0.63	0.08-2.39	0.002	0.0004-0.003	0.03	0.03-0.05
			Bark of *Larix sibirica*			
B	1.99	0.25-3.4	0.02	0.002-0.031	0.37	0.3-0.44
R	0.97	0.25-1.7	0.02	0.002-0.05	0.37	0.26-0.56
S	4.49	1.5-10.8	0.05	0.005-0.12	0.42	0.21-0.99
I	5.68	0.8-10.6	0.06	0.008-0.11	0.62	0.58-0.67
			Bark of *Bétula platyphýlla*			
B	0.04	0.02-0.06	0.0004	0.0003-0.0004	0.06	0.04-0.07
R	0.02	0.01-0.06	0.0004	0.00004-0.0008	0.07	0.04-0.1
S	0.21	0.01-1.6	0.002	0.0001-0.006	0.05	0.03-0.1
I	0.13	0.01-0.5	0.001	0.0003-0.004	0.05	0.04-0.06

Note: B—background areas. *Land-use zones of the city:* R—natural recreation, S–settling (residential), and I—industrial.

The Pb/Mn ratio characterizes the relationship between the technogenic and biophilous elements, i.e., the level of anthropogenic load. An optimum value of this ratio for undisturbed terrestrial vegetation was taken equal to the ratio of the clarkes (natural abundances of Pb and Mn) [41]. It is equal to 0.006, which attests to the low portion of technogenic elements in the TE composition of the plants which are not involved in the physiological processes. The excess of Pb in plants inhibits respiration and suppresses photosynthesis because of the disturbance in electron transport [30].

The assimilating organs of birch in all the land-use zones of Zakamensk are not subjected to technogenic pressure, because average values of the Pb/Mn ratio are low (0.0004–0.006). A local anomaly with the maximum of 0.032 was found in the residential zone near the Lenin street, the major highway in the city, which is the source of Pb. Minimum Pb/Mn values in the needles of larch were found in the background (0.0005) and natural recreation (0.002) zones, and maximum values were found in the residential zone with the average Pb/Mn ratio of 0.04 and the local maximum of 0.339; these values exceed those in the unpolluted terrestrial vegetation by 7 and 56 times, respectively [42].

Perennial organs usually accumulate higher amounts of pollutants [5]. The Pb/Mn ratio indicates that an increase in technogenic loads is accompanied by the rise in the concentration of Pb. However, in the bark of birch in all the land-use zones, the Pb/Mn ratio is no higher than that in the unpolluted vegetation and ranges from 0.0004 to 0.002. In contrast, in the bark of larch, the average Pb/Mn ratio is minimal (0.02) on the background plots and increases in the industrial zone up to 0.06 with local peaks of up to 0.119, which is 20 times higher than the normal value.

A comparison of data on Pb concentrations in the trees growing in different land-use zones of Zakamensk with the maximum permissible Pb concentration shows that the bark of larch is subjected to the highest technogenic Pb load: average Pb concentrations in it vary from 5.8 to 9.9 mg/kg rising up to 25 and 19 mg/kg in local anomalies within the residential and industrial zones, respectively. The first anomaly is found in the center of the city, near the Modonkul River, where a ford operates in the summer. The source of Pb is probably related to motor oil, gasoline, and car exhausts. In the second anomaly, the accumulation of Pb is related to the emissions from the "Foundry" Smelter and thermal power plant. Active accumulation of Pb in the bark of larch is favored by dust precipitation and retention on the rough surface of the bark. In the needles of larch, the highest concentrations of Pb (3.3 mg/kg with local peaks up to 9.1 mg/kg) are observed in the residential zone. The organs of birch are characterized by generally low concentrations of Pb in all the zones, except for the residential zone, where Pb concentration in the leaves averages 1.5 mg/kg.

Thus, coniferous trees accumulate Pb more actively than small-leaved (birch) trees. This can be explained by the morphological specificity of the assimilating organs, i.e., by the presence of waxy cuticle firmly fixing TEs on larch needles [20–21]; and also by the pretreatment of the samples—the washing of birch leaves with distilled water to remove dust particles—can also be the reason for the lower concentration of Pb in the birch leaves.

The Cu/Zn ratio characterizes the supply of these biogenic metals participating in the synthesis of enzymes. The optimum ratio in the nonpolluted terrestrial vegetation is 0.27 [17]. In the assimilating and perennial organs of birch trees from different land-use zones of Zakamensk, the average Cu/Zn ratio varies from 0.03 to 0.07 attesting to the high concentrations of Zn (Table 1). This ratio in larch needles is close to the optimum (0.28–0.31). The most significant imbalance of Zn and Cu (Cu/Zn = 0.36) is observed in larch needles in the residential zone. In the bark of larch, minimum values of the Cu/Zn ratio (0.37) are disclosed in the background and natural recreation landscapes; they exceed the optimum by 1.4 times. Maximum values (0.62) are typical of the residential zone. In the local maximum in the industrial zone, this ratio increases up to 0.99.

A significant positive correlation between Cu and Zn concentrations in the assimilating organs of larch and birch trees was discovered (correlation coefficients $r = 0.65$, and $r = 0.88$, respectively). This fact contradicts previously published data about antagonism of Cu and Zn, as their uptake by the roots is controlled by the same mechanism, and one element can inhibit the uptake of the other [30, 43]. Such a discrepancy can be explained by the fact that, in our study, the Cu/Zn values are close to the optimum for nonpolluted terrestrial vegetation and that the uptake of these elements by the leaves and needles from the atmosphere plays a significant role in their total absorption. For perennial organs of birch and larch, such a correlation is weak ($r = 0.56$, and $r = 0.57$).

Copper belongs to biophilous elements; it is always present in soils, plants, and animal tissues and participates in various metabolic processes. It enters the composition of plastocyanin that is involved in electron-transfer processes between photosystem II and photosystem I; it is also present in Cu-containing proteins and enzymes that catalyze oxidation of ascorbic acid and biphenols, and hydroxylation of monophenols [36]. The critical phytotoxic concentration of Cu reaches 20 mg/kg of dry phytomass [34]. In the studied tree species of Zakamensk, such a concentration has not been exceeded. The concentration of Cu in the assimilating and perennial organs varies from 3.2 to 9.4 mg/kg; the local maximum (15 mg/kg) was determined in larch needles in the industrial zone.

Zinc participates in redox processes in plant tissues; it stabilizes air-exchange processes and affects photosynthesis; it is included in the composition of tryptophan (an amino acid) [36]. The phytotoxic concentration of Zn is 300 mg/kg of dry phytomass. With an increase in the anthropogenic load, the concentration of Zn in the organs of the trees increases. Minimum values were found in the background and natural recreation areas, and maximum values – in the industrial zone. Birch is a more active concentrator of Zn in comparison with larch: average Zn concentrations in the leaves and bark of birch were within 78.5–182 and 99–132 mg/kg of dry phytomass, respectively; in the needles and bark of larch, they were within 11.5–19.5 and 7.2–15.8 mg/kg of dry phytomass, respectively. Taking into account the fact that birch leaves were preliminary washed with distilled water, it can be supposed that the main input of Zn into the leaves was from the Zn-contaminated soils. Similar results were obtained in other studies of the TE composition of tree species in Europe [44–47].

5. Conclusions

Biogeochemical studies performed in Zakamensk (Buryat Republic) showed that the trees of this city are subjected to high anthropogenic loads. The resulting imbalance of elements in the trees is characterized by the total index of the biogeochemical transformation Z_v calculated for the assimilating and perennial organs. The changes in the TE composition of larch needles are specified by intense accumulation of Cr, W, V, Pb, Bi, Mo, Sb, Ni, Cd, Co, and Sn; birch leaves concentrate W, V, Mo, Bi, Pb, Sb, Cd, and Co. The highest values of Z_v index are typical of larch needles, which are explained by the fixation of precipitating dust particles containing TEs by the wax film on needle surfaces. The accumulation of TEs in the assimilating organs of larch and birch is more intense than that in the perennial organs.

Two stable biogeochemical anomalies were revealed in the center and in the north of the city. In the central anomaly, Z_v values for larch needles and birch leaves reach 205–320 and 71–100, respectively. HMMs enter the plants from tailings, thermal power plant, and the "Foundry" Smelter. In the northern anomaly, Z_v is up to 260 for larch needles and 72 for birch leaves. This anomaly is located in the residential zone; it is contaminated with HMMs blown out from the Modonkul technogenic sand deposit.

Despite the high anthropogenic load, the organs of the trees preserve their capacity to take up cationic elements (Cu, Zn, Sr, Cd, Ba) from the soil, which is generally typical of humid environments with acid soils, in which these elements are mobile. Anion-forming ore elements W and Mo and Bi as the major accessory are also actively absorbed from the soils. Tailings serve as the main sources of soil and plant pollution with TEs. The surface of tailings remains barren, and they are characterized by good aeration and increased water permeability. As a result, the material of tailings is subjected to wind and water erosion and to chemical weathering under the impact of sulfuric acid. These processes enhance the lateral migration of TEs in the soil cover and their accumulation by the tree species.

The ecological state of the trees in the urban environment was judged from the Fe/Mn and Pb/Mn ratios. Their values attest to the disturbance of photosynthetic processes in the trees, especially in the residential zone of the city. The highest values of the Fe/Mn (4.7) and Pb/Mn (0.042) ratios were found in larch needles. The high Fe/Mn ratio is explained by the deficit of Mn in the environment and the active bioconcentration of ferrous oxides that may induce chlorosis and the stress state of the trees. The high Pb/Mn ratio is due to the excessive content of Pb resulting in the inhibition of respiration and suppression of photosynthesis. The Cu/Zn ratio in the needles and bark of *Larix sibirica* in the city is within 0.28–0.31, i.e., it is close to the optimum for the unpolluted terrestrial vegetation. In the leaves and bark of *Bétula platyphýlla*, the Cu/Zn ratio is 0.03–0.07, which attests to an imbalance in the concentrations of these elements.

A comparative analysis of different organs of the trees showed that the needles and bark of *Larix sibirica* can be used as sensitive indicators of the environmental pollution with HMMs during the growing season and long-term periods. This is favored by the presence of a wax film on the needle surface ensuring fixation of dust particles with the pollutants precipitating from the atmosphere and by the rougher surface of the larch bark, which ensures more active biological uptake of TEs from the atmosphere.

Acknowledgements
This study was carried out in the framework of the research supported by the Russian Science Foundation (project № 14-27-00083).

References

[1] Markert B 1993 Instrumental analysis of plants *Plants as Biomonitors. Indicators for Heavy Metals in Terrestrial Environment* ed B Markert (Weinheim: VCH Verlagsgesellschaft) pp 65–103

[2] Bargagli R 1998 Trace Elements in Terrestrial Plants: An Ecophysiological Approach to Biomonitoring and Biorecovery (Berlin: Springer-Verlag) p 324

[3] Weiss P, Offenthaler I, Öhlinger R and Wimmer J 2003 Higher plants as accumulative bioindicators *Bioindicators & Biomonitors, Principles, Concepts and Applications* vol 6, ed B Markert, et al. (Amsterdam: Elsevier) pp 465–500

[4] Fujiwara F G, Gómez D R, Dawidowski L, Perelman P and Faggi A 2011 Metals associated with airborne particulate matter in road dust and tree bark collected in a megacity (Buenos Aires, Argentina) *Ecol. Indic.* **11** 240–7

[5] Shcherbenko T A, Koptsik G N, Groenenberg B-J, Lukina N V and Livantsova S Y 2008 Uptake of nutrients and heavy metals by pine trees under atmospheric pollution *Moscow Univ. Soil Sci. Bull.* **63** 51–9

[6] Koptsik G N, Koptsik S V, and Omlid D 1999 Transformation of the elemental composition of plants in forest biogeocenoses of the northern taiga under the impact of aerial pollution *Vestn.*

Mosk. Univ., Ser. 17: Pochvovedenie **1** 37–49.
[7] Nieminen T M, Derome J and Saarsalmi A 2004 The applicability of needle chemistry for diagnosing heavy metal toxicity to trees *Water. Air. Soil Pollut.* **157** 269–79
[8] Lindeberg J 2004 X-ray Based Tree Ring Analyses *Acta Univ. Agric. Sueciae Sivestria* 1–25
[9] Anon 2011 *Explanatory note on realization of the target republican program "Environmental safety in the Buryat Republic in 2009–2011 and up to 2017"* (Ulan-Ude)
[10] Smirnova O K and Plyusnin A M 2013 *Dzhida Ore Region (Environmental Problems)* (Ulan-Ude: Izd. Buryat. Nauchn. Ts. SO RAN) p 181
[11] Nogina N A 1964 *Transbaikal Soils* (Moscow: Science Publ.) p 312
[12] Jones J B J and Case V W 1990 Sampling, handling, and analyzing plant tissue samples *Soil testing and plant analysis* (Madison. Wisconsin: SSSA book series) pp 389–247
[13] Kosheleva N E, Makarova M G and Novikova O V 2005 Heavy metals in the leaves of tree species in urban landscapes *Vestn. Mosk. Univ., Ser. 5: Geografiya* **3** 74–81
[14] Saet Yu E, Revich B A, Yanin E P, Smirnova R S, Basharkevich I L, Onishchenko T L, Pavolva L N, Trefilova N Ya, Achkasov A I, and Sarkisyan S Sh 1990 *Environmental Geochemistry* (Moscow: Nedra) p 335
[15] Ufimtseva M D and Terekhina N V 2005 *Phyptoindication of the state of urban geosystems in Saint Petersburg* (St. Petersburg: Nauka) p 339
[16] Kasimov N S, Bityulova V R, Kislov A V, Kosheleva N E, Nikiforova E M, Malkhazova D M and Shartova N V 2012 Problems of the ecogeochemistry of large cities *Okhrana i razvedka nedr* **7** 8–13
[17] Elpat'evskii P V and Arzhanova V S 1990 *Geochemistry of Landscapes and Technogenesis* (Moscow: Nauka) p 196
[18] Novikova O V and Kosheleva N E 2007 Ecogeochemical assessment of the state of arboreal vegetation in Quito (Ecuador) *Vestn. Mosk. Univ. Ser. 5: Geografiya **1** 43–8*
[19] Kosheleva N E, Timofeev I V, Kasimov N S, Kisselyova T M, Alekseenko A V and Sorokina O I 2016 Trace Element Composition of Poplar in Mongolian Cities *Biogenic-Abiogenic Interactions in Natural and Anthropogenic Systems* Lecture Notes in Earth System Sciences ed V. Frank-Kamenetskaya, et al. (Switzerland: Springer International Publishing) pp 165–78
[20] Aznar J C, Richer-Laflèche M, Bégin C and Bégin Y 2009 Lead Exclusion and Copper Translocation in Black Spruce Needles *Water. Air. Soil Pollut.* **203** 139–45
[21] Gandois L and Probst A 2012 Localisation and mobility of trace metal in silver fir needles *Chemosphere* **87** 204–10
[22] Lin Z Q, Schuepp P H, Schemenauer R S and Kennedy G G 1995 Trace metal contamination in and on Balsam fir (*Abies balsamea* (L) Mill.) Foliage in southern Quebec, Canada *Water, Air, Soil Pollut.* **81** 175–91
[23] Müller C and Riederer M 2005 Plant Surface Properties in Chemical Ecology *J. Chem. Ecol.* **31** 2621–51
[24] Raitio H, Tuovinen J P and Anttila P 1995 Relation between sulphur concentrations in the Scots pine needles and the air in northernmost Europe *Water, Air, Soil Pollut.* **85** 1361–6
[25] Rautio P and Huttunen S 2003 Total vs. internal element concentrations in Scots pine needles along a sulphur and metal pollution gradient *Environ. Pollut.* **122** 273–89
[26] Trimbacher C and Weiss P 2004 Norway Spruce: A Novel Method using Surface Characteristics and Heavy Metal Concentrations of Needles for a Large-Scale Monitoring Survey in Austria *Water, Air, Soil Pollut.* **152** 363–86
[27] Perel'man A I 1989 *Geochemistry* (Moscow: Vysshaya Shkola) p 528
[28] Esenzholova A Zh and Panin M S 2012 Bioindication potential of the leaves of trees and shrubs in Temirtau *Vestn. Tomsk. Gos. Univ. Biologiya* **3(19)** 160–8
[29] Opekunova M G 2013 *Diagnostics of the Technogenic Transformation of Landscapes on the Basis of Bioindication: Author's abstract of doctoral (Geogr.) dissertation* (St. Petersburg: St.

Petersburg State University)
[30] Kabata P A 2011 *Trace Elements in Soils and Plants. Fourth Edition* (Boca Raton: CRC Press) p 548
[31] Kopylova L V 2010 Accumulation of iron and manganese in the leaves of trees in technogenic areas of Trans-Baikal region *Izvest. Samarskogo Nauchn. Ts. RAN* **12** 709–12
[32] Lyanguzova I V 2010 *Tolerance of the Components of Forest Ecosystems in the Russian North towards Aerotechnogenic Pollution: Author's abstract of doctoral (Biol.) dissertation* (St. Petersburg: V L Komarov Botanical Institute)
[33] Kopylova L V 2011 Specific features of the uptake of heavy metals by tree species in the urban environment *Vestn. Irkutsk. Gos. Univ.* **44(3)** 91–9
[34] Kazantsev I V 2008 *Ecological assessment of the impact of the rail transport on the contents of heavy metals in soils and plants of the right-of-way: Author's abstract of Cand. Sci. (Biol.) dissertation* (Samara)
[35] Bashkin V N and Kasimov N S 2004 *Biogeochemistry* (Moscow: Nauchnyi mir) p 648
[36] Kulagin A and Shagieva Yu A 2005 *Arboreal plants and the Biological Conservation of Industrial Pollutants* (Moscow: Nauka) p 190
[37] Giniyatullin R Kh, Kulagin A Yu, Batalov A A, and Salikhova R N 1997 *Concentrations of Some Metals in the Aboveground Organs of Betula pendula under Conditions of Industrial Contamination*
[38] Kopylova L V 2002 Assessment of the level of soil contamination with heavy metals and the intensity of their uptake by arboreal plants *Uchenye zap. Zabaikal'skogo Gos. Univ.* **1(42)** 70–5
[39] Kopylova L V and Yakimova E P 2013 Specificity of the accumulation of metals by arboreal plants in the urban environment *Uchenye zap. Zabaikal'skogo Gos. Univ. Ser. Estestven. nauki* **1(48)** 74–81
[40] Alvarez T M C, Leal A and Martínez L R 1980 Iron-manganese interaction and its relation to boron levels in tomato plants *Plant Soil* **55** 377–88
[41] Dobrovol'skii V V 2003 *Fundamentals of Biogeochemistry* (Moscow: Izd. Ts. "Akademiya")
[42] Vega F A, Covelo E F, Vazquez J J and Andrade L 2007 Influence of mineral and organic components on copper, lead, and zinc sorption by acid soils *J. Environ. Sci. Heal. Part A* **42** 2167–73
[43] Graham R D 1981 Absorption of copper by plant roots, *in Copper in Soils and Plants* ed J F Loneragan, A D Robson and R D Graham (New York: Academic Press) p 141
[44] Brekken A and Steinnes E 2004 Seasonal concentrations of cadmium and zinc in native pasture plants: consequences for grazing animals *Sci. Total Environ.* **326** 181–95
[45] Kopponen P, Utriainen M, Lukkari K, Suntioinen S, Kärenlampi L and Kärenlampi S 2001 Clonal differences in copper and zinc tolerance of birch in metal-supplemented soils *Environ. Pollut.* **112** 89–97
[46] Marguí E, Queralt I, Carvalho M L and Hidalgo M 2007 Assessment of metal availability to vegetation (*Betula pendula*) in Pb-Zn ore concentrate residues with different features *Environ. Pollut.* **145** 179–84
[47] Steinnes E, Lukina N, Nikonov V, Aamlid D and Røyset O 2000 A Gradient Study of 34 Elements in the Vicinity of a Copper-Nickel Smelter in the Kola Peninsula *Environ. Monit. Assess.* **60** 71–88

Photosynthetic Physiological Characteristics of *Gazania rigens* L. Under Drought Stress

T T Gao[1], S W Zheng[1], X H Zhou[1], D X Wang[1] and X P Lu[1, 2]

[1] Key Laboratory of Architecture and Urban Environment, Gold Mantis School of Architecture, Soochow University, Soochow, Jiangsu 215123, P.R. China

E–mail: longzs@suda.edu.cn

Abstract. To investigate the responses of photosynthetic physiological characteristics of *Gazania rigens* L. to drought stress, the changes of three cultivars ('Xingbai', 'XH' and 'Hongwen') photosynthetic values under drought stress were determined via LI-6400 portable photosynthesis analyzer (LI-COR, USA), and the relationships between photosynthesis and drought resistance of each cultivar were analyzed. The results showed that, three cultivars net photosynthetic rate (*Pn*), stomatal conductance (*Gs*), transpiration rate (*Tr*), and light use efficiency (*LUE*) value indicated the trend of decreasing gradually and there existed significant reduction in *Pn* and *Gs* values. There were extremely negative significant correlations between drought stress treatment days and *Pn*, *Gs*, *Tr*, water use efficiency (*WUE*) and *LUE* values. A small amount of leaves began to turn soft and yellow after drought stress treatment for 10 days, but they could recover to grow quickly after rehydration. The *Pn* values of 'Hongwen' decreased quickly and changed in a large range, so it had a poor resistance; the *Pn* values of 'Xingbai' decreased slowly and changed in a small range, so it had a strong resistance; while the changes of their hybrids –'XH' were between its parents. This research would provide a theoretical basis for gazania resistance cultivar breeding and application in landscape.

1. Introduction

Gazania (*G. rigens* L.), with another name – treasure flower, belonging to Asteraceae family and Gazania spp., is a perennial herb and known for its beautiful flower shape and horticultural appeal. With the characteristics of evergreen, flowering in three seasons, dwarf, interesting flower shape, gazania can be used as potted plant, ground cover plant, flower border and so on [1]. Now it has been widely used in landscaping of the world.

Water is a vital factor that limits plant distribution and growth, and photosynthesis is directly influenced by water [2]. Photosynthesis is one of the main factors determining plants' biomass. Only when plants keep high photosynthesis under stress can plants maintain its survival and development. Moreover, the ability to resistance is reflected by the values of photosynthesis [3]. Photosynthesis and the changes of the plant itself response the plant growth condition and stress resistance. The leaves of plants will close its stoma when subjected to drought stress, which leads to decrease photosynthesis rate [4]. *Pn* and *Gs* values of shrubs and herbs decrease with the reduction of water, which is harmful for plant growth [5].

The photosynthetic physiological characteristics of plants in the Asteraceae family under drought stress were studied by many scholars. Research on Dahlia photosynthetic characteristics showed *Pn*, *Tr*, *Gs*, *WUE* values decreased with the prolonged drought stress, while *Ci* values increased first and then decreased [6]. Compared with the control group, water stress decreased stomatal density on the surface of the leaf, as well as the *Gs* values and moisture loss; water stress reduced leaf values of *Pn*, *Tr*, *Gs* and other physiological indexes of *Flaveria bidentis* [7]. Influence of different degree of drought stress on the photosynthetic physiological characteristics and water use efficiency have been researched, which indicated ground-cover chrysanthemum belonged to cover plants with high drought resistance, and it was suitable to plant ground-cover chrysanthemum in the arid areas [8].

Deficit irrigation can improve plants' tolerance to stress. Previous studies on different gazania cultivars' drought resistance are mainly focused on physiological responses to drought stress, such as polyamines, antioxidant enzymes, isozymes, abscisic acid (ABA) and so on [9], while the study of photosynthetic characteristics under drought stress was not involved. Photosynthesis is one of the main performances of plant heterosis, and there exists heterobeltiosis in the hybrids photosynthesis capability, especially under the water stress condition [10]. In order to provide a theoretical basis for gazania resistance cultivar breeding and landscape plants low carbon maintenance, 'Xingbai', 'Hongwen' and their hybrid 'XH' were chosen as experimental materials. Photosynthetic physiological values of three cultivars under drought stress were determined via portable photosynthesis analyzer.

2. Materials and methods

2.1. Materials

'Xingbai', 'Hongwen' and their hybrid 'XH' were chosen as experimental materials. They were propagated in early May 2013 by cutting and these cuttings were transplanted to plastic pots after 30 days growing. Each pot had three holes and the pot volume is $40\times30\times15$ cm^3 with uniform substrate (Liangtu common soil for gardening, China) in it. There were 6 pots and every pot had 3 cultivars, 9 plants in total, with each cultivar having three plants. Then they were watered and fertilized with unified management for future experiment.

2.2. Methods

2.2.1. Experimental design. These pots were soaked fully in water at 3:00 PM for 20 minutes to get enough water on the day before the experiment. The treatments were as follows: these plants were in drought stress without watering in the whole experiment period. The artificial drought stress device frame was made of $200\times100\times50$ cm^3 aluminum alloy, the top of which was covered with transparent glass as other 4 sides of the device usually was open. These sides were covered only in this case that it was rainy. In this way, it could not only simulate natural environment as much as possible to let wind and light in freedom but also prevent the interference caused by rainfall. Each treatment had three repeats and the first day of the experiment was chosen as control. Relative photosynthetic indicators were recorded respectively on the 1st, 4th, 7th, 10th, 13th and 16th day of drought stress experiment.

2.2.2. Setting and determination of photosynthetic gas exchange parameters. The second position leaves were chosen as experimental materials in each treatment, and photosynthetic characteristics like *Pn*, *Gs*, intercellular CO_2 concentration (*Ci*), and *Tr* were determined via LI-6400 portable photosynthesis analyzer (LI-COR, USA). The parameters of the analyzer in this experiment are as follows: red and blue light inside the photosynthetic system was used to provide continuous and stable illumination; leaf chamber was 2×3 cm in size and light intensity was 1500 $\mu mol{\cdot}m^{-2}s^{-1}$; flow rate of sample chamber was 500 $\mu mol{\cdot}s^{-1}$. During determination, the atmospheric CO_2 concentration was about 400 $\mu mol{\cdot}mol^{-1}$, and the test time was 8:30-11:30 AM with the experimental temperature controlled at 25~27 ℃.

2.3. Data processing method

Use SPSS 17.0 (Statistical program for social sciences) and Microsoft Excel 2003 to analyze statics and map. *WUE= Pn / Tr, LUE= Pn / Light intensity* to calculate *WUE* and *LUE* values [11]. And the data in this paper was mean values which were derived from three repeats.

3. Results

3.1. Effects of drought stress on gazania Pn

Compared with the 1st day experimental values, 'Xingbai', 'XH' and 'Hongwen' *Pn* showed the trend of decreasing gradually with the prolonged drought stress (Figure 1), and *Pn* values decreased significantly from the days 10 to 13. Under the same treatment days, the values of 'XH' from days 1 to 10 were lower than its parents 'Xingbai' and 'Hongwen', while the values of 'XH' from days 10 to 16 were between its parents; the values of 'Hongwen' from days 1 to 10 were higher than 'Xingbai' and 'XH', while it decreased greatly than 'Xingbai' and 'XH' from days 10 to 13 and the value was 10.51 $\mu mol{\cdot}m^{-2}{\cdot}s^{-1}$; 'Xingbai', 'XH' and 'Hongwen' decreased 14.31 $\mu mol{\cdot}m^{-2}{\cdot}s^{-1}$, 13.38 $\mu mol{\cdot}m^{-2}{\cdot}s^{-1}$, 17.02 $\mu mol{\cdot}m^{-2}{\cdot}s^{-1}$ respectively compared with the 1st day of experiment.

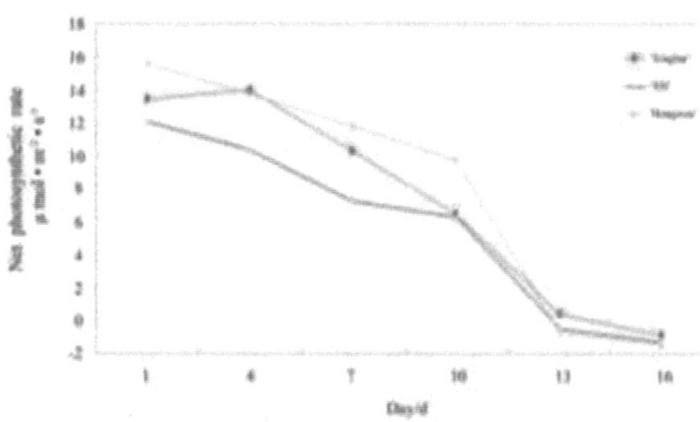

Figure 1. Effects of drought stress on gazania net photosynthetic rate.

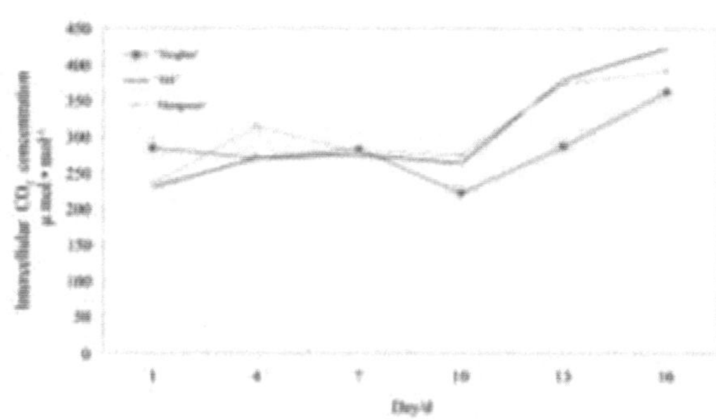

Figure 2. Effects of drought stress on gazania intercellular CO2 concentration.

3.2. Effects of drought stress on gazania Gs and Tr

With the prolonged drought stress, 'Xingbai', 'XH' and 'Hongwen' *Gs* showed the trend of decreasing gradually (Table 1). Under the same treatment circumstance, the values of 'XH' were between its parents' from days 4 to 13, and it didn't reach to significant level; the values of 'XH' were a slightly higher than its parents on the 16th day treatment; the three cultivars values reached to the lowest point on the 16th day treatment. What's more, 'Xingbai', 'XH' and 'Hongwen' decreased 185.45 $mmol \cdot m^{-2} \cdot s^{-1}$, 135.61 $mmol \cdot m^{-2} \cdot s^{-1}$, 258.85 $mmol \cdot m^{-2} \cdot s^{-1}$ respectively compared with the 1st day value. However, the changes of three cultivars' *Tr* values were not the same. 'Xingbai' *Tr* showed the trend of decreased gradually while values of 'XH' and 'Hongwen' decreased firstly, then increased and decreased finally once again. Compared with the 1st day value, treatment values of the 16th day of the three cultivars reached to the lowest point, 'Xingbai', 'XH' and 'Hongwen' decreased 2.29 $mmol \cdot m^{-2} \cdot s^{-1}$, 2.17 $mmol \cdot m^{-2} \cdot s^{-1}$, 3.63 $mmol \cdot m^{-2} \cdot s^{-1}$ respectively. It was obvious that the values of 'XH' were between or lower than its parents' under the same treatment.

Table 1. Stomatal conductance and transpiration rate of gazania under drought stress.

Treatment	*Gs* ($mmol \cdot m^{-2} \cdot s^{-1}$)			*Tr* ($mmol \cdot m^{-2} \cdot s^{-1}$)		
Days	'Xingbai'	'XH'	'Hongwen'	'Xingbai'	'XH'	'Hongwen'
1 d	214.22a	171.67a	272.96a	3.00a	2.84a	3.92a
4 d	133.46b	134.06b	224.07b	1.94b	1.71c	2.65c
7 d	97.14c	112.28b	173.95c	1.46bc	1.32e	2.45d
10 d	88.42d	114.11b	144.16d	1.34bc	2.08b	3.23b
13 d	64.06e	79.03c	123.58e	1.13bc	1.63d	2.26e
16 d	28.77f	36.06d	14.11f	0.71c	0.67f	0.29f

3.3. Effects of drought stress on gazania Ci

The changes of 'Xingbai' *Ci* was complex under drought stress, but the three cultivars showed the trend of increased firstly and then decreased, finally increased once again in the whole (Figure 2). The values of 16th day treatment were significantly higher than others' and values of 'Xingbai', 'XH' and 'Hongwen' increased 77.87 $\mu mol \cdot mol^{-1}$, 192.77 $\mu mol \cdot mol^{-1}$, 156.26 $\mu mol \cdot mol^{-1}$ respectively compared with the 1st day treatment. The values of 'XH' *Ci* were between or lower than its parents' under the same treatment from days 1 to 10, while 'XH' values were higher than its parents from treatment days 10 to 16.

3.4. Effects of drought stress on gazania WUE

'Xingbai', 'XH' and 'Hongwen' *WUE* values increased first and then decreased with the prolonged drought stress (Figure 3). The values of three cultivars reached to the highest point under drought stress for four days. 'Xingbai', 'XH' and 'Hongwen' decreased 4.91 $\mu mol \cdot mmol^{-1}$, 6.31 $\mu mol \cdot mmol^{-1}$, 8.93 $\mu mol \cdot mmol^{-1}$ respectively compared with the 1st day treatment. The values of 'XH' were between its parents' under the same treatment.

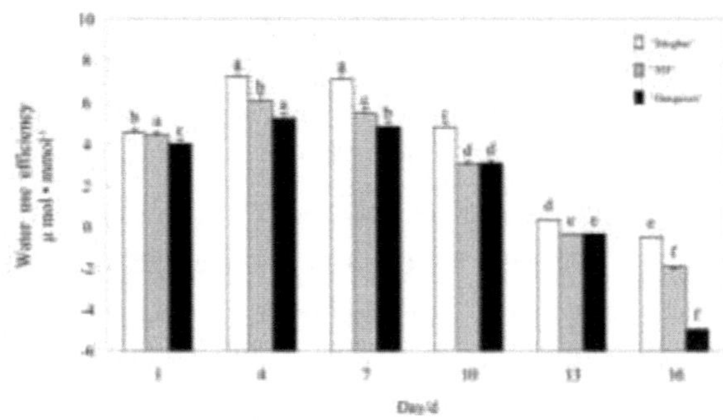

Figure 3. Effects of drought stress on gazania water use efficiency.

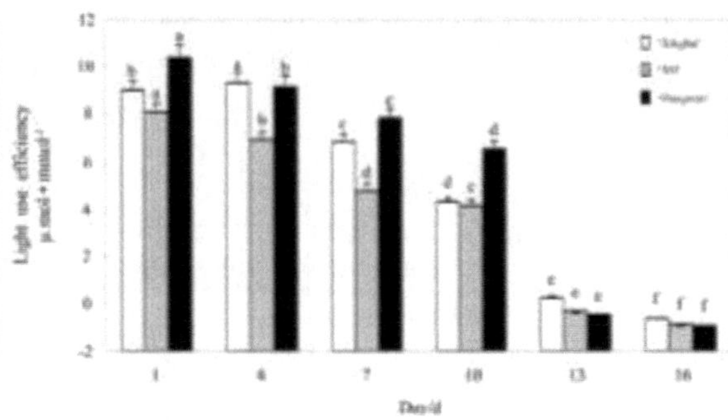

Figure 4. Effects of drought stress on gazania light use efficiency.

3.5. Effects of drought stress on gazania LUE

With the prolonged drought stress, 'Xingbai' *LUE* value increased on the day 4th treatment when the other three cultivars decreased gradually in general (Figure 4). 'Xingbai', 'XH' and 'Hongwen' decreased 9.54 μmol·mmol^{-1}, 8.92 μmol·mmol^{-1}, 11.35 μmol·mmol^{-1} respectively compared with the 1st day treatment. The values of 'XH' were between or lower than its parents under the same treatment. The three cultivars *LUE* values decreased significantly than others' and they decreased 4.00 μmol·mmol^{-1}, 3.84 μmol·mmol^{-1}, 6.01 μmol·mmol^{-1} respectively from days 10 to 13.

3.6. Correlationships of gazania photosynthetic indicators under drought stress

There existed a certain correlations among gazania photosynthetic indicators under drought stress (Table 2). The three cultivars *Pn, Gs, Tr, WUE, LUE* were extremely negative significance with drought days. There was no correlation between 'Xingbai' *Ci* and treatment days, while there existed extremely positive significance between 'XH', 'Hongwen' *Ci* and treatment days.

Table 2. Correlation of photosynthetic characteristics under drought stress.

		Day	*Pn*	*Gs*	*Ci*	*Tr*	*WUE*
'Xingbai'	*Pn*	-0.966**					
	Gs	-0.954**	0.853**				
	Ci	0.448	-0.480*	-0.375			
	Tr	-0.828**	0.729**	0.880**	-0.292		
	WUE	-0.776**	0.897**	0.567*	-0.612**	0.459	
	LUE	-0.966**	1.000**	0.853**	-0.480*	0.729**	0.897**
'XH'	*Pn*	-0.972**					
	Gs	-0.940**	0.924**				
	Ci	0.898**	-0.948**	-0.919**			
	Tr	-0.754**	0.725**	0.853**	-0.789**		
	WUE	-0.877**	0.915**	0.814**	-0.900**	0.514*	
	LUE	-0.972**	1.000**	0.924**	-0.948**	0.725**	0.915**
'Hongwen'	*Pn*	-0.944**					
	Gs	-0.960**	0.880**				

IOP Conf. Series: Earth and Environmental Science **41** (2016) 012027 doi:10.1088/1755-1315/41/1/012027

Ci	0.839**	-0.910**	-0.807**			
Tr	-0.810**	0.788**	0.882**	-0.868**		
WUE	-0.859**	0.909**	0.873**	-0.815**	0.825**	
LUE	-0.944**	1.000**	0.880**	-0.910**	0.788**	0.909**

** significant differences at P< 0.01 level; * significant differences at P< 0.05 level

4. Discussions

Drought is the main abiotic stress factors limiting plant growth and high recuperative capacity of plants from drought damages is critical for plant survival during drought stress. The impact of drought on the plants photosynthesis and photosynthetic organs will strengthen with the increasing extent of drought stress. Drought stress reduce plant's photosynthesis, these results are consistent with it [12].

The three cultivars *Pn*, *Gs* and *LUE* decreased gradually while there was a short increasing time for *Ci, Tr* and *WUE*. The changes of *Pn* values were mainly influenced by two factors: stomata limit caused by partial stomata closing, and non-stomata limit caused by photosynthetic activity decreasing in mesophyll cells [13]. *Pn* value decreased while *Ci* value increased gradually from days 1 to 7, which indicated the reason that why *Pn* value was decreased was not the change of *Gs*, because the changed *Gs* would lead CO_2 concentration to decrease. It was that non-stomata limit, such as increasing resistance of stomatal diffusion in mesophyll cells, decreasing of CO_2 solubility, reducing affinity of Rubisco enzyme on CO_2, decreasing of RuBP regeneration or decreasing of key components stability for photosynthetic apparatus, hindering the usage of CO_2 caused the accumulation of intracellular CO_2 [14]. *Pn* value decreased and *Gs*, *Ci* decreased as well from treatment days 7 to 10, which indicated that the decrease of *Gs* limited the entrance of CO_2 - stomata limit. *Ci* increased gradually from days 10 to 16, which manifested that the main reason causing *Pn* decreasing was non-stomata limit. Therefore, gazania *Pn* value is mainly influenced by stomata and non-stomata limits under drought stress. *WUE* is a stable indicator to evaluate the relation between carbon fixation and water consumption [15], and it can show water usage level meanwhile it limits the adaptability to stress [16].

Three cultivars *Tr* decreased gradually in general with the prolonged drought stress, which was the same as *Gs* changed. It is probably because that the decrease of *Gs*, *Tr* reduce water loss in plant body. *WUE* is determined by *Pn* and *Tr*. Moreover, *WUE* value increased first in this experiment from treatment days 1 to 4. *WUE* value showed the trend of decreasing gradually with the prolonged drought stress and this might be a kind of adaptability to drought stress. Meanwhile, the biggest value of *WUE* occurred on the 4th day treatment, which revealed that it is not necessarily the highest water use efficiency occurs in high field water-holding capacity condition. Moderate water deficit will improve the plant water utilization rate.

5. Conclusion

There exists extremely significant changes between *Pn*, *Gs*, *Tr*, *WUE*, *LUE* and treatment days, so it can be an indicator to evaluate the photosynthetic characteristics response of gazania to drought stress. According to the values and the general trends of different measured variables under drought stress, we can select cultivar(s) with strong resistance and high ornamental value to enrich the landscape architecture.

References

[1] Goldblatt P, Manning J and Bond P 2000 *Cape plants:a conspectus of the Cape flora of South Africa* (Pretoria: National Botanical Institute)

[2] Wang N N, Hu Z H and Shen Y B 2011 *Acta Botanica Boreali-Occidentalia Sinica* **31** 0101

[3] Liu J, Yang J Y, Zhang P J, Song X M and Zhang L J 2011 *Chinese J. Grassland* **33** 75

[4] You Y, Sun Y B, Lv F B, Zhu G F and Cao J X 2011 *Chinese J. of Trop. Crops* **32** 1245

[5] Niu S L, Jiang G M, Gao L M, Li Y G and Liu M Z 2003 *Acta phytoecological scinica* **27** 318

[6] Fan S L, Yuan Z H, Feng L J, Wang X H, Ding X M, Zheng H L and Jia X H 2011 *Chinese Agric. Sci. Bull.* **27** 119

[7] Wang X Y 2011 *Response of Flaveria bidentis to Soil Water Stress* (Hebei China: Agric. Univ. of Hebei Press) p 04

[8] Shi L R, Wang Y P, Liu G R and Li M Z 2011 *J. of Henan Agric. Sci.* **40** 119

[9] Zhou X H, Wang W, Dai B, Wu Y Q and Lu X P 2013 *Northern Hort.* **09** 67

[10] Li B, Zhan Y G, Xu Y G, Zeng F S, Zhang G Q 2012 *Acta Botanica Boreali-Occidentalia Sinica* **32** 2312

[11] Chen K, Hu G Q , Keutgen N, Janssens M J and Lenz F 1999 *Scientia Hort.* **81** 43

[12] Cao H, Xu X F, Han Z H, Wang X W and Guo T Q 2004 *Acta Hort. Sinica* **31** 285

[13] Jones H G 1998 *J. Exp. Bot.* **49** 378

[14] Allen D J and Ort D R 2001 *Trends Plant Sci.* **6** 36

[15] Chen T X, Hao W J, Chen X B and Lu X S 2009 *Chinese J. Grassland* **31** 41

[16] Ray I M, Townsend M S and Muncy C M 1999 *Crop Sci.* **39** 1088

IOP Conf. Series: Earth and Environmental Science **41** (2016) 012028 doi:10.1088/1755-1315/41/1/012028

Nutrient removal of a floating plant system receiving low-pollution wastewater: Effects of plant species and influent concentration

J J Duan[1], J N Zhao[1], L H Xue[1] and L Z Yang[1, 2]

[1]Institute of Agricultural Resources and Environment, Jiangsu Academy of Agricultural Sciences, Nanjing 210014, China

Email: lzyang@issas.ac.cn

Abstract. Plant floating bed was adopted in this study to compare the purification effect of four plant species (*Oenanthe javanica*, *Ipomoea aquatica*, *Hydrocotyle vulgaris*, and *Iris sibirica*) receiving high and low treated domestic sewage. The experiment was conducted for eight months during the low temperature season. The results indicated that the average removal rates of TN and NH_4^+-N in *I. aquatica* floating bed were relatively high both under high and low influent concentration during the first stage of the experiment. During the second stage, *H. vulgaris* showed the best performance for nitrogen treatment, and the average removal rates of TN were 70.7% and 87.7% under high and low influent concentration, while the average removal rates of NH_4^+-N were as high as 98.9% and 98.9%, accordingly. Moreover, *H. vulgaris* contributed most for plant assimilation to nitrogen removal among different plant floating systems. It was also found that the existence of hydrophytes effectively controlled the rise of water pH value and algae growth and reproduction, which helped to improve the aquatic environment. The results provide engineering parameters for the future design of an ecological remediation technology for low-pollution wastewater purification.

1. Introduction

As the rapid development of the economy, the problem of agricultural non-point pollution is getting increasingly serious. Runoffs from the farmland, containing nitrogen, phosphorus and other nutrients, go into ditches firstly. Ditches are the bond of farmland and rivers or lakes. In order to reduce the risk of farmland nutrients into water bodies, Yang set forward an innovative ecological ditch including a variety of plants planted in the side wall and bottom, and other auxiliary engineering facilities like interception dam and interception box [1]. The promising ecological ditch could promote the sedimentation of particulate materials and intercept nitrogen and phosphorus in drainage, which is an effective ecological treatment technology for the control of non-point pollution [1, 2]. However, the stereoscopic space of the ecological ditch was not fully utilized, and the upper space of it could be used efficiently.

Floating beds were often used on the surface of water for polluted river water, eutrophic water and domestic wastewater treatment [3–5]. It can be used in the ecological ditch to purify the upper space of the drainage. Little hydrophytes, such as *Oenanthe javanica*, *Ipomoea aquatica*, *Iris sibirica*, were appropriate to grow on floating bed system.

Floating bed systems exhibit high removal rates of nitrogen and phosphorus during the growing season of the plant [6, 7]. However, when cold weather arrives, the performance of floating beds is often getting worse because of the weakness of the macrophytes. Few macrophytes, like *O. javanica* and *I. sibirica*, could grow well in cold weather, and had purification effect on polluted river water [8, 9]. But the performance of these plants for nutrient removal at different influent concentrations was not fully studied.

In the present study, we investigated the purification effect of four plant species at two influent concentrations.

2. Materials and methods

2.1. Plant material

O. javanica, I. aquatica, H. vulgaris and *I. sibirica* were selected as the plant materials which were commonly planted in the Yangtze River Delta. In this study, hydrophytes except *H. vulgaris* were got from a garden centre in Suqian, Jiangsu province, China; *H. vulgaris* was obtained from a pond around Tai Lake in Yixing, Wuxi city, China. The plantlets were washed with distilled water and preincubated in experimental water for about 3 weeks. When the plantlets were about 16 cm high, uniform and healthy ones were chosen for the experiment.

2.2. Experimental water

In this study, treated domestic wastewater (DW), which was pumped from a septic tank in the Jiangsu Academy of Agricultural Science (Nanjing, China), was used as the experimental water. After ultraviolet sterilizing, DW was diluted to high or low concentration by tap water before use (namely, H groups and L groups, respectively). The wastewater used in the experiment was a kind of low-pollution wastewater (LPW), which met the Chinese Farm Irrigation Water Standard [10]. The characteristics of the experimental water were exhibited in Table 1.

Table 1. Characteristics of the domestic wastewater during the experiment.

Treatment	TN ($mg•L^{-1}$)	NH_4^+-N ($mg•L^{-1}$)	NO_3^--N ($mg•L^{-1}$)	TP ($mg•L^{-1}$)	COD[a] ($mg•L^{-1}$)	BOD_5[b] ($mg•L^{-1}$)	FC[c] /100 mL	C:N[d]
H groups	20.25 ±7.23	17.23 ±5.95	1.19 ±1.49	1.62 ±0.89	40.0 ±17.3	14.19 ±1.34	4900	2.0
L groups	10.90 ±6.18	9.90 ±7.00	1.15 ±0.79	0.94 ±0.63	27.4 ±10.8	10.74 ±1.14	3400	2.5

Note: H groups and L groups represented treatment groups receiving domestic wastewater of high and low concentration.
[a] Chemical oxygen demand
[b] 5-day biochemical oxygen demand
[c] Fecal Coliform
[d] C and N were measured as COD and TN in C:N ratio

2.3. Design of the floating plant system

Floating bed was used to receive DW, which was made of a 20 mm diameter PVC pipe (65.0 cm length × 50.0 cm width) with grilling (4.0 cm × 4.0 cm) binding on it. Plastic tanks (internal dimensions of 74.8 cm × 52.0 cm × 48.5 cm) were used as the container.

O. javanica, I. aquatica, and *I. sibirica* were planted on the floating bed receiving high and low concentration LPW to form six treatment groups (H1, H2, H3, L1, L2, and L3); Control groups were set receiving the two kinds of LPW (Control 1 and Control 2). Each treatment had three replicates. *I. aquatica* was harvested at the end of December, and then *H. vulgaris* was planted in H2 from January to May. Other treatments remained unchanged.

Apart from *H. vulgaris*, hydrophyte seedlings were planted into the nutritive cups on the holes of the grilling and were held stably with a sponge. *H. vulgaris* of a considerable weight with other hydrophytes, was placed directly on the surface of the floating bed (Figure 1). Aerator was added in each tanks to provide enough dissolved oxygen to the system. The plastic tanks were filed with 145 L of DW at the beginning of the experiment. The experiment lasted from October 2, 2014, to May 9, 2015.

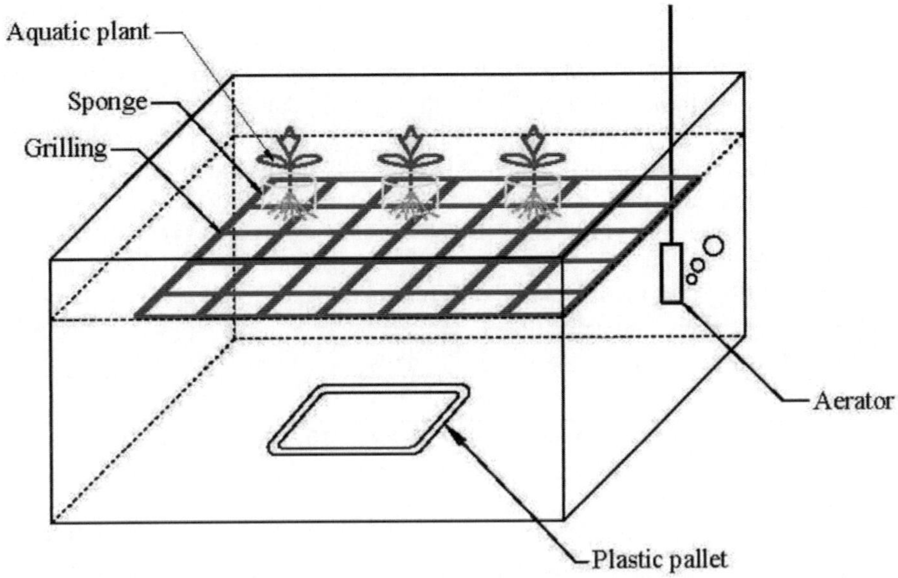

Figure 1. Composition of the floating plant system used in the present study.

2.4. Sampling and analysis

About 50 mL of water samples in the plastic tanks were collected every week at 9:00 am, stored in a refrigerator of 4℃, and measured immediately. Flow injection analysis was used to analyse TN, TP, NH_4^+-N, and NO_3^--N on a Skalar San++ System (Skalar Co., The Netherlands). 2 mL of water samples was added into a tube before measuring Chemical oxygen demand, and COD analyser (DR1010 COD, HACH, China) was used. Temperature, pH, and dissolved oxygen (DO) were monitored using a multi-parameter water quality monitor (HI9828100, HANNA) before water sampling. Chlorophyll a in water was measured with ethanol-non-grinding method [11].

Plant height and biomass were measured before and after the experiment. The relative growth rate (*RGR*) of the plant was calculated based on W_{ti} and W_{to}, defined as:

$$RGR\ (\%) = 100 \times (W_{ti} - W_{t0}) / W_{t0} \qquad (1)$$

where W_{ti} is the average fresh weight of plants on day i of the experiment and W_{t0} is the initial average fresh weight of the plants.

At the end of the study, plants were harvested and removed from the floating plant system. Plant samples were prepared and analysed for N concentration via the Kjeldahl method [12]. N accumulations were estimated by multiplying the N concentration by the plant biomass [13].

Nitrate, free amino acid, and soluble sugar of the leaves were analysed according to standard methods [14].

2.5. Statistical analysis

One-way analysis of variance (ANOVA) was performed to test the effects of different hydrophytes and DW concentrations on plant growth situation and nitrogen removal. Statistical analyses were conducted with SPSS13.0. Differences were considered statistically significant when $p < 0.05$.

3. Results and discussion

3.1. Changes in water quality

In the present study, pH values of plant floating beds were lower than in control groups, and pH values in H groups were lower than in L groups during the growth period of *I. aquatica* and *H. vulgaris* (Table 2). DO concentrations were higher during the growth period of *I. aquatica*. Except for H1, DO concentrations in other treatments met the grade one of the *Surface Water Quality Standard* (GB 3838-2002). The relatively low DO values in plant floating beds than control groups might ascribed to plant's residues consuming oxygen through decomposition in low temperature season.

As one of the water quality indexes, DO concentration reflexes the pollution degree of water bodies and physical, chemical and biological characteristics [5]. Hydrophytes can transport O_2 to the root through the internal aerenchyma, and providing oxygen to the underground part [15]. In addition, biofilm can be formed on the surface of the plant roots, providing enough oxygen to heterotrophic bacteria and nitrobacteria [16].

Table 2. Physical parameters in different treatment groups.

Growth stages	Treatment group	pH	Water temperature(℃)	DO(mg•L^{-1})
the growth period of *I. aquatica*	H1	7.84 (6.90-9.71)	16.72 (7.70-23.60)	6.92 (0.13-13.80)
	H2	7.96 (6.86-9.70)	16.56 (7.40-22.70)	7.57 (0.10-13.56)
	H3	8.21 (6.57-9.64)	16.72 (7.70-23.20)	8.27 (0.08-16.6)
	Control 1	8.40 (6.92-10.39)	16.66 (7.80-23.10)	9.12 (0.04-16.00)
	L1	8.57 (7.26-9.83)	17.42 (7.70-23.20)	9.06 (0.83-14.50)
	L2	8.53 (7.19-10.33)	16.96 (7.80-23.3)	9.65 (1.32-1.95)
	L3	8.54 (7.33-9.72)	16.98 (7.70-23.20)	9.36 (1.78-15.30)
	Control 2	9.02 (7.26-10.76)	17.07 (8.20-22.80)	12.66 (2.15-22.42)
the growth period of *H. vulgaris*	H1	7.48 (6.86-8.55)	15.12 (6.10-23.28)	3.95 (0.13-12.37)
	H2	7.63 (6.45-8.97)	14.91 (5.80-22.90)	4.28 (3.18-13.47)
	H3	8.23 (7.09-9.50)	15.15 (6.10-24.17)	4.38 (2.03-13.29)
	Control 1	9.04 (7.41-11.27)	15.06 (6.00-25.80)	4.72 (0.55-21.23)
	L1	8.69 (6.99-10.70)	15.49 (6.10-23.06)	4.51 (0.79-15.61)
	L2	8.10 (6.42-10.32)	15.22 (6.10-23.32)	4.61 (1.46-15.76)
	L3	8.42 (7.22-10.15)	15.41 (6.10-24.08)	4.47 (0.77-15.89)
	Control 2	10.11 (7.29-11.70)	15.42 (6.10-25.80)	5.96 (1.80-26.87)

Note: Data in the above table are average value (minimal value-maximal value).

The nitrogen removal of plant floating beds did not have obvious advantage compared to control groups during the growth period of I. aquatica(Figure 2-4). The average removal rates of TN and NH_4^+-N in I. aquatica floating bed were relatively high both under high and low concentration of LPW. With the raise of the temperature in spring, the removal efficiency of Nitrogen increased in plant floating beds (Figure 5-7). Among the hydrophytes in the experiment, H. vulgaris showed the best performance. The average removal rates of TN were 70.7% and 87.7% for high and low LPW treatment in H. vulgaris floating beds, while the average removal rates of NH_4^+-N were as high as 98.9% and 98.9%, accordingly.

The chlorophyll a concentrations in plant floating beds were significantly lower than in control groups (Table 8), showing higher algae growth without hydrophytes in the low pollution aquatic water.

Bu and Xu reported that *Canna indica* and *Accords calamus* and other two floating bed plants all had the algae inhibiting effect, which confirmed to our study [5].

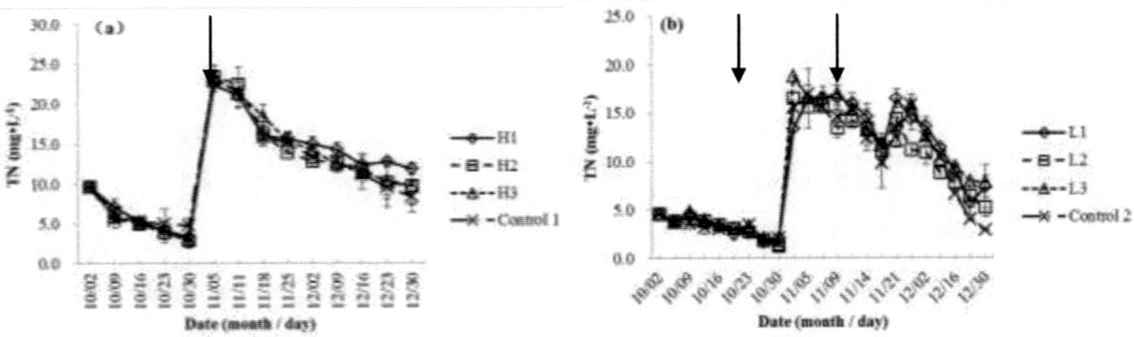

Figure 2. Changes of TN concentration in H groups (a) and L groups during the growth of *I. aquatica*.

Note: arrows represent water exchanging. The same below.

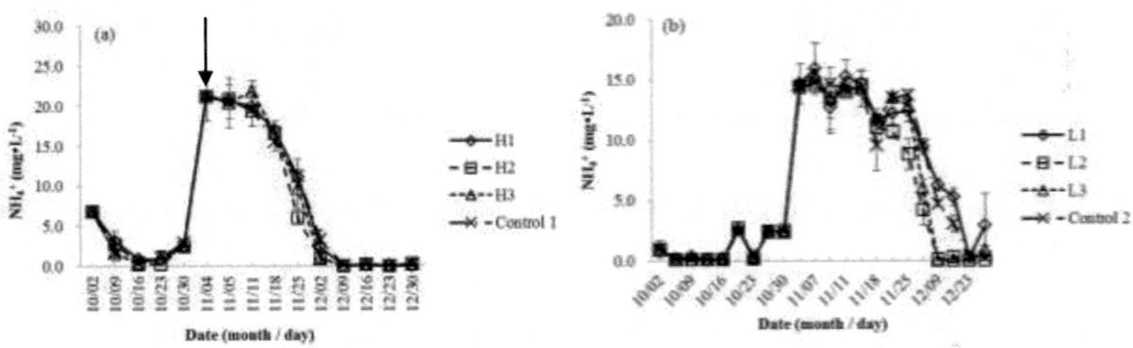

Figure 3. Changes of NH_4^+-N concentration in H groups (a) and L groups during the growth of *I. aquatica.*

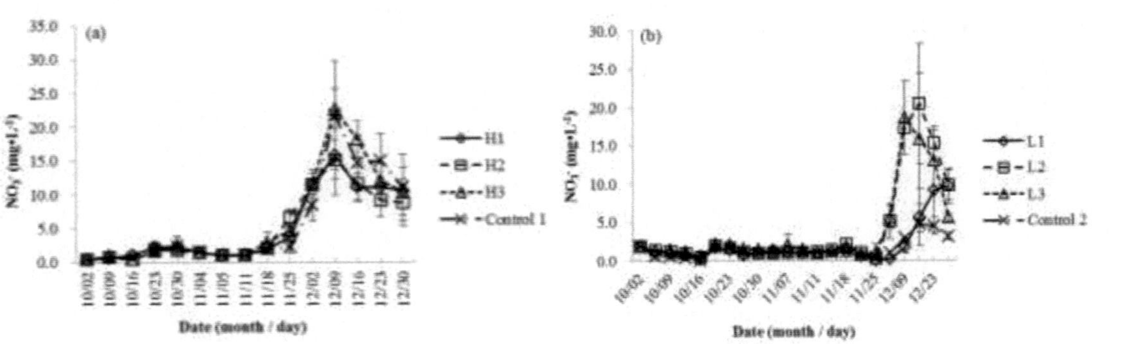

Figure 4. Changes of NO_3^--N concentration in H groups (a) and L groups during the growth of *I. aquatica.*

IOP Conf. Series: Earth and Environmental Science **41** (2016) 012028 doi:10.1088/1755-1315/41/1/012028

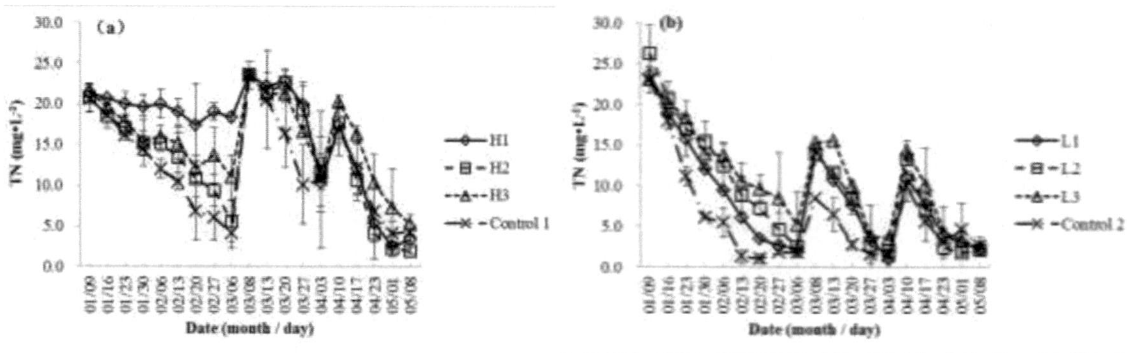

Figure 5. Changes of TN concentration in H groups (a) and L groups during the growth of *H. vulgaris*.

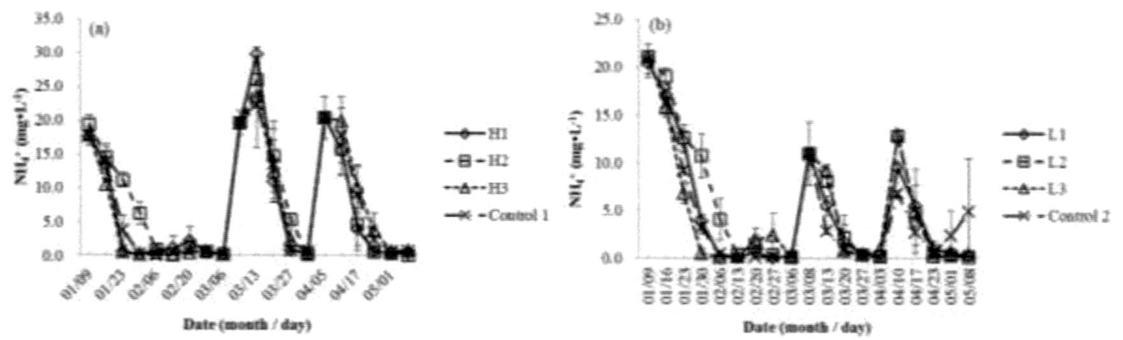

Figure 6. Changes of NH_4^+-N concentration in H groups (a) and L groups during the growth of *H. vulgaris*.

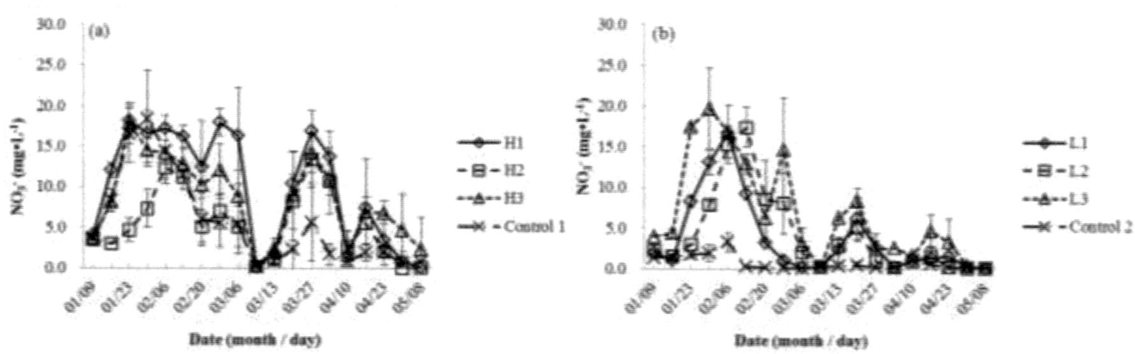

Figure 7. Changes of NO_3^--N concentration in H groups (a) and L groups during the growth of *H. vulgaris*.

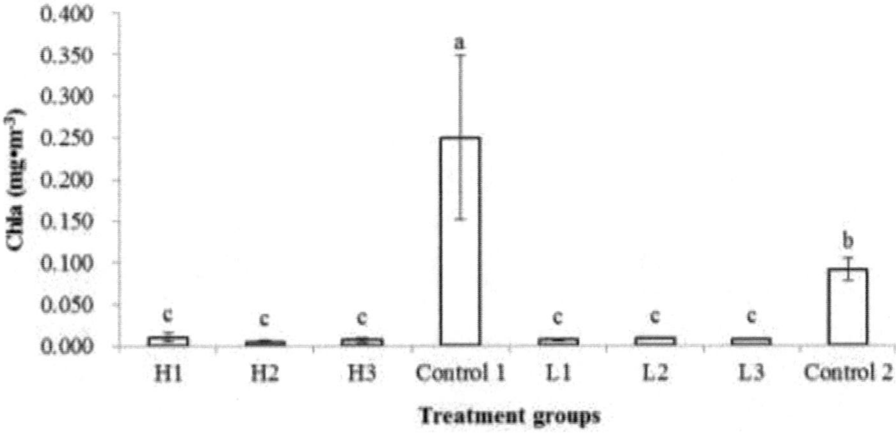

Figure 8. The concentration of chlorophyll a in the outflow of each treatment group.

3.2. Plant growth and nutrient accumulation

Although the biomass and average plant height of *I. sibirica* were larger, the same indexes were not the highest among the four plants at harvest. The biomass of *H. vulgaris* was the highest under high or low concentration of LPW treatments, and *O. javanica* and *I. aquatica* were larger in H treatment groups. At the same time, the RGR of *H. vulgaris* was the highest under both of the influent concentration; and the average root length of *I. sibirica* was the highest (Table 3). TN assimilation of *H. vulgaris* was the highest of all during the experiment, and the ratio of plant assimilation to TN input were 60.1% and 108.0% in H2 and L2 groups during the growth of *H. vulgaris*, respectively (Tabel 4).

As a edible plant, *I. aquatica* favorites high temperature and humid environment, and distributed widely in China and India [17]. *I. aquatica* has a high economic value, compared with other hydrophytes. It has been used for heavy metals and organic pollutants adsorption, and agricultural wastewater treatment [17-19]. In the present study, the removal rates of TN and NH_4^+-N in *I. aquatica* floating bed was relatively high receiving high and low LPW. Free amino acids and soluble sugar were higher in *I. aquatica* than in *O. javanica* both in H groups and L groups at harvest by quality determination (Table 5). The NO_3^--N in stem and leaf were 296.7 and 173.8 mg·kg^{-1} in *O. javanica*, and the corresponding values were 188.0 and 187.3 mg·kg^{-1} in *I. aquatica.* The NO_3^--N concentrations were at a low level among leafy vegetables which can be safe to eat [20, 21].

Table 3. Growth situation of plants.

Treatment groups	Plants	Initial planting		Harvesting			
		Biomass (g)	Average plant height (cm)	Biomass (g)	Average plant height (cm)	Average root length (cm)	Relative growth rate RGR (%)
H 1	*O. javanica*	7.0±0.2	15.5±1.4	153.0±27.3	64.2±9.9	17.2±5.2	2085.7
H 2	*I. aquatica*	7.7±0.2	17.1±2.1	44.3±7.4	18.9±4.2	17.3±4.8	475.3
	H. vulgaris	6.5±0.0	3.5±3.1	398.3±35.5	51.1±7.5	13.5±2.3	6027.7
H 3	*I. sibirica*	21.5±0.3	29.3±5.8	89.0±19.4	30.4±4.7	21.4±7.0	314.0
L 1	*O. javanica*	7.0±0.1	16.3±1.4	48.1±9.4	30.2±15.5	20.6±4.5	587.1
L 2	*I. aquatica*	7.9±0.0	17.9±1.8	27.5±0.7	15.4±2.4	13.1±4.5	248.1
	H. vulgaris	6.5±0.0	4.7±5.1	353.5±31.2	52.9±5.9	13.4±2.3	5338.5
L3	*I. sibirica*	22.5±1.2	27.0±8.3	98.7±22.3	30.4±5.7	23.6±4.7	338.7

Table 4. Contribution of plant assimilation to TN removal in different plant floating system.

Treatment group	Input amouts of TN ($g•m^{-2}$)	Removal amounts of TN ($g•m^{-2}$)	TN assimilation ($g•m^{-2}$)	Ratio of assimilation to input (%)	Ratio of assimilation to removal (%)
H 1	50.27	27.08	9.85	19.6	36.4
H 2-*I. aquatica*	16.32	10.14	2.81	17.20	27.68
H 2-*H. vulgaris*	33.96	24.50	23.45	69.05	95.70
H 3	50.27	35.15	3.41	6.8	9.7
L 1	37.39	22.72	2.70	7.23	11.9
L 2-*I. aquatica*	18.70	8.11	1.17	6.25	14.40
L 2-*H. vulgaris*	18.69	15.45	20.18	107.98	130.61
L 3	37.39	19.51	3.17	8.5	16.3

Table 5. Quality determination of *O. javanica* and *I. aquatica.*

Treatment group	Plant	Tissue	NO_3^--N ($mg•kg^{-1}$)	Free amino acids ($mg•kg^{-1}$)	Soluble sugar (%)
H 1	*O. javanica*	leaf	173.8±57.1	322.7±17.1	1.56±0.63
		stem	296.7±153.6	324.1±116.5	0.94±0.69
L 1	*O. javanica*	leaf	-	294.8±46.6	1.30±0.35
		stem	-	280.5±67.7	1.20±0.45
H 2	*I. aquatica*	leaf	187.3±24.4	1060.0±210.2	2.61±1.44
		stem	188.0±53.4	1258.3±141.0	1.99±0.98
L 2	*I. aquatica*	leaf	-	-	-
		stem	70.1±10.8	1176.7±152.6	3.21±1.00

4. Conclusions

A floating Plant system was adopted to investigate purification effect of four plant species at high and low influent concentrations in this study. *O. javanica* and *I. aquatica* had both the nitrogen absorption effect and economic benefits, and were compared with common *H. vulgaris* and *I. sibirica*. Results showed that the average removal rates of TN and NH_4^+-N in *I. aquatica* floating bed were relatively high both under high and low concentration of LPW. During the growth of *H. vulgaris.*, *H. vulgaris* showed the best performance for nitrogen treatment, and nitrogen assimilation of it was the highest of all in the experiment. Quality determination illustrated that the NO_3^--N concentrations of stem and leaf in *O. javanica* and *I. aquatica* were at a low to medium level among leafy vegetables which can be safe to eat. It was also found that the existence of hydrophytes effectively controlled the rise of water pH value and algae growth and reproduction, which helped to improve the aquatic environment.

Acknowledgements

This work was financially supported by the Jiangsu Agriculture Science and Technology Innovation Fund (No. CX(15)1004), the Special Fund for Agro-Scientific Research in the Public Interest (No. 201503106)and the National Natural Science Funds of China (No. 41501324).

References

[1] Yang L, Zhou X, Wang J, Wang D, Shi W and Shi L 2005 Ecological ditch system with

interception function and its effects on controlling farmland non-point pollution *Chinese Journal of Ecology* **24** 1371

[2] Wang Y, Wang J, Li W, Bo L and Yang L 2010 Initial exploration of mechanism of ecological ditch intercepting nitrogen and phosphorus in drainage from farmland *Journal of Ecology and Rural Environment* **26** 586

[3] Zhang Z, Feng M and Yang L 2007 Nitrogen and phosphorus removal and N_2O emission from domestic sewage in floating-bed plant systems *Acta Ecologica Sinica* **27** 4333

[4] Guo Y, Liu Y, Zeng G, Hu X, Li X, Huang D, Liu Y and Yin Y 2014 A restoration-promoting integrated floating bed and its experimental performance in eutrophication remediation *J. Environ. Sci.* **26** 1090

[5] Bu F and Xu X 2013 Planted floating bed performance in treatment of eutrophic river water *Environ. Monit. Assess.* **185** 9651

[6] Zhu L, Li Z and Ketola T 2011 Biomass accumulations and nutrient uptake of plants cultivated on artificial floating beds in chinas rural area *Ecol. Eng.* **37** 1460

[7] Zhang Q, Achal V, Xu Y and Xiang W 2014 Aquaculture wastewater quality improvement by water spinach (Ipomoea aquatica Forsskal) floating bed and ecological benefit assessment in ecological agriculture district *Aquac. Eng.* **60** 48

[8] Xin Z, Li X, Yan Z, Zhou Y, Jia Y, Tang Y, Guo W and Sun Y 2011 Purification effects of different stubble height Oenanthe javanica (Blume) DC floating bed systems in winter in two successive experimental periods *Chinese J. Ecol.* **30** 2745

[9] Gao J, Wang W, Guo X and Zhu S 2014 Nutrient removal capability and growth characteristics of Iris sibirica in subsurface vertical flow constructed wetlands in winter *Ecol. Eng.* **70** 351

[10] Duan J, Feng Y, Yu Y, He S, Xue L and Yang L 2016 Differences in the treatment efficiency of a cold-resistant floating bed plant receiving two types of low-pollution wastewater *Environ. Monit. Assess.* **188** 283

[11] Yang Y, Xia M, Yang J, Wang T and Ma W 2011 The research of methods of measuring chlorophyll a in water *Environmental Monitoring in China* **27** 24

[12] Lu R 2000 *Soil agricultural chemical analysis method* (Beijing: China Agricultural Science and Technology Press) p 308

[13] Hu G, Zhou M, Hou H, Zhu X and Zhang W 2010 An ecological floating-bed made from dredged lake sludge for purification of eutrophic water *Ecol. Eng.* **36** 1448

[14] Zhang Z and Qu W 2003 *Experiment guide for plant physiology* (Beijing: Higher Education Press) p 118

[15] Tanner C 2001 Plants as ecosystem engineers in subsurface-flow treatment wetlands *Water Sci. Technol.* **44** 9

[16] Orth H, Lertpocasombut K and Wilderer P 1987 Wastewater treatment for industrial estates in Southeast Asia using water hyacinths *Water Sci. Technol.***19** 85

[17] Zhang Q, Achal V, Xu Y and Xiang W 2014 Aquaculture wastewater quality improvement by water spinach (Ipomoea aquatica Forsskal) floating bed and ecological benefit assessment in ecological agriculture district *Aquac. Eng.* **60** 48

[18] Yang U, Yoon S, Chung J, Kim Y, Park K, Park T and Shim S 2012 Water spinach (Ipomoea aquatic Forsk) reduced the absorption of heavy metals in an in vitro bio-mimicking model system *Food Chem. Toxicol.* **50** 3862

[19] Chen J, Wang K, Chen H, Lu C, Huang L, Li H, Peng T and Chang S 2010 Phytoremediation of Cr (III) by Ipomonea aquatica (water spinach) from water in the presence of EDTA and chloride: Effects of Cr speciation *Bioresour. Technol.* **101** 3033

[20] Shen M, Zhai B, Dong H and Li J 1982 Studies on nitrate accumulation in vegetable crops *Acta Horticulturae Sinica* **9** 41

[21] Wang C, Zong Z, Li S and Chen B 2002 Nitrate accumulation in vegetables and its residual in vegetable fields *Environmental Science* **23** 79

Influences of biochar addition on vegetable soil nitrogen balance and pH buffering capacity

Y Yu[1, 2], AO Odindo[2], L Xue[1] and L Yang[1,3]

[1]Institute of Agricultural Resources and Environment, Jiangsu Academy of Agricultural Sciences, Nanjing, Jiangsu, 210014, China

[2]University of KwaZulu-Natal, Scottsville, 3209, South Africa

E-mail: lzyang@issas.ac.cn

Abstract. Leaching is a major path for chemical nitrogen fertilizer loss from in vegetable soil, which would destroy soil pH buffering capacity soil and result in acidification. It has been a common phenomenon in Tai Lake Region, China. However, few study focused on the change soil pH buffering capacity, especially the effect of soil amendment on pH buffering capacity. In this study, a pot experiment was conducted to research the effects of biochar addition to a vegetable soil on nitrogen leaching and pH buffering capacity with pakchoi (*B.chinensis L.*) growth as the experimental crop. The results showed that biochar could significantly increase the pakchoi nitrogen utilization efficiency, decrease 48%-65% nitrogen loss from leaching under the urea continuous applied condition. Biochar also could effectively maintain the content of soil organic matter and base cations. Therefore, it rose up soil pH buffering capacity by 9.4%-36.8% and significantly slowed down acidification rate. It was suggested that 1%-2% addition ratio was recommended from this study when used as similar soil condition.

1. Introduction

Nitrogen leaching and soil acidification have been the most important threatens for vegetable system [1-2]. According to previous studies, utilization efficiency of present nitrogen fertilizer was only 33 % in northern China and 18 % in southern China [3-4], and the rest of them left in soil, released [H^+] in the process of nitrification and lost from leaching [5]. Additionally, base cations were inclined to loss with leachate contributed to further acidification happening [6]. Regulation nitrogen in vegetable soil will be the key to decrease leaching and mitigate soil acidification.

Biochar, a form of activated carbon, has been used experimentally to solve numerous soil problems in recent years. But studies have indicated that biochar will increase the risk of NO_3^--N leaching and [H^+] concentration by improving soil nitrification [7]. On the other hand, the liming effect of biochar may induce a neutral trend[8] and the adsorption may decrease available nitrogen in soil[9-10], which could change nitrogen transformation and utilization efficiency. The effects of biochar addition on soil leaching and pH buffering capacity would be determined by multi factors. The differences of biochar addition ratios, soil types, fertilizer rates all make the results complicate and various [11] when referring to soil leaching and pH buffering capacity.

We conducted the experiment with soil collected from vegetable field in Taihu Lake region. The field in that region had grown paddy for hundreds of years but some of them was used for vegetable growing in recent 10 year, which brought serious leaching loss and soil acidification due to increasing fertilizer input. It was set eight treatments with two nitrogen fertilizer rates and four biochar addition ratios to find the responds of nitrogen leaching on biochar addition, determine the effect of soil pH buffering capacity and nitrogen distributions in different treatments based on biochar recovery method. The results could be used as guide for vegetable soil amendment with biochar in Taihu Lake region.

2. Materials and methods

2.1. Soil, plant and biochar descriptions

Soil was collected from the top 0-20 cm in a vegetable field (31°17′N, 119°54′E) after harvest in Yixing County, Taihu lake region, Jiangsu Province, South-eastern China. This region has a sub-tropical monsoon climate with an average annual rainfall of 1,177 mm. The field was converted from a paddy field and had a 7-year vegetable plantation history, with 3-4 seasons of leafy vegetables per year. Urea and compound fertilizer were commonly applied at a total rate of 800 kg N/ha per year. The soil is paddy soil, with a pH (H2O) 5.65; an organic matter content 17.9 g/kg; total nitrogen concentrations of 1.87 g/kg, respectively; and mineral nitrogen, phosphorus and potassium concentrations78.8, 61.0 and 70.1 mg/kg, respectively.

Experiment was carried out with pakchoi (Brassica chinensis L.) as vegetable plant.Biochar (BC) was produced from rice straw using a biogas-energy slow pyrolysis system under oxygen-limited conditions. The temperature was increased to 450 °C then maintained for over 5 hours. The properties in this study are as follows: total nitrogen and carbon concentrations were 12.5 and 503.1 g/kg, respectively; pH 8.9; and BET 7.4 m^2/g.

2.2. Experiment design

Experiment was conducted with each pot (28 cm height, 15 cm width and 10cm depth) filled with 2 kg of air-dried soil. Eight treatments were included two nitrogen rates (110 kg N/kg and 0 kg N/kg) and four biochar addition ratios (0% w/w, 1% w/w, 2% w/w and 5% w/w) with six replicates. Chemical N fertilizer (110 mg N/kg) was added to pots in the U and U+BC treatments (P0+U, P0+U+BC, P+U and P+U+BC) at the 14th day following seed germination. P_2O_5 (85 mg /kg) and K_2O (60 mg /kg) were applied as base fertilizers to each pot at the beginning of each growing season.

Biochar was taken turn to wash with 0.5 mol/L HCl, 0.1mol/L NaOH solution and deionized water, and then dried to content weight at 80℃. Some of biochar divided and packed into nylon mesh bags (0.15 mm, 1 g/bag biochar) that allowed water, nutrients and microbes but not soil or biochar particles to pass through. Three biochar packages were buried into the soil in each pot and the rest was mixed with soil till the biochar weight was enough to the ratio of the treatments 15 days before planting starting. One gram of biochar was added to each pot at the end of each growing season to compensate the sampling loss.

Experiment lasted three growing seasons, from Sep. 2014- Apr. 2015 in greenhouse of Jiangsu Academy of Agricultural Sciences. Pak choi was sowed into soil. The day pakchoi germinating was set as the 1st day in each growing season. The quality of seedling was controlled at 30 per pot by thinning at the 7th day after germination. And it was harvest at the 35th day after germination.

2.3. Sampling and measurements

After pakchoi was harvested, the plant biomass was determined by weighing, samples were oven-dried at 70°C until a constant weight was achieved and then filtered through a 0.149-mm sieve the total N (TN) analysis (Flash EA 1112 series; Thermo Finnigan, Elk Grove Village, IL). Plant N accumulation (mg N per pot) = dry biomass of whole plant (g) × total N content (g/g).

One soil composite sample was collected from each pot by mixed five cores sampling. Mineral nitrogen concentrations (NH_4^+-N、NO_3^--N) , soil organic matter, exchangeable K, Na, Ca, Mg and Al were determined by methods from Soil Agricultural Chemical Analysis Method.

The pH buffer capacities were determined by titrating 50 g of each soil sample in sealable polyethylene bags with H_2SO_4 and $CaCO_3$ at rates of 0, 1×10^{-2}, 2×10^{-2}, 4×10^{-2}, 6×10^{-2} and 8×10^{-2}

mol/L mol (H^+ or $1/2CO_3^{-2}$)/L/kg. The rates of $CaCO_3$ were added to the soils as a suspension in distilled water [12]. After amendments were added, they were mixed with the soil. Distilled water was added to each soil/amendment mixture to reach approximately field capacity. The bags were sealed and stored at room temperature for 1 month. The soils were then air-dried and ground for determination of pH (1:2.5, soil:water). Regression analysis was used to determine the buffer capacities within the linear range of the titration curve: pHBC=1/a ×10, where, pHBC=pH buffer capacity of the soil at the end of experimental period (10-3 mol H+/kg/pH), 1=one unit pH range used in calculation, a=slope of linear.

The acidification rate is defined as the rate of acid addition that needs to be neutralized in order to maintain a constant pH: AR= ΔpH× pHBC× BD×Vol/ 10^6, where, AR=acidification rate (10^3 mol H^+ /ha), ΔpH=change in pH over the experimental period (pH unit), pHBC=pH buffer capacity of the soil at the end of experimental period (10^{-3}mol H+ $kg^{-1}pH^{-1}$), BD=bulk density of the soil (kg/m^3), Vol=volume of soil per unit area (m^3/ha), BD×Vol=2.4×10^6 kg/ha in the soil type.

2.4. Leaching

Over irrigation was utilized to simulate rainfall and to collect leachate, as well as to continuously record the volume. It was set to happen four times each growing season: before sowing, 10th, 20th, and 30th days after germination. The field capacity was 46%, and therefore irrigation water requirement was set as the differences between 50% w/w soil weight and current soil moisture content. Water was irrigated by several times to avoid flooding. Samples were collect six hours after irrigation, and nitrogen concentration and volume were measured for nitrogen leaching loss calculation.

2.5. Satistical analysis

Nitrogen distribution and leaching were showed by means from six replicates in three growing seasons, calculated in pots. However, data related to soil buffering capacity and base cations was acquired from samples collected at the end of experiment. Relative significance of different treatments was accomplished by multivariable analysis. Significant differences among means were determined by Duncan's multiple range tests at the 5% level. Figures were accomplished by Microsoft Excel 2010.

3. Results and discussion

3.1. Mineral nitrogen distribution under different biochar ratios

As showed in figure 1, 4%-22% mineral nitrogen from fertilizer was absorbed by biochar from soil, and the nitrogen content absorbed increased with the rate of biochar addition. Biochar addition improved pakchoi nitrogen accumulation by 95%, 101%, 71% at 1% biochar addition, 2% biochar addition, 5% biochar addition, respectively. Nitrogen residual rate reached 46% of fertilizer input when without biochar addition. All biochar addition could reduce nitrogen residual, but the decreased residual content was not in proportional to biochar addition rates.

3.2. Effect of biochar addition on soil nitrogen leaching

The biochar addition could reduce leachate volume for its porosity. More biochar addition showed better volume-reduce function. As showed in table 1, the reduction of leachate volume under 5% biochar addition was 35% - 60% in three growing season, the corresponding values were 28%-44% and 28%-44% under 1% biochar addition and 2% biochar addition, respectively.

The nitrogen concentration of leachate was 10-27 mg/L in treatments without chemical nitrogen fertilizer, and it increased to 72-100 mg/L in treatments after nitrogen fertilizer. Averagely, biochar addition decreased the nitrogen concentration of leachate by 33.3%. The reduction of nitrogen concentration of leachate showed no significant differences among different biochar addition ratios. Therefore, nitrogen leaching loss was reduced by 48.3%, 53.7% and 65.0% in 1%, 2% and 5% biochar addition treatments, respectively.

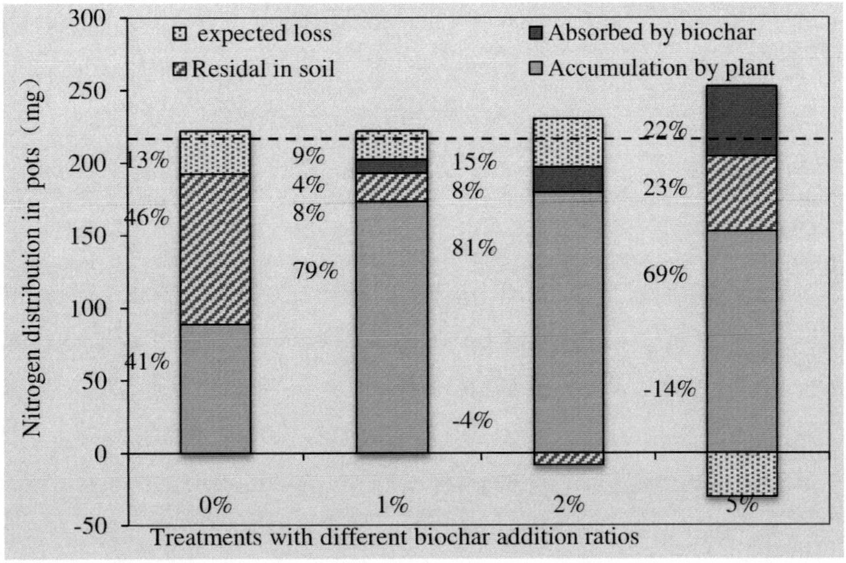

Figure 1. The mineral nitrogen distribution under different biochar ratios. Dotted line presented the nitrogen content (220 mg per pot) from chemical nitrogen fertilizer input.

Table 1. Effect of biochar addition on soil leaching.

Biochar addition ratios and treatments		Leachate volume (mL/season)	Nitrogen concentration (mg/L)	Nitrogen leaching loss (mg/season)
0	No nitrogen fertilizer	308.26±38.41a	16.29±3.05a	5.07±2.62a
	Nitrogen fertilizer		63.74±7.84a	19.77±2.97a
1%	No nitrogen fertilizer	229.38±37.27b	15.74±2.72a	3.50±0.92a
	Nitrogen fertilizer		43.50±11.56b	10.23±1.97b
2%	No nitrogen fertilizer	192.41±37.00bc	16.19±3.65a	2.96±0.96a
	Nitrogen fertilizer		48.77±14.40ab	9.15±3.27b
5%	No nitrogen fertilizer	152.90±29.87c	16.46±4.08a	2.12±1.28a
	Nitrogen fertilizer		45.30±13.57b	6.93±4.11b

3.3. Effect of biochar addition on soil pH buffering capacity

No significance change of pH presented in treatments without nitrogen fertilizer after three growing season compared to original soil (Table 2). It indicated that plant selective adsorption in this study did not contributed to soil acidification significantly. Furthermore, biochar addition did not enhance soil pH in study period. However, chemical nitrogen fertilizer input without biochar reduced both soil pH 0.44 units and buffering capacity. Biochar addition increased soil pH buffering capacity in treatments without nitrogen fertilizer by 21.6%, 24.3% and 36.8% at 1%, 2% and 5% addition ratios, respectively. The corresponding ratios with nitrogen fertilizer were 9.4%, 20.2% and 27.2%, respectively. Moreover, soil acidification rate was affected by nitrogen fertilizer in the absence of biochar addition. Biochar addition decreased soil acidification rate by 37.1%, 80.4% and 17.0% at 1%, 2% and 5% addition ratios when compared to treatment without biochar addition.

IOP Conf. Series: Earth and Environmental Science **41** (2016) 012029 doi:10.1088/1755-1315/41/1/012029

Table 2. Changes of soil pH, buffer capacities and acidification rates under different treatments.

Biochar addition ratios and treatments		pH	pHBC (10^{-3} mol/kg)	AR (10^3 mol/hm^2)
0	No nitrogen fertilizer	5.78±0.12 a	8.56±0.24bc	-2.64±2.46c
	Nitrogen fertilizer	5.21±0.10 c	7.45±0.58d	7.80±1.43 a
1%	No nitrogen fertilizer	5.64±0.14 a	10.42±0.87a	0.24±3.50bc
	Nitrogen fertilizer	5.40±0.16bc	8.15±0.33cd	4.91±3.34ab
2%	No nitrogen fertilizer	5.82±0.09a	10.65±0.18a	-4.23±2.26c
	Nitrogen fertilizer	5.57±0.12ab	8.95±1.09bc	1.53±2.58abc
5%	No nitrogen fertilizer	5.61±0.11ab	11.72±0.10a	1.12±2.95bc
	Nitrogen fertilizer	5.37±0.14bc	9.48±0.33b	6.47±3.18ab
Original soil			5.65±0.05a	9.01±0.39bc

Soil organic matter content reduced 9.4%-18.2% by planting three growing seasons, but biochar addition mitigated the reduction (Table 3). The same trends were also found in contents of base cations. Contents of exchangeable K^+、Na^+ and Mg^{2+} decreased 35.8%-69.0%, 61.1%-63.0% and 43.7%-51.6% in treatments without biochar addition. Exchangeable K^+ was the most affected by nitrogen fertilizer in especial. However, biochar maintained contents exchangeable K^+, Na^+ and Mg^{2+}, even increased exchangeable K^+ content 34.7%. Otherwise, contents of exchangeable $Ca^{2+}$and Al^{3+} seemed no significant responses to nitrogen fertilizer or biochar addition.

It could be demonstrated that the effect of biochar on soil pH buffer capacities was not direct but multipath. And differences between addition ratios were not always significant. However, biochar effectively reduced nitrogen leaching (Table 1) and maintain the content of soil organic matter and base cations(Table 3), which mitigating the acidification caused by the invalidation of first-level buffering system [13] and the increasing of CEC induced by addition of biochar[14].

Table 3. Changes of contents of soil organic matter and base cations under different treatments.

Biochar addition ratios and treatments		Organic matter (g/kg)	Content of base cations (10^{-3} mol/kg)				
			K^+	Na^+	Ca^{2+}	Mg^{2+}	Al^{3+}
0	No nitrogen fertilizer	14.65±1.44c	4.02±1.26c	3.15±0.37b	52.57±4.11a	9.97±0.72c	3.66±0.63a
	Nitrogen fertilizer	16.21±1.69bc	1.94±0.18d	3.31±0.34b	27.79±0.65c	8.57±0.50c	3.24±0.19a
1%	No nitrogen fertilizer	17.42±2.03bc	5.95±0.76b	9.03±0.42a	40.11±2.19ab	14.71±3.87b	3.90±0.21a
	Nitrogen fertilizer	18.09±1.99abc	6.16±0.89b	7.68±0.63a	42.24±2.13ab	19.79±4.08a	3.56±0.67a
2%	No nitrogen fertilizer	18.14±1.51abc	6.87±0.87b	9.38±0.82a	45.92±3.15a	15.02±0.80b	3.76±0.16a
	Nitrogen fertilizer	21.00±1.80a	6.11±0.68b	8.93±0.99a	46.25±5.79a	20.39±2.62a	3.94±0.58a
5%	No nitrogen fertilizer	17.57±2.02bc	8.43±1.00a	8.47±0.91a	37.75±3.68b	17.31±0.62b	3.54±0.21a
	Nitrogen fertilizer	19.46±2.03ab	4.87±0.78c	8.16±0.95a	40.33±5.34ab	17.63±3.25b	3.57±0.33a
Original soil		17.90±1.87a	6.26±1.04b	8.51±0.69a	44.50±4.88ab	17.70±2.64b	3.59±0.26a

4. Conclusion

Biochar addition had significant effects on soil nitrogen distribution balance. Nitrogen utilization efficiency increased from 41% to 81% with 2% biochar addition, and residual nitrogen was decreased from 46% to 8%. Biochar addition reduced soil nitrogen leaching loss 48%-65% at different addition ratios. The reduction mechanism showed different under different biochar addition ratios. Reduction was induced to low nitrogen concentration in leachate at 1% and 2% biochar addition ratios, while contributed to decreased leaching volume at 5% biochar addition ratio. Biochar mitigated soil pH decrease though it decreased 0.44 units by nitrogen fertilizer without biochar. Meanwhile, it increased soil pH buffering capacity by 22%-37%, and reduced acidification rate by 80%-170% by maintaining the contents of soil organic matter and base cations.

Acknowledgements

Funding for this work was supported by a project grant from the National Science Foundation for Distinguished Young Scholars of China (Grand No. 41501320), and the Special Fund for Agro-scientific Research in the Public Interest (Grand No. 201503106).

References

[1] Ti C, Luo Y and Yan X 2015 Characteristics of nitrogen balance in open-air and greenhouse vegetable cropping systems of China *Environ. Sci. Pollut. R.* **23** 18508-18

[2] Duan L, Huang Y and Hao J 2004 Vegetation uptake of nitrogen and base cations in China and its role in soil acidification *Sci. Total Environ.* **330** 187-98

[3] Song X, Zhao C, Wang X and Lia J 2009 Study of nitrate leaching and nitrogen fate under intensive vegetable production pattern in northern China *C. R. Biol.* **332** 385–92

[4] Min J, Zhao X, Shi W, Xing G and Zhu Z 2011 Nitrogen balance and loss in a greenhouse vegetable system in Southeastern China *Pedosphere* **21** 464–72

[5] Shi W, Yao J and Yan F 2009 Vegetable cultivation under greenhouse conditions leads to rapid accumulation of nutrients, acidification and salinity of soils and groundwater contamination in South-Eastern China *Nutr. Cycl. Agroecosys.* **83** 73-84

[6] Barak P, Jobe B, Krueger A, Peterson L and Laird D 1997 Effects of long-term soil acidification due to nitrogen fertilizer inputs in Wisconsin *Plant Soil* **197** 61-9

[7] Ball P, Mackenzi M, Deluca T and Montana W 2010 Wildfire and charcoal enhance nitrification and ammonium-oxidizing bacterial abundance in dry montane forest soils *J. Environ. Qual.* **39** 1243-53

[8] Lehmann J, Da Silva Jr J, Steiner C, Nehls T, Zech W and Glaser B 2003 Nutrient availability and leaching in an archaeological Anthrosol and a Ferralsol of the Central Amazon basin: fertilizer, manure and charcoal amendments *Plant Soil* **249** 343-57

[9] Clough T and Condron L 2010 Biochar and the nitrogen cycle: Introduction *J. Environ. Qual.* **39** 1218-23

[10] Berglund L, DeLuca T and Zackrisson O 2004 Activated carbon amendments of soil alters nitrification rates in Scots pine forests *Soil Biol. Biochem.* **36** 2067–73

[11] Laird D, Fleming P, Davis D, Wang B, Horton R and Karlen D 2010 Impact of biochar amendments on the quality of a typical Midwestern agricultural soil *Geoderma* **158** 443-9

[12] Tarkalson D, Payero J, Hergert G and Cassman K 2006 Acidification of soil in a dry land winter wheat-sorghum/corn-fallow rotation in the semiarid U.S. great plains *Plant Soil* **283** 367-79

[13] Liao B and Dai Z 1991 Soil buffering capacity to acid precipitation and weathering characteristics of soil minerals *Acta Sci. Circum.* **11** 425-31

[14] Xu R, Zhao A, Yuan J H and Jiang J 2012 pH buffering capacity of acid soils from tropical and subtropical regions of China as influenced by incorporation of crop straw biochars *J. Soil Sediment* **2** 494-502

Movement and transformation of potassium in fertiliser micro-sites in latosol

Y D He[1,2], X B Cui[3], L X Wang[1,2],Y X Liu[1,2], T Jing[1,2] and B Z Wang[1,2,4]

[1] Haikou Experimental Station, Chinese Academy of Tropical Agricultural Sciences, Haikou 570102, China

[2] Hainan Key Laboratory of Banana Genetic Improvement, Haikou, 570102, China

[3] Environment and Plant Protection College, Hainan University, Haikou 570228, China

E-mail: 13807615366@163.com

Abstract: A soil column method was used to analyse the characteristics of the movement and transformation of potassium (K) in a latosol. The internal temperature was maintained at 28 to 30 °C. The distances moved by the added K were 50 to 100 mm after 7 and 28 d, respectively. Potassium moved faster within 7 d during the incubation time of 28 d. The concentrations of water-extractable K, and exchangeable K, had significant negative linear relationships with distance from fertiliser placement in the region of K movement. In incubation time, water-extractable K and exchangeable K had significant effects on the concentration distributions for fertiliser micro-sites, but had no significant effects on non-exchangeable K. Most of K was still in an available form at fertiliser micro-sites after moving into the soil except that 4.22% to 11.19% of all applied K was fixed by soil.

1. Introduction

Potassium (K) is one of the three essential factors in plant growing. The amount of potassium necessary for plant growth is matched with nitrogen, and higher than phosphorus [1]. A deficiency of potassium causes a severe reduction in crop yield, and crop quality, and reduces the effectiveness of other inputs. In recent years, crops have removed more K from the soil with the expansion in the number of varieties of high-yield crops, increases in multiple cropping indices, and the development of more intensive agriculture. This causes increasing potassium deficient soil areas to appear: soil potassium depletion is becoming increasingly serious [2]. Potassium fertiliser application is an important way of alleviating this soil potassium deficit. China, the world's top consumer of potash fertiliser, is deficient in raw potash resources. Potash supply is realised by imports given China's less developed potash industry [3]. Plants can attain their full potential by improving the utilisation of potassium in this kind of situation. So an accurate assessment of the supply of potassium to different soils, and guidelines for rational fertilisation, are becoming more important. In order to improve utilization efficiency of K fertilisers, the investigation not only the amounts of different K fractions that other fertilisers may have on the K balance in soils, but also the movement and transformation of K in soils were required.

Most researchers have focused on leaching and transporting potassium in field experiments. However, there is little research available which examines the diffusion and transformation of potassium in fertiliser micro-sites. Dong Yuanyan studied the transport and diffusive coefficients of phosphate and potassium in red soils by using a soil column method, which did not involve the effectiveness and distribution of K, when KH_2PO_4 is applied [4]. The movement and transformation of K in fertiliser micro-sites in fluvo-aquatic soil and red earth were investigated through an incubation experiment using a soil column by Zhenyu Du and Jianmin Zhou when KCl was applied [5,6]. There have been no reports on the diffusion and transformation of potassium in latosols in Hainan, China, to date. To provide support for scientific management and to guide a rational use of fertiliser, for this study try to provide novel insight into the effects of the application of low (0.2 g) and high (0.5 g) concentrations of K_2SO4 on the K^+ movement and transformation profile at 7 and 28 d in a latosol, by using a laboratory experiment in soil column.

2. Materials and methods

2.1. The condition of test point

The soil (latosol) used in this study was calcareous, and belonged to those Ustic Argosoils according to the Chinese soil classification system, and was approximately equivalent to a luvisol in the FAOUNESCO system. The soil sample was obtained from the top 15 cm of a cultivated layer in a field located near Chengmai city, Hainan Province, China (N 19° 56′ 22.71′′, E 109° 54′ 06.33′′). The soil texture was silt loam, and some selected chemical and physical properties index were as follows: organic matter content 17.4 g/kg, pH 5.42, hydrolysed nitrogen 62.7 mg/kg, effective phosphorus 16.69 mg/kg, and available potassium 11 mg/kg. After removing visible plant residues, stones, and other material, when the soil was air-dried, ground to pass a 0.9mm aperture sieve of soil. The K fertiliser used in this study was an analytically pure grade of K_2SO_4, which had been the soil before being used in these experiments.

2.2. Methods

Laboratory soil column experiments were conducted to evaluate the movement and transformation of potassium in fertiliser micro-sites. The wax columns (Patent No: zl201320728057.4), 150 mm in height with a cylindrical cavity have an internal diameter of 50 mm were used for soil incubation [7]. It was use that jelly melted and mixed to make the wax column was 2:1(Two parts of paraffin wax,1 part petroleum). Two pieces of filter paper were used to close the bottom of the wax column: the 360 g of soil was packed into each column to obtain a final bulk density of 1.22 g/cm^3. One filter paper (approximately 49 mm in diameter) was placed on the soil surface to separate the soil from the fertiliser, and to ensure that the fertiliser solution could be evenly spread over the soil. The packed soil columns were where placed vertically on the fine sand disc, and become wetted by capillary rise to a moisture content of 401 g/kg. After abandoning a filter papers covering the bottom, the top and bottom of the packed wax blocks prevent moisture lose that were covered with plastic film. The packed wax blocks were allowed to equilibrate for 48 h at 25 °C before K treatments were applied.

The K fertiliser treatments were 0.2 g and 0.5 g of K_2SO_4, respectively. The K fertilisers were added at a rate of 0.09 g and 0.224 g *per* column, respectively. Which were equivalent to the addition of 55 kg K_2O hm^{-2} and 138 kg K_2O hm^{-2} in field, respectively. Along one side of crop rows spaced 45 cm apart. This application device was intended to simulate field conditions similar to those near fertiliser granules or banding. A soil column with application of K was used as the control trial. Different incubation times (1 w and 4 w) were also used for two different K concentration treatments, respectively. Three replicates were used in a completely randomised design for experiment. The wax blocks were covered, then incubated vertically condition at 28 °C. After 1 w or 4 w, the filter paper on the soil cylinder was removed for K analysis to evaluate the amount of K retained. It use for blade to cut the cylinder into 30 slices, 5 mm thick.

Soil samples of each slice were analysed to find the concentrations of different potassium, and soil moisture.

Water-soluble K (WSK) was determined as follow: 1.8 g of soil, which was equivalent to 1 g of water-free soil, and 10 mL distilled water were added to 50 mL centrifuge tube and shaken on a reciprocating shaker at 25 °C ,and 30 min. The mixture was centrifuged for 10 min, and the suspension was separated and the K concentration in the solution was determined by flame spectrometry.

Exchangeable K (EK) was extracted from 1.8 g of soil, 10ml of 1 M ammonium acetate solution at pH 7.0 in 50 mL centrifuge tubes and shaken at 25 °C for 1 h [8]. The mixture was centrifuged for 10 min, the suspension separated and the K^+ concentration in the solution was determined by flame spectrometry. it is difference that between WSK and the NH_4OAc-extracted for K Exchangeable calculated.

Non-exchangeable K (NEK) was determined as follow: 1.8 g of soil, and 10 mL of 1M HNO_3 were mixed in glass beaker (which was covered with a watch glass, size 250 mL). The mixture was boiled for 10 min, filtered, and then made up to 100 mL. The K^+ concentration was determined by flame spectrometry. NEK was calculated by subtracting the K extracted with NH_4OAc from that extracted with HNO_3 method [9].

All results presented were the means of three replicates. The statistical analysis of the data was based on MS-Excel™ 2003 and SAS 9.1 software.

3. Results and discussion

3.1. Movement of K and water-soluble K in fertiliser micro-sites in a latosol

Water-soluble K is the available potassium in a soil solution that is immediately available to plants and is potentially subject to leaching. The amount of WSK should dependent on the equilibrium among the

IOP Conf. Series: Earth and Environmental Science **41** (2016) 012030 doi:10.1088/1755-1315/41/1/012030

different forms of K in the soil and the application, or removal of K from the soil. Keeping the balance of WSK should be dependent on soil properties such as pH, CEC, clay mineralogy, and could be influenced by the alteration of ion concentrations in soil solution and the total concentration of soluble anions.

The depths of K movement can be slightly changed by the concentration of fertiliser applied: here they were 50 mm (for 0.2 g K_2SO_4) and 55 mm (for 0.5 g K_2SO_4) after 1 w, and 80 mm (for 0.2 g K_2SO_4) and 100 mm (for 0.5 g K_2SO_4) after 4 w (Fig. 1). The migration rate of K in earlier stages was faster than in later stages. This was because the potassium content in earlier stage was higher. When the soil water content was 150 g/kg and its bulk density was 1.60 g cm^{-3}, the distances moved by the added K were 18 to 34 mm after three to 14 d in red soil, according to Dong *et al.* [4] The movement and transformation of K in fertiliser micro-sites on a red soil were studied through an incubation experiment using a soil column. When the soil water content was 370 $g{\cdot}kg^{-1}$ and bulk density was 1122 g cm^{-3}, the distances moved by the added K were 60 to 110 mm after 7 d and 28 d [5]. There were large differences between the results in the two different experiments, possibly because the physico-chemical properties of the soil were different.

The concentration of WSK contained in soil section gradually decreased with increasing distance from the fertiliser application site (Figure 1). It showed a negative linear relationship between the contents of water-soluble K and distance (Table 1). The concentration of WSK, cultivated for 1 w, was higher than that when cultivated for 4 w at 0 to 30 mm distance in the soil column. At distances greater than 30 mm, the results were the opposite. This showed that the WSK was migrating downwards at between 1 w and 4 w.

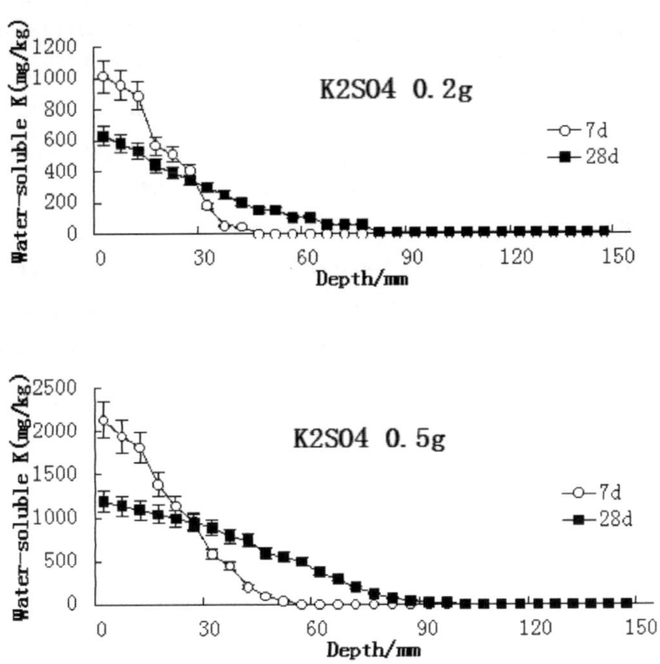

Figure 1. The water-extracted potassium content in the soil sample.

To analyse the effects of fertilisation and cultivation time on the water-soluble potassium content distribution, we used a linear best-fit between the water-soluble potassium contents produced by the four treatments and distance: analysis of variance and multiple comparisons were then used to find the coefficients of the regression equation. The results showed that water-soluble potassium content decreased with distance from the point of fertilisation under high fertiliser doses; meanwhile, the rate of decline (b_1) was significantly increased (Table 1). For the same amount of fertilisation, the rate of decline at 7 d was bigger than that at 28 d. The intercept (b_0) and regression coefficient (b_1) of the four regression equations were significantly different. This showed that the amount of fertiliser and time affected the distribution of water-soluble potassium content at fertiliser micro-sites.

Table 1. The relationship between the water-soluble K (y, mg/kg) and the depth (x, mm) was determined by one-dimensional linear regression ($y = b_0 + b_1x$).

Amount of K_2SO_4 (g)	Incubation time (w)	Regression coefficient		
		b_0	b_1	R^2
0.2	1	1160.13	-25.38	0.963**
0.2	4	610.5	-7.94	0.949**
0.5	1	2344.78	-45.65	0.980**
0.5	4	1314.56	-14.01	0.986**

Notes: Different lower case letters indicate significant liming effects at a 0.05 probability level; ** stands for very significant ($P < 0.01$).

3.2. *The content distribution of exchangeable K at fertiliser micro-sites in a latosol*

Exchangeable K is the total K adsorbed by the organic matter, or clay minerals that can easily be exchanged with other cations. The distribution pattern of exchangeable K is the same as that of water-soluble K. From Figure 2, it was seen that the exchangeable K content gradually decreased with increasing distance. It showed z negative linear relationship between the exchangeable K contents and distance (Table 2). The results showed that exchangeable K content increased as the distance from the point of fertilisation decreased: the rate of decline (b_1) was also significantly increased: when the amount of fertiliser was the same, the rate of decline at 7 d was bigger than that at 28 d. The intercept (b_0) and regression coefficient (b_1) of the four regression equations were markedly different. This showed that the amount of fertiliser and application time on the distribution of exchangeable K content were significantly affected at such fertiliser micro-sites.

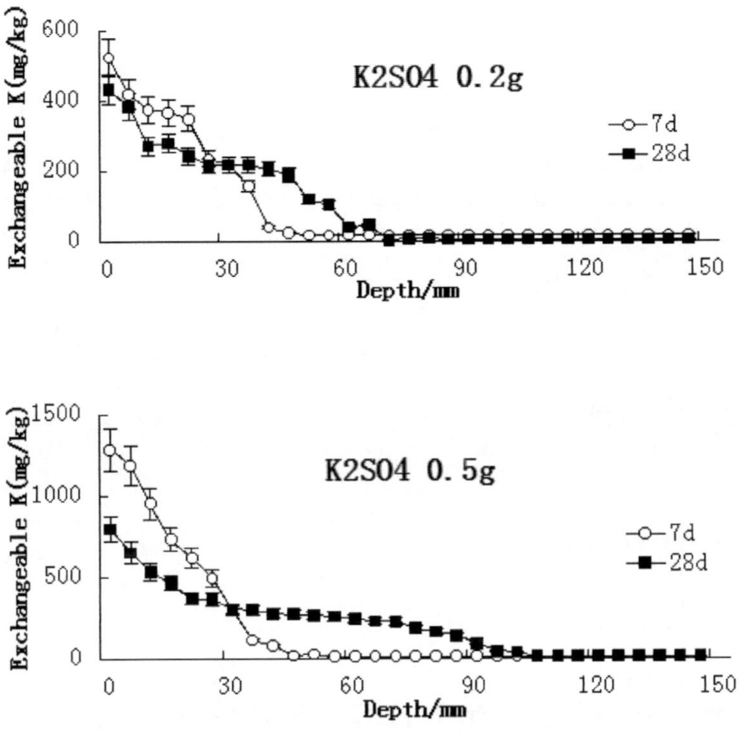

Figure 2. The exchangeable K content in the soil sample.

Table 2. The relationship between the exchangeable K (y, mg/kg) and depth (x, mm) was determined by one-dimensional linear regression ($y = b_0 + b_1x$).

Amount of K_2SO_4 (g)	Incubation time (w)	Regression coefficient		
		b_0	b_1	R^2
0.2	1	563.69	-10.67	0.965**
0.2	4	403.69	-5.14	0.940**
0.5	1	1353.84	-27.56	0.958**
0.5	4	606.78	-5.69	0.832**

Notes: Different lower case letters indicate significant limiting effects at a 0.05 probability level; ** represents very significant ($P < 0.01$).

3.3. The content distribution of non-exchangeable K at fertiliser micro-sites in a latosoll

Figure 3 showed that the non-exchangeable K contents gradually decreased with soil depth. There was no significant difference in transfer volume of non-exchangeable K under different treatment times at these fertiliser micro-sites. Rate of fertiliser addition was another important factor affecting potassium fixation in latosols. The NEK contents of treatments subjected to the addition of added 0.5 g K_2SO_4, under the same conditions, was higher than that when adding 0.2 g. The latosol had a lower ability to fix potassium (albeit with local, micro-regional differences therein).

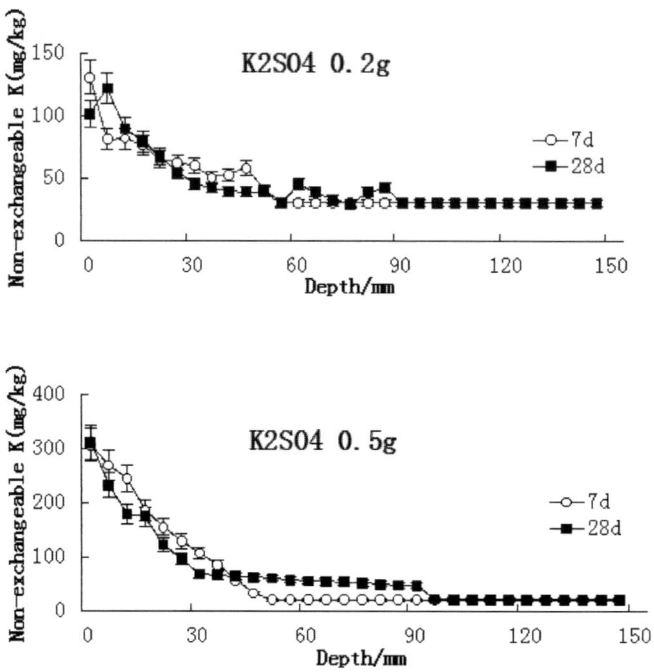

Figure 3. The non-exchangeable K content in the soil samples.

3.4. The migration of different forms of K at fertiliser micro-sites in a latosol

When 0.2 g of K was applied, the migration distance was 50mm after 1 w. The average forward speed was 7.14 mm/d. The distance moved by the K was 80 mm after 4 w. The average forward speed was 1.43 mm/d in the final 3 weeks. The average forward speed in the first week was over five times faster than in the final three weeks. The migration distance of K, when applying 0.5 g, was 55 mm after 1 w. The average speed of its movement was 7.86 mm/d. The migration distance of K was 100 mm after 4 w and its average forward speed was 2.14 mm/d in the final three weeks. The average speed in the first week was over four times faster than the final three weeks, thus it can be seen that potassium moved faster within the first 7 d of the 28 d incubation period.

From Table 4, the recovery rate of K was between 97.1% and 101.1% due to possible errors and over-extraction of chemically bound K in the soil's structure. Recycled K was used in lower amounts than would otherwise have been the case.

Table 3. The migration of K fertilisers at fertiliser micro-sites in a latosol.

Amount of K_2SO_4 (g)	Incubation time (d)	Migration amount (mg)				
		WSK	EK	NEK	available K	TK
0.2	7d	55.45b	30.43a	3.78a	85.88a	89.66a
0.2	28d	50.79d	34.79a	5.17a	85.57a	90.75a
0.5	7d	128.75a	68.23b	17.65b	196.98b	217.42b
0.5	28d	130.71c	71.66b	25.51b	202.37b	227.88b

IOP Conf. Series: Earth and Environmental Science **41** (2016) 012030 doi:10.1088/1755-1315/41/1/012030

[a] Different lower case letters indicate significant limiting effects at a 0.05 probability level.

One week later, 34.21 mg (38.16%) and 88.67 mg (40.78%) of K was transformed into other forms with K doses of 0.2 g and 0.5 g, respectively. After 4 w, there were 39.96 mg (44.03%) and 97.17 mg (42.64%) of K, respectively, which had transformed into other forms thus proving that the rate of movement of K and the inversion quantity and transformation rate were significantly improved by extending the diffusion time. Along with the rise of fertiliser dose, the rate of movement of K and the inversion quantity were significantly increased (albeit the rule governing the transformation rate was not obvious).

4. Conclusions

Fertiliser has a significant impact on the various forms of potassium migration volume. The internal temperatures remained at between 28 and 30 °C. The distances moved by the added K were 50 to 100 mm after 7 d and 28 d. Potassium moved faster within the first 7 d of the 28 d incubation time. The average forward speed in the first week was over four to five times faster than in the final three weeks.

The balance between WSK and EK finished instantaneously, but the balance between EK and NEK finished slowly [10]. Most of the K was still in an available form (89.24% to 95.78%) at these fertiliser micro-sites after moving into the soil except that 4.22% to 11.19% of the applied K was fixed by the soil. About 40% of the K was transformed into other forms in all treatments.

EK and WSK showed similar distribution patterns. The concentrations had significant negative linear relationships with distance from fertiliser placement in the region of K movement.

Both K application amount, and incubation time, had significant effects on the concentration distributions of WSK and EK in fertiliser micro-sites. With extended diffusion time, the rate of movement of K, the inversion quantity, and the transformation rate were significantly improved. Along with increased dose, the rate of movement of K and the inversion quantity were significantly improved (albeit the changing trend of the rate of transformation was not obvious).

Acknowledgment

This work was financially supported by the Natural Science Foundation of Hainan Province (Grant No. 20153110) and the Ministry of Agriculture's Tropical Agro-technique Extension and Systems of Construction Project (15RZNJ-16).

References

[1] Kering M K, Butler T J, Biermacher J T, Jagadeesh Mosali, John A and Guretzky 2013 Effect of potassium and nitrogen fertilizer on switchgrass productivity and nutrient removal rates under two harvest systems on a low potassium soil *Bioenergy Res* **6** 329-35

[2] Zhan L P, Li X K, Lu J W, Cong R H,Wang J,Wang Z and Liao Z W 2013 Potassium Supply Capacity and Potassium Dynamics in Different Types of Soils under Paddy Rice-ryegrass Rotation *ACTA Pedologica sinica* **50** 592-9

[3] Zhang Y C, Li L, Shi X L, Liu S X, Gao Q and Lou Y J 2013 Progress on researches of Soil Potassium Forms and Effectiveness *Journal of Jilin Agricultural Science* **38** 52-4

[4] Dong Y Y 1995 Transport and Diffusive Coefficient of Phosphate and Potassium in Ren Soil Journal of Huazhong *Agricultural University* **14** 247-52

[5] Du Z Y and Zhou J M 2005 Movement of Potassium in Fertilizer Microsites in Red Earth *Acta Pedologica Sinica* **42** 1036-9

[6] Du Z Y, Zhou J M, Wang H Y, Chen X Q and Wang Q H 2009 Movement and transformation of potassium in in a cinnamon soil as affected by humic acid *Journal of Soil and Water Conservation* **23** 202-5

[7] Du Z Y, Wang Q H, Liu F C, Ma H L, Ma B Y and Malhi S S 2013 Movement of Phosphorus in a Calcareous Soil as Affected by Humic Acid *Pedosphere* **23** 229-35

[8] Mouas B Mouas B,Pierre B,Malika Boudiaf Naït Kaci, Arezki Derridj and Bruce Velde 2013 Potassium status in bulk and rhizospheric soils of olive groves in North Algeria *Geoderma* **197** 161-8

[9] Lu R K 1999 *Analytic Methods for Soil and Agro-chemistry* (Beijing: Chinese Agricultural Press) pp 27-31 (in Chinese)

[10] Huo N and Wang H T 2014 Research Progress of Soil Potassium Availability on Soil Moisrure *Journal of SHanxi Agricultural Sciences* **42** 1330-4

IOP Conf. Series: Earth and Environmental Science **41** (2016) 012031 doi:10.1088/1755-1315/41/1/012031

The complete mitochondrial genome of the *bagarius yarrelli* from honghe river

M Du[1, 3], C J Zhou[2], B Z Niu[1], Y H Liu[1, 3], N Li[1],J L Ai[1] and G L Xu[1]

[1]College of Life science and technology, Key Lab for Quality, Efficient cultivation and Security Control of Crops in Colleges and University of Yunnan province, Honghe University, Mengzi, 661199, Yunnan Province, P.R. China

[2]College of Fisheries, Henan Normal University, Xinxiang, 453007, Henan Province, P.R. China

E-mail: kidliu1968@126.com; du2005min@126.com

Abstract. The total length of mitochondrial DNA sequence of the Bagarius yarrelli from the Honghe river of China is determined in this paper. The total length of the circular molecule is 16524 base pair which denoted a similar gene order to that of the other bony fishes, which include a non-coding control region, a replicated origin, two ribosome RNA (rRNA) genes, 22 transfer RNA (tRNA) genes as well as 13 protein-coding genes. Its whole base constitution is 31.4% for A, 26.9% for C, 15.7% for G and 26.0% for T, with an A+T bias of 57.4%. Those mitochondrial data would contribute to further study molecular evolution and population genetics of this species.

1. Introduction

The whole mitochondrial genome DNA is a closed-circular molecule, which was found in most multicellular animals [1], for example *Cranoglanis bouderius* [2], grass carp[3], *Octopus minor*[4], *Ovis aries*[5], *Alligator sinensis*[6]. Mitochondrial DNA has applied for study on phylogeny for its maternally inherited [7].

Yellow sisorid-catfish (*Bagarius yarralli*)[8-9] is inhibited in yunnan province, China, e.g., Honghe River, Lanchangjiang River, Lujiang River. It is benthic carnivorous fish [10] and has succeeded in artificial propagation[11].

2. Materials and methods

B. yarralli Sample was collected from the Honghe River at hekou county,Yunnan province, China. The muscle tissues were cut to preserve in absolute ethanol for DNA extraction through Phenol-chloroform method [12].

Eighteen primers specific for B. yarralli were designed to produce and sequence the 17 overlapping fragments of the whole mitochondrial DNA genome according to the sequence of B. yarralli from Lanchangjiang River,which the GenBank accession number is JQ026260 or NC_021606, as well as the other fishes of Sisoridae contains Glyptothorax trilineatus(NC_021608), Oreoglanis macropterus (NC_021607),Pareuchiloglanis gracilicaudata(NC_021603) (Table 1),of which Primer Baya14Fand Baya14R was from Niu et al.[13], primer Baya1F and Baya1R, Baya16F and Baya16R, Baya17F and

IOP Conf. Series: Earth and Environmental Science **41** (2016) 012031 doi:10.1088/1755-1315/41/1/012031

Baya17R, Baya18F and Baya18R were from Du et al.[14], primer Baya5F and Baya5R, Baya6F and Baya6R, Baya7F and Baya7R, Baya9F and Baya9R were from Du et al. [15].

Polymorphism Chain Reaction was performed on a EDC810 thermal cycler (Eastwin Life Sciences, Inc.) with the 50μl PCR reaction volume and 30 cycles. Tiangen gel extraction kits (Tiangen biochemical science and technology Beijing co., LTD) was used to purified incised purpose fragment and then cloned into the PMD18T-vector (TaKaRa) via top10 competent cell and sequenced through a 3770 DNA sequencer (ABI PRISM) according to the operation manual.

3. Results and discussion

Here, we report the overall length of mitochondrial genome of B. yarralli at Honghe River with the GenBank accession number KT983411.The whole length of the mtDNA genome was 15624 bp, comprising of 16S ribosomal RNA gene and 12S ribosomal RNA gene (16S rRNA and 12S rRNA), thirteen genes which have protein-coding charactistics, twenty-two genes were associated with transfer RNA genes (tRNA) and a non-coding control region and a replication locus of L strand and you can found the information at Table 1. Its whole base constitution of the heavy chain was adenine for 31.4%, Cytosine for 26.9%, Guanine for 15.7% and Thymine for 26.0%, while A+T>G+C. A large proportion of this B. yarralli mitDNA genes were located in heavy strand while 8 transfer RNA (transfer RNA-Gln, transfer RNA-Ala, transfer RNA-Asn, transfer RNA-Cys, transfer RNA-Tyr, transfer RNA-Ser(UCN), transfer RNA-Glu, transfer RNA-Pro) and NADH dehydrogenase subunit 6(ND6) genes were located in light strand. The genome information was illuminated through compared with the whole mitochondrial sequence within the bamboo rat (Rhizomys pruinosus, KC789518), Glyptothor fukiensis fukiensis(JQ917224) and the B. yarralli from Lanchangjiang River. The structural organization of the mtDNA gene are analogical to the vertebrate mitDNA genome [16-18].

Table 1. Sequence primers in the analyses of this Bagarius yarrelli mitDNA genome.

Primers name for L strand	Sequence(5'→3')	annealing temperature (°C)	Length(bp)	location L strand
Baya1F 1805	CCTTCTCCAAGCCTACGTGTA		54	824
Baya1R	GATAGAAACTGACCTGGATTGCT			
Baya2F 2462	CAATCCCCTTTCAGAGYCCA		54	809
Baya2R	GTTTGTGCRACTGCTCGKA			
Baya3F 2897	GGTTCAAATCCTCCTCTTAAT		54	1000
Baya3R	TGGYCCTTGGGCATTCAGGCAC			
Baya4F 3814	GTGCCTGAATGCCCAAGGRCCA		54	1371
Baya4R	TGRGTTGCATTCAGRAGATGTGGGGT			
Baya5F 5116	GAGTGAAAAYCTCCTAGTCYCT		54	1340
Baya5R	GTTTCTCATTTAATWGAKCCTCC			
Baya6F 6348	TAGACACCCGAGCATAYTTYACA		54	1172

Baya6R	GAAGGYCARGTTTTCGTAGTC		
Baya7F 7202	GCCTCCCCYRTAATAGAAGAA	54	1208
Baya7R	GACGGCCAGGCCYATATTTA		
Baya8F 8076	CCTGAAACTGACCATGRCACT	54	894
Baya8R	TGTGCCTTCTCGTACRATGTCT		
Baya9F 8756	TACAAGAAAACGTATAATGGCC	54	1258
Baya9R	CCGAAATCAGAGGTCTTRTRT		
Baya10F 9765	GATCAGCACGACTCCCCTT	54	1244
Baya10R	CTTCTACGTGGGCTTTDGGT		
Baya 11F 10357	CCAACCATCATRCTATTYCCA	54	1444
Baya 11R	AGGCCTCTTTCGGTGGAYAA		
Baya 12F 11436	AACACCACATAYGARCGAAC	54	1500
Baya 12R	CTTGAKGTTGAGAATGCTACGATT		
Baya 13F 12648	CMGCAATAGAAGGCCCMACC	54	1445
Baya 13R	TGGAATGTTRGTRGTTTTTGC		
Baya 14F 13755	TYTAAACAGCCCGAAGCGCMC	54	1016
Baya 14R	GCACCTCAGAAKGATATTTGWCCYC		
Baya 15F 14422	AACCACCGTTGTAAYTCAACTAT	54	1200
Baya 15R	CTCCGGATTACAAGACCGGTGCT		
Baya 16F 15264	CACACATCHAAACAACAAGGHA	54	1400
Baya 16R	GGTGTGGCRTAGCAAGGCGTC		
Baya 17F 113	TTATACATGCAAGTATCCGCA	54	1000
Baya 17R	TAAGCTAGCTGTTTRGCTCAGG		
Baya 18F 895	TTGAATTAGGCTCTGAGACG	54	1000
Baya 18R	TACACGTAGGCTTGGAGAAGG		

L indicates light strands while H denote heavy strands, respectively

IUB codes was used to labeled the mixed bases:R denote A or G; Y denote C or T; K denote G or T; M denote A or C; S denote G or C; W denote A or T.

With the exception of NADH1 and COX1 genes were started from GTG, the other eleven genes which encoded protein were initiated by ATG. The NADH1,COX1,ATP8,ATP6,NADH4L and NADH5 were shared the complete termination codon TAA, while the NADH2,COX2,COX3, NADH3, NADH4 and Cytb genes were stopped by incomplete termination codon T and NADH6 gene has the complete termination codon of TAG. The length of 16S ribosome RNA and 12S ribosome RNA were 1668bp while 954bp respectively, which were situated between the transfer RNA-Leu(UUR) and transfer RNA-Phe genes were separated by the transfer RNA-Val gene. This speculated 22 tRNA genes were dispersed among the genes which encoded protein; ribosome RNA genes and D-Loop region with vary from 67 to 75bp in length. 7 segments of gene overlap from one to eight bp while nine intergenic spacer sections of gene from 1 to 15bp were found in the entire genome. The length of the control region was 895bp and was situated between transfer RNA-Pro and transfer RNA-Phe gene, of which nucleotide ingredient was rich A+T content of 66.5% (Table 2).

Table 2. Characteristics of the mitochondrial genome of Bagarius yarrelli from Honghe River of china.

Locus	From	To	Length(bp)	Start codon	Stop codon	Space(+) Overlap(-)	Strand
tRNA-Phe	1	69	69			0	H
12S rRNA	70	1023	954			0	H
tRNA-Val	1024	1095	72			0	H
16S rRNA	1095	2762	1668			0	H
tRNA-Leu(UUR)	2763	2837	75			0	H
NADH1	2838	3812	975	GTG	TAA	0	H
tRNA-Ile	3818	3889	72			+6	H
tRNA-Gln	3959	3889	71			-1	L
tRNA-Met	3959	4028	70			-1	H
NADH2	4029	5073	1045	ATG	T--	0	H
tRNA-Trp	5074	5146	73			0	H
tRNA-Ala	5217	5149	69			+2	L
tRNA-Asn	5291	5219	73			+1	L
Rep-origin	5292	5322	31			0	L
tRNA-Cys	5389	5323	67			0	L
tRNA-Tyr	5460	5390	71			0	

L							
COI	5462	7012	1551	GTG	TAA	+1	H
tRNA-Ser(UCN)	7083	7013	71			0	L
tRNA-Asp	7088	7159	72			+4	H
COII	7175	7865	691	ATG	T--	+15	H
tRNA-Lys	7866	7940	75			0	H
ATP8	7942	8109	168	ATG	TAA	+1	H
ATP6	8100	8783	684	ATG	TAA	-8	H
COIII	8783	9566	784	ATG	T--	-1	H
tRNA-Gly	9567	9639	73			0	H
NADH3	9640	9985	346	ATG	T--	0	H
tRNA-Arg	9986	10056	71			0	H
NADH4L	10049	10353	297	ATG	TAA	-6	H
NADH4	10347	11727	1381	ATG	T--	-5	H
tRNA-His	11728	11797	70			0	H
tRNA-Ser(AGY)	11798	11865	68			0	H
tRNA-Leu(CUN)	11871	11943	73			+5	H
NADH5	11944	13764	1821	ATG	TAA	0	H
NADH6	13761	14276	516	ATG	TAG	-4	L
tRNA-Glu	14345	14277	69			0	L
Cyt b	14348	15488	1141	ATG	T--	+2	H

tRNA-Thr H	15489	15560	72	0
tRNA-Pro L	15629	15559	71	-2
D-loop H	15630	16524	895	0

Notes: tRNA, transfer RNA;12 S rRNA and 16 S rRNA, 12S ribosomal RNA and 16 S ribosomal RNA; NADH1–6, 4 L, NADH dehydrogenase subunits 1–6, 4 L; COI–III, cytochrome c oxidase subunits I–III;ATP 6 and 8, ATPase subunits 6 and ATPase subunits 8; bp, base pair(s); cyt b, cytochrome b.

The Maximum likelihood tree (by Mega 5.0 software) were established on account of the whole mitDNA genome nucleotide information of this B. yarrelli from the Lanchangjiang river (JQ026260)and the Honghe river in this study(Bagarius yarrelli-hh) , Danio rerio(AC024175.3), as well as the other 11 species of fish under the kimura 2-parameter model and the 1000 bootstrap method. The phylogenetic tree denoted that Glaridoglanis andersonii(NC_021600),Pseudecheneis sulcata(JQ026259), Exostoma labiatum(NC_021601),Glyptothorax fukiensis, Glyptothorax trilineatus(NC_021608) and Cranoglanis bouderius(NC_008280) clustered into one branch. Euchiloglanis kishinouyei(NC_021598),Oreoglanis macropterus, Pareuchiloglanis gracilicaudata, Pseudecheneis yunnanensis(NC_021604) formed into a clade while the Bagarius yarrelli lc from the Lanchangjiang river and Bagarius yarrelli hh from the Honghe river clustered into together.

Acknowledgement
This work is supported by the National Natural Science Foundation of China (no. 31360638), the young academic leaders of reserve talent project in Yunnan Province(no. 2015HB059), the Science Research Fund of Yunnan Provincial Department of Education (no.ZD2013009), the innovative practice teaching reform project in Honghe University (no.CXSJ1340), and the young academic leaders of reserve talent project in Honghe University (no. 2014HB0203), Honghe University doctoral special project (no. 14bs11).

Declaration of interest
The authors declare that there is no conflict of interest.

References
[1] Nakao M, Yokoyama N, Sako Y, Fukunaga M and Ito A 2002 The complete mitochondrial DNA sequence of the cestode Echinococcus multilocularis (Cyclophyllidea: Taeniidae) *Mitochondrion* **1** 497 – 509
[2] Peng Z G, Wang J and He S P 2006 The complete mitochondrial genome of the helmet catfish Cranoglanis bouderius (Siluriformes: Cranoglanididae) and the phylogeny of otophysan fishes *Gene* **376** 290–7
[3] Wang C H, Chen Q, Lu G Q, Xu J W, Yang Q L and Li S F 2008 Complete mitochondrial genome of the grass carp (Ctenopharyngodon idella, Teleostei): Insight into its phylogenic position within Cyprinidae *Gene* **424** 96–101
[4] Cheng R, Zheng X, Lin X, Yang J and Li Q 2012 Determination of the complete mitochondrial DNA sequence of Octopus minor *Mol. Biol. Rep.* **39** 3461-70
[5] Hiendleder S, Lewalski H, Wassmuth R and Janke A 1998 The Complete Mitochondrial DNA Sequence of the Domestic Sheep (Ovis aries) and Comparison with the Other Major Ovine Haplotype *J. Mol. Evol.* 7(4) 441-8
[6] Wu X B,Wang Y Q, Zhou K Y, Zhu W Q, Nie J S and Wang C L 2003 Complete mitochondrial DNA sequence of Chinese alligator,Alligator sinensis, and phylogeny of crocodiles *Chinese*

Sci. Bull **19** 2050-4

[7] Ma H Y, Ma C Y and Ma L B 2011 Population genetic diversity of mud crab (Scylla paramamosain) in Hainan Island of China based on mitochondrial DNA *Biochem Syst. Ecol.* **39** 434-44

[8] Cheng Q T and Zheng B S 1987 Systematic synopsis of Chinese fishes (Beijing: Science Press) pp 218-9

[9] Chu X L and Chen Y R 1989 Fish fauna of Yunnan (Beijing: Science Press) pp 194-5

[10] Leng Y, Tian S K, Liu Y T, Qiu J R, Xue C J, Yi Y and Li X S 2011 Preliminary studies on the feeding habits of wild Bagarius yarralli(sykes) *Modern Agricultural Science and Technology* **19** 329-30

[11] Xue C J, Zhang Z X, Ma J Y, Sun Y, Luo Y X and Li X S 2012 Preliminary studies on artificial propagation and embryonic development of Bagarius yarralli *Journal of Hydroecology* **33**(5) 54-6

[12] Taggart J B, Hynes R A and Ferguson A 1992 A simplified protocol for routine total DNA isolation from salmonid fishes *Journal of Fish Biology* **40** 963-5

[13] Niu B Z, Du M, Liu Y H, Bai L and Ai J L 2016 Cloning and polymorphism analysis of mitochondrial DNA ND6 gene of Bagarius yarrelli *Jiangsu Agricultural Sciences* **44** (4) 62-5

[14] Du M, Niu B Z, Jia M Y, Liu Y H and AI J L 2017 Gene Cloning and Polymorphism Analysis of Mitochondrial 12S rRNA and 16S rRNA of Bagarius yarrelli *Journal of southwest university(natural science edition)* **38**

[15] Du M, Niu B Z, Wang T T and Liu Y H 2016 Study on the Polymorphism and Phylogeny of Cytochrome Oxidase Gene of Bagarius yarrelli *Journal of shanghai ocean university* **25**(3) 337-43

[16] Ma Z H, Yang X F, Zhang X J and Yang R B 2013 Complete mitochondrial genome of the freshwater goby Odontobutis potamophila (Perciformes: Odontobutidae) *Mitochondrial DNA* 24

[17] Zhao F, Zhang T Z, Su J P, Nevo E and Lin G H 2014 Mitochondrial genome of bamboo rat Rhizomys pruinosus *Mitochondrial DNA* **25**(5) 381

[18] Zhou C J, Wang D Q, Yu M L and He S P 2012 The complete mitochondrial genome of Glyptothorax fukiensis fukiensis (Teleostei, Siluriformes: Sisoridae) *Mitochondrial DNA* **23**(6) 414-6

IOP Conf. Series: Earth and Environmental Science **41** (2016) 012032 doi:10.1088/1755-1315/41/1/012032

The impacts of climate change on water resources in the Second Songhua River Basin, China

J S Yu[1], X L Yao[1], W C Sun[1,2,3] and Z J Li[1]

[1]College of Water Sciences, Beijing Normal University, No.19 Xinjiekouwai Street, Beijing 100875, China

[2]Joint Center for Global Change Studies, Beijing 100875, China

E-mail: sunny@bnu.edu.cn

Abstract. The Northeast China Plain is one of the main grain growing regions in China. Due to the high latitude and black soil ecological system, the crop growth in there is vulnerable to climate change, which makes it important to evaluate the influences of climate change on water resources. In this study, the influences of climate change on water resources of a typical basin in northeast China, the Second Songhua River Basin were assessed using the SWAT model. Ensemble downscaled output from sixteen GCMs for A1B emission scenario in 2050s was adopted as the regional climate scenario and was used to drive SWAT model to predict hydrological changes. The prediction shows that mean evapotranspiration of whole basin increases in most time of a year. Stream flow of Fuyu gauging station downstream this basin exhibits a decrease trend from April to June, November and December, and increases in the remaining period of the year. It is indicated that water resources may not be sufficient in spring for irrigation and the possibility of flood in summer may increases, indicating countermeasures should be made to ensure agricultural water use and prevent possible damages of flood on crop.

1. Introduction

Climate change is now a leading issue on the environmental and socioeconomic agenda worldwide [1]. The Soil and Water Conservation Society (SWCS) [2] stated that, under climate change, there is clear potential for increased risks of environmental consequences. Since the climate system interacts with the hydrological cycle, the impact of climate change on hydrology has become a growing concern [1, 3]. Xu et al. [4] indicated that the impact of climate change on water resources and the potential implications for water resource management have been the focus of a multitude of scientific investigations over the past two decades. Jiang et al. [5] argued that changes in the availability of local and regional water resources are one of the most important and immediate effects.

The report of the IPCC (Intergovernmental Panel on Climate Change) (2013) concluded that the speed of climate change is faster than the estimation in the past and the increase of the global mean atmospheric temperature will likely be more than 1.5 and 2.0 °C by the end of this century (compared with 1850-1900). This increase process will not end by 2100, and the increase of temperature only might be controlled within 2.0 °C with the most strict emission scenario. As to the precipitation, generally, the contrast in precipitation between wet and dry regions and between wet and dry seasons will increase. Changes in the global hydrological cycle in response to climate change over the 21st century will not be uniform [6], and need to carry out further research. Increased temperatures in winter and spring will affect the precipitation phase, and, as a consequence, the snow/precipitation

IOP Conf. Series: Earth and Environmental Science **41** (2016) 012032 doi:10.1088/1755-1315/41/1/012032

ratio and the volume of water stored in snow cover will be changed [7]. Therefore, the hydrology of rivers in the northern hemisphere is sensitive to climate change [7]. Meanwhile, there still remain some uncertainties regarding the responses of the hydrological cycle to climate change, such as uncertainties in future climate projections and in the response of hydrological processes [8].

Also, the impact of climate change on vegetation growth is significant, especially the influence on crop growth, water consumption and grain production [9]. The northeast of China is the main grain producing region, and as such, the influence of climate change on the hydrological cycle, water resources, and crop growth in this area deserve more attention. Crop responses to rising temperatures are nonlinear, and may be different below and above threshold points [10]. Changes in precipitation variability and air humidity affect water consumption and the degree of water stress, resulting in changes in transpiration rates at leaf level [11].

To date, there have been many studies spanning various disciplines documenting climatic predictions around the word. However, there are still some limits associated with these predictions [12]. For example, some previous studies have predicted future climate scenarios simply from analysis of historical data and their trends [13]. This method blatantly ignores the uncertainty in climate change and the increased frequency of abnormal climates event in recent decades, so historical data and their trends are unreliable, and cannot represent a realistic picture of the current and future situations [14].

Climate scenario model takes account of this climate uncertainty and has been considered as an effective means to predict future climate [15-17]. GCMs (General Circulation Models) are based on the fundamental laws of physics. These models consider the content of greenhouse effect in the atmosphere. Meanwhile, it is also recognized that different climate model provide different projection. So it is recommended to consider several GCMs to obtain the better projection when undertaking any study of impacts on the hydrological regime [18]. GCMs are also subject to several limitations, in particular the limited spatial detail of the relatively coarse grid of a GCM, and their consequent inadequacy to model small scale variability. Gratifyingly is that a variety of methods, such as dynamical downscaling and statistical downscaling, have been developed to create finer spatial resolution data sets [19]. Therefore, at present, GCMs provide us with the most reliable and robust methods for assessing the response of the climate system to changes in forcing [20], and have been widely used in many studies [3, 21, 22].

Located in the cold temperate zones, Northeast China Plain is a sensitive area of climate changes [23]. Its high latitude and black soil ecosystem also serve to exacerbate its sensitivity, and the changes in climate are obvious [24, 25]. In the context of warming, the climate in there warmed significantly [26], and the frequency of severe dryness is increased [27]. Consecutive dry days, rainstorms and floods occurred frequently in the 1990s [28]. The precipitation extremes over 1-5 days increased significantly [29]. Several studies showed that the cloud amount in northeast China had decreased strongly since 1950 [30], thus generating more events of serious drought [27]. As the major agricultural region in northeast China, studies about the impact of climate change on water resources in the Northeast China Plain are very significant and will provide information to promote protection of soil hydrology and ecology [14]. That is, because the variability in climate in the Northeast China Plain plays a vital role in the aspect of ecological and agricultural water requirements [31], it's necessary to analyse the impact of climate change on water resources in the future in there. To date, the water resources prediction researches are not so numerous in this region, and the future climate prediction method is lack of reliability in the existing researches. Coupling research of climate prediction and hydrological simulation and their uncertainty analysis are lack of in-depth study. Therefore, the coupling prediction of water resource with mechanism model in the context of warming in this basin is urgent.

In this study, we set up a semi-distributed hydrological model, and analysed the future climate in the Second Songhua River Basin in the Northeast China Plain from 2040 to 2069 from sixteen GCMs. The influence of climate change on hydrological cycle and water resources in this basin has been evaluated by using the semi-distributed hydrological model with this future climate. Then, some countermeasures for agriculture addressing climate change in this basin have been proposed. This

IOP Conf. Series: Earth and Environmental Science **41** (2016) 012032 doi:10.1088/1755-1315/41/1/012032

analysis would allow us to evaluate and draw conclusion about the influence of climate change on water resources.

2. Study area

The Second Songhua River Basin, located between 124.62°-128.8°E，41.75°-45.4°N in the Northeast China Plain. This river originates from Tianchi Lake in the Changbai Mountain and travels 958km from southeast to northwest draining a total area of 73400 km^2, and flows through 26 counties in Jilin Province (Figure 1). The 90- m resolution digital elevation map shows that the basin elevation ranges from 121 m to 2729 m and decreases from the southeast to the northwest. The Second Songhua River Basin is mainly covered by black soil in northeast China, and has become one of the most important regions for cereal grain production. In this basin, the main types in the southeast mountains are the dark brown earth and albic bleached soil, and the main types in the northwest plain are chernozem, meadow soil and black soil. To date, many valleys and tablelands between the Changbai Mountains have been reclaimed for cultivation so far. The main land use types in this basin are forest and agricultural land, and rice is the main crop (Figure 2, Table 1). Forest mostly situates in the southeast mountains, which is the headstream of Songhua River, and paddy land mostly situates in the northwest plain.

The region has a mid-latitude monsoon climate, and the amount of precipitation decreases from southeast to northwest. Precipitation mainly occurs in summer, and is accompanied by high temperatures [14]. According to the long-term history data, the average annual precipitation ranges from 400 mm to 600 mm, 70% of which occurs in the period from June to September. Daily average temperatures are below freezing from late November to early March of the following year. According to the statistics data of gauging station in the river downstream outlet, the observed average monthly stream flow from July to September is greater than 500 cubic meters per second.

According to the statistical analysis of meteorological disasters on crops during recent four decades in Jilin Province, the rice is sensitive to the impact of flood and the maize is more sensitive to the drought than to the flood [32]. Through the quantitative evaluation of the meteorological disaster from 1971-2008, the influence of drought on agricultural production has become more serious over time. Thus the massive crop in this basin makes it sensitive to climate change. In their study of climate change and extreme climate events in northeast China, Zhao [33] indicated that there was a significant increasing trend in the annual mean temperature in northeast China, which, over the last 40 years, was higher than both the global and national rate. In the Jilin Province, the impact of drought on the agriculture is larger than the sum of flood, wind and hailstone and cold impact, and the drought impact increased seriously [34]. In the analysis of characteristic of drought and flood evolution from 1950-2010, Yang et al. [35] indicated that the drought has happened 30 times and the flood 25 times in Jilin Province. The drought increased obviously especially from the 1970s, and the flood frequency has the highest in1980s and then has a small decrease. Jilin province undergoes the characteristic of drought and flood coexists currently. As the major agricultural region with black soil ecosystem and high latitude in northeast China, studies about the impact of climate change on water resources in this area are very significant and will provide information to promote protection of soil hydrology and ecology [13].

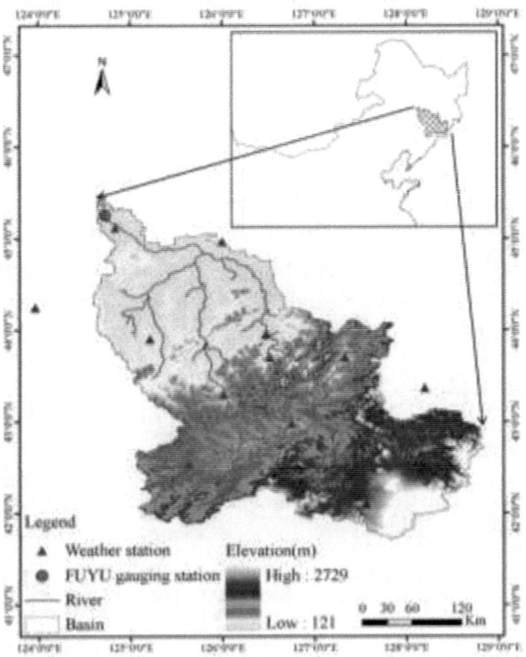

Figure 1. Geographic information of the Second Songhua River Basin and its stream flow gauging station and weather stations.

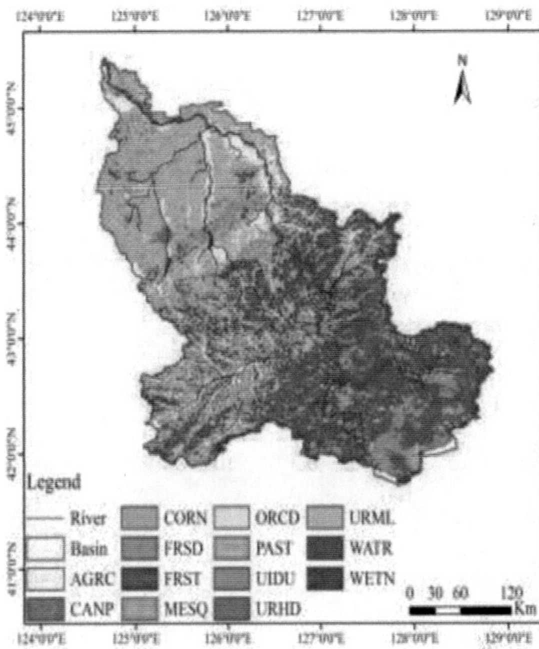

Figure 2. Types of land use in the Second Songhua River Basin.

Table 1. The definition of land use type ID and the area proportion.

ID	Ratio (%)	Area (km^2)	Definition
FRST	41.3	30328.9	Forest-Mixed
CORN	32.6	23928.4	Agricultural Land-dry land
AGRC	8.8	6429.8	Agricultural Land-paddy field
FRSD	4.4	3251.6	Forest-Deciduous
URML	3.8	2759.8	Residential-Med/Low Density
MESQ	2.5	1842.3	Shrub
PAST	2.0	1475.3	Pasture
WATR	1.7	1247.8	Water
ORCD	0.9	638.6	Orchard
URHD	0.8	565.2	Residential-High Density
WETN	0.6	469.8	Wetlands-Non-Forested
CANP	0.6	433.1	Bare land
UIDU	0.1	36.7	Industrial

3. Hydrological model

3.1. Model description

Hydrological model is a useful tool in analysing the hydrological processes in large basin. Physical model represents the underlying hydrological and land surface processes in greater detail than conceptual or statistical models [36]. SWAT model is a physically-based model to evaluate watershed hydrology and water developed by the United States Department of Agriculture (USDA) [37]. Studies have demonstrated that it performs especially well when simulating hydrological, environmental, and biological processes and it has been applied in many parts of the world for a range of climate conditions and resolutions [38-40].

Daily weather data have to be input into the model. Climate change scenarios for precipitation and temperature can be easily manipulated as the weather input files. After calibration and validation of its parameters, the SWAT model was used in this study to evaluate and understand the impacts of climate change on hydrologic cycle and water resources. The Nash-Sutcliffe (E_{NS}) coefficient of efficiency [41] and the Deterministic Coefficient (R^2) were used as the objective functions for optimizing the model performance. They are defined as:

$$E_{NS} = 1 - \frac{\sum_{i=1}^{n} (Q_{m,i} - Q_{s,i})^2}{\sum_{i=1}^{n} (Q_{m,i} - Q_{ma})^2} \tag{1}$$

$$R^2 = \frac{\left[\sum_{i=1}^{n}(Q_{m,i} - Q_{ma})(Q_{s,i} - Q_{sa})\right]^2}{\sum_{i=1}^{n}(Q_{m,i} - Q_{ma})^2 \sum_{i=1}^{n}(Q_{s,i} - Q_{sa})^2} \tag{2}$$

Where $Q_{m,i}$ (m^3s^{-1}) and $Q_{s,i}$ (m^3s^{-1}) are the observed and simulated stream flows for a time step i, Q_{ma} (m^3s^{-1}) is the average observed stream flow, and Q_{sa} (m^3s^{-1}) is the average observed stream flow for the whole simulation period.

Calibration and validation are performed in SWAT Calibration and Uncertainty Programs (SWAT-CUP), which is a computer program for calibration of SWAT model. In this study, the sensitivity analysis and uncertainty analysis of parameters, calibration and validation are performed in SWAT-CUP. In order to do these, the results of SWAT model and observed data are required input into the SWAT-CUP.

3.2. Uncertainty analysis

In the application of hydrological model with GCMs coupling model, the uncertainties in predicting water resources include climate modelling uncertainty and the uncertainty associated with the hydrological model. There are many existing studies that report climatic projections and uncertainty for different emission scenarios [7, 42] and downscaling methodologies [43, 44]. Parallel research activities, such as those discussed by Semmler et al. [43] have focused on generating an ensemble of climate simulations based on different GCMs and multiple future climate scenarios. In that study, combination of different GCMs was used to evaluate the projected changes in stream flow in Ireland. To date the use of an ensemble median (or mean) is an effective means to improve the outcome of climate simulations that is often better than any individual future climate projection [45]. In order to reduce the uncertainty, this study used the ensemble average projected temperature and precipitation from 16 GCMs for A1B emission scenario (medium CO_2 emission, and leveling-off of emission in the mid-21st century) downscaled to a resolution of 0.5 °C for the period from 2040 to 2069. These GCMs result are created by using fine resolution observed data and statistical downscaling method to generate the climate scenario at fine spatial scale. This ensemble GCMs for A1B emission scenario with downscaling method can reduce the uncertainty associated with the climate change modelling.

Parameter is the other main source of uncertainty in hydrological model. The questions of how to identify the sensitive parameters and how to quantify the uncertainty of these parameters have been widely investigated [46].

Three soil parameters chosen for this study are the available water content (*AWC*), bulk density (*BD*), and soil saturated hydraulic conductivity (*Ks*). They can all be obtained by calibrating the hydrological model based on the optimization algorithm [47]. However, if these significant soil parameters take part in the calibration may increase the "Equifinality" effect (same simulations with different parameter values), which can increase the uncertainty of hydrological model [48]. And the values obtained by calibration ignored the physical significance of parameters. Through comparative analysis of different computation methods, this study calculated these parameters using the following soil transfer functions, which have been widely used [49, 50].

$$AWC = \theta_{33} - \theta_{1500} \quad (3)$$

$$BD = (1-\theta_S) \times 2.65 \quad (4)$$

$$Ks = 1930 \times (\theta_S - \theta_{33})(3-\lambda)$$
$$\lambda = 1/B$$
$$B = \frac{[\ln(1500) - \ln(33)]}{[\ln(\theta_{33}) - \ln(\theta_{1500})]} \quad (5)$$

Where Θ_{33} and Θ_{1500} are the water contents at capillary pressures of 33 and 1500 kPa, Θ_s is the saturated soil moisture.

This study adopted the General Likelihood Uncertainty Estimation (GLUE) method to analyse the parameter uncertainty of SWAT model, and chose the E_{NS} as the likelihood objective function. The parameter uncertainty was assessed by E_{NS}, R^2, the posterior distribution of the parameter cumulative probability, *P*_factor and *R*-factor. In Figure 3, x-axis is the value range of parameter, and y-axis is the cumulative probability. In the simulation that used the calculated parameters, the posterior distributions of the other parameters were closer to the prior distribution, which indicated that calculated soil parameters reduced the impacts of other parameters on the posterior distribution. In Table 2, the *P*_factor and *R*-factor were raised by Abbaspour et al. [51] as the indexes of model uncertainty by analysing the cumulative probability distribution output variables in the 2.5 % and 97.5 % quintiles. The *P*_factor indicates the probability that the measured values are located within the confidence interval (95 PPU), while the R-factor indicates the width of the confidence interval. When the *P*_factor is close to 1 and the R-factor is close to 0, the uncertainty is less. In this study, the confidence interval (95 PPU) contained more measured values with calculated soil parameters, and the R^2 and E_{NS} results were generally higher than those simulated with calibrated soil parameters. Thus, the simulation obtained from calculation is closer to the actual hydrological regime of the river. As a result, the high sensitive soil parameters (AWC, BD, Ks) obtained from this calculation method instead of calibration can obviously reduce the uncertainty of hydrological model.

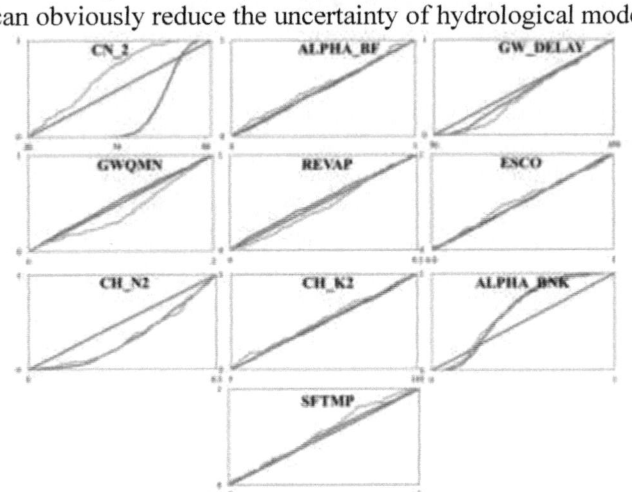

Figure 3. Posterior cumulative probability distributions of the other ten parameters in two conditions (—priori cumulative probability distribution;—simulation with calculated soil parameters;— simulation with validated soil parameters).

IOP Conf. Series: Earth and Environmental Science **41** (2016) 012032 doi:10.1088/1755-1315/41/1/012032

Table 2. Simulation results of two methods to obtain soil parameters.

Situation	*P*_factor	*R*-factor	R^2	E_{NS}
soil parameters obtained from calculation	0.7	0.9	0.69	0.67
soil parameters obtained from validation	0.61	0.87	0.64	0.58

3.3. Model setup and calibration

To set up the SWAT model, the observed stream flow data are needed to calibrate the SWAT model parameters. Due to limitations in the observed data, this study only used the observed stream flow data from one hydrological gauging station, the Fuyu station (45.17N, 124.8E, the mouth of the river basin). The data at the Fuyu station can be used to represent the complete hydrological process of the whole basin, so it is appropriate to use data from this station for the calibration. Daily observed data for the period from 1980-1986 and 2006-2011 were used. After the uncertainty analysis about these three soil parameters, calculating by soil transfer functions can reduce the uncertainty of hydrological model. Then, there are seven other parameters need to be calibrated, as listed in Table 3. These parameters of each watershed were obtained through SWAT-CUP based on the observed stream flow at the Fuyu gauging station.

Table 3. Model parameter description.

Parameter	Description	Unit	Range		Value
v__CN2.mgt	The runoff curve number	-	20	90	80.555
v__ALPHA_BF.gw	Base flow alpha factor	days	0	1	0.79664
v__GW_DELAY.gw	Groundwater delay time	days	30	450	253.33
v__GWQMN.gw	Threshold depth of water in the shallow aquifer required	mm H_2O	0	2	1.48305
v__GW_REVAP.gw	Groundwater "revap" coefficient	-	0	0.2	0.19834
v__ESCO.hru	Soil evaporation compensation factor.	-	0.01	1	0.9850
v__CH_N2.rte	Manning's "n" value for the main channel	-	0	0.3	0.15494
v__CH_K2.rte	Effective hydraulic conductivity in main channel alluvium	mm/hr	5	130	97.822
v__ALPHA_BNK.rte	Base flow alpha factor for bank storage	days	0	1	0.20515
v__SFTMP.bsn	Snowfall temperature	°C	-5	5	-2.1091

Meteorological data drive the hydrological model, and should include at least daily precipitation and daily max/min temperature data. For this study, we used daily meteorological data from 14 weather stations (Table 4), obtained from the Chinese Meteorological Data Sharing Service System. The observation period is from 1980 to 2011, and the data format is daily. Other input data for the SWAT model included a 90 m resolution Digital Elevation Map, a 90 m resolution land cover map, a 1:10 00000 resolution soil map, and soil texture information. All the input data were fully quality controlled and their properties and sources are tabulated in Table 5. The SWAT model can fully use the spatial variation of these 14 weather stations to simulate the hydrological process of each sub basin. Based on the observed stream flow data from the Fuyu station, the calibration and validation periods were from 1980-1986 and from 2006-2011, respectively.

Table 4. The information of weather station used in this paper.

Station name	Latitude	Longitude	Elevation
Qianguoerluosi	45.12	124.83	134.70
Changling	44.25	123.97	189.30

Sanchahe	44.97	126.00	196.60
Changchun	43.90	125.22	236.80
Yantongshan	43.30	126.02	271.60
Yongji	43.70	126.52	229.50
Jinlin	43.95	126.47	183.40
Jiaohe	43.70	127.33	295.00
Dunhua	43.37	128.20	523.70
Meihekou	42.53	125.63	339.90
Huadian	42.98	126.75	263.30
Jinyu	42.35	126.82	549.20
Donggang	42.10	127.57	774.20
Songjiang	42.53	128.25	591.40

Table 5. Data used in SWAT modelling in the Second Songhua River Basin.

Description	Remark
Digital Elevation Map (DEM)	From GLOBE data ,90 arc second resolution, Available at: http://www.ngdc.noaa.gov/mgg/topo/globe.html
Land cover map	Interpreted by remote sensing data,90 arc second resolution
Soil map	From China Soil Scientific Database,1:10 00000 resolution, Available at http://www.soil.csdb.cn/
Soil texture	From China Soil Scientific Database, Available at http://www.soil.csdb.cn/
Daily precipitation	From China Meteorological Date Sharing Service System, Available http://cdc.cma.gov.cn/home.do
Daily max/min temperature	From China Meteorological Data Sharing Service System, Available at: http://cdc.cma.gov.cn/home.do
Monthly precipitation, max/min temperature, wind speed, solar radiation, relative humidity	From China Meteorological Data Sharing Service System, Available at: http://cdc.cma.gov.cn/home.do
Daily observed stream flow	From China hydrological yearbook

After calibration and validation, these seven parameters were then used in SWAT model to simulate the hydrological processes across the whole basin in reference period and future scenario.

4. Future scenario of climate change

The future regional mean precipitation and maximum and minimum temperatures projected by the ensemble of 16 GCMs under an A1B emission scenario were derived for the Second Songhua River Basin for 2050s (2040-2069) from the Climate Wizard dataset (available at: http://www.climatewizard.org/index.html) . The data at finer spatial scale in this dataset are created by downscaling techniques and are revised by past observed data [52]. In this dataset, the relative change in monthly precipitation and the absolute change in the mean temperature were downscaled to a resolution of 50 km using the two-step bias correction spatial downscaling method. This method is simple and inexpensive, and can be repeatedly rerun to generate large ensemble of daily precipitation series at the point or catchment scale, which has been applied extensively at regional, continental, and global scales [42, 52, 53]. The future climate prediction of the 16 GCMs is provided as the monthly relative change and absolute change. The percentage changes in the monthly mean precipitation between the future scenario (2050s) and the reference period were used to scale the daily precipitation data for the 2050s. The absolute change in the monthly mean temperature and the reference period data were directly used as input data for the 2050s. The reference period data were the observed daily data of 14 weather stations from 1980 to 2011, which represented the current climate situation across the whole basin. The future scenario data obtained from the relative and absolute changes of 16 GCMs in the 2050s reflected the future climate situation across the whole basin. The hydrological regime differences between these simulations of reference period and future scenario are considered as the impact of climate change on hydrological processes.

The average relative changes in the monthly mean precipitation and the absolute change in the monthly mean temperature of the whole basin between the reference period and the future climate scenario in the 2050s, respectively, are shown in Figure 4. With the exception of October, the precipitation will increase throughout the year. For the period from December to February, precipitation is predicted to increase significantly (30 %-33.7 %). Similarly, the temperature increase is greater in winter than in summer. The monthly mean temperature is predicted to increase in all of the twelve months, with an average increase of 2.93 °C. These changes were in general agreement with the IPCC report (2014). The increased temperatures in the winter and spring seasons will accelerate the speed of snow melting and will result in increased heat availability. The changes in precipitation and temperature will have significant effects on plant growth and ecosystem health in the Second Songhua River Basin.

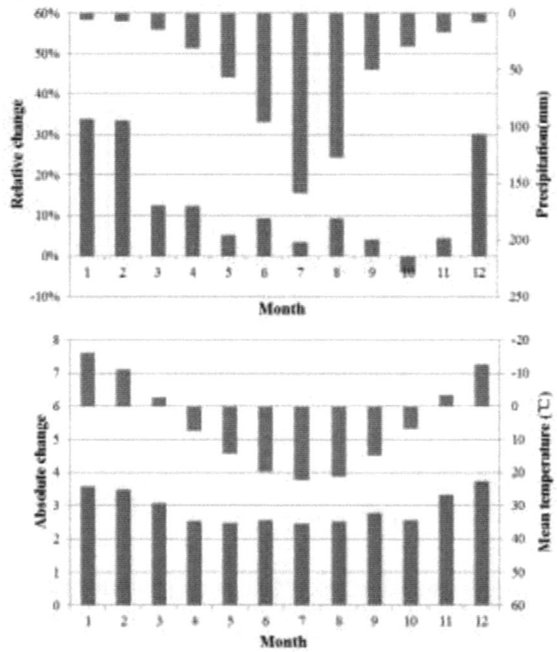

Figure 4. Basin averaged changes (red columns) of monthly mean precipitation and temperature between baseline (blue columns) and future climate scenario in 2050s.

5. Result and discussion

5.1. The performance of SWAT model

Stream flow data collected at the Fuyu gauging station downstream of this basin reflects the river discharge of the whole basin and was used to calibrate and validate the model. The daily observed and simulated stream flow for the Fuyu gauging station is shown in Figure 5. The simulation of calibration period shows better agreement with the observed data than the simulation of validation. As the impacts of climate change will be explored at the monthly scale at Fuyu, the monthly simulated stream flow was examined and is demonstrated in Figure 6. The monthly simulation both in calibration and validation is better than the daily one. According to the analysis, the average relative error of yearly total volume between observed stream flow and simulated stream flow is 0.298, and the average relative error of peak value is 0.215. The daily and monthly model performance for the calibration and validation period is tabulated in Table 6, which indicates that the simulation of stream flow by SWAT model is reasonable. The good performance at both the daily and monthly scale demonstrates that the SWAT model is an appropriate choice for the impact assessment being planned.

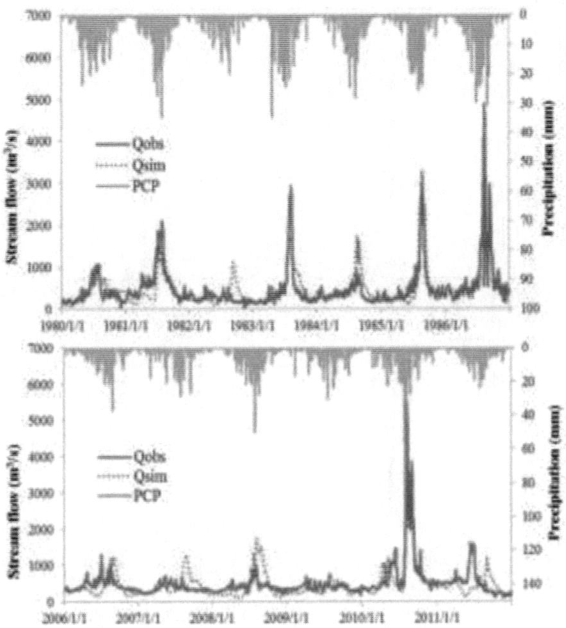

Figure 5. Daily stream flow simulations at Fuyu station (Qobs: observed stream flow; Qsim: simulated stream flow; PCP: precipitation).

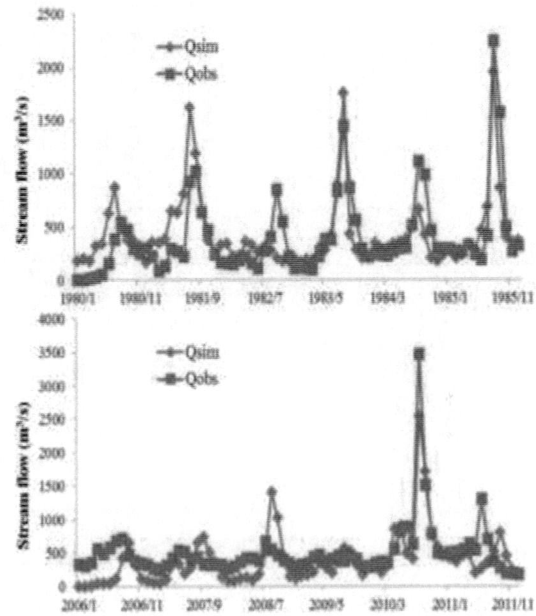

Figure 6. Comparison between monthly averaged simulated and observed stream flow at Fuyu station (Qobs: observed stream flow; Qsim: simulated stream flow).

Table 6. The model performances for calibration and validation periods.

Discharge Type	Process	Simulation Period	R^2	E_{NS}
Daily	calibration	1980-1986	0.7	0.69
	validation	2006-2011	0.5	0.46
Monthly	calibration	1980-1986	0.72	0.67
	validation	2006-2011	0.56	0.48

5.2. Hydrological response to climate change

The daily meteorological data for the 2050s A1B scenario as the input data for the SWAT model to simulate hydrological processes in the future. The impact of climate change on the hydrological regime in the Second Songhua River Basin is illustrated by comparing the ET of the whole basin and the stream flow data at the Fuyu gauging station between the reference period and 2050s. The monthly relative changes in ET and stream flow are shown in Figure 7. As regards the relative change, in most months, the ET is predicted to increase in the 2050s, with particularly obvious increases from December to February. The absolute change differs from the relative change; the largest absolute change is predicted to occur in May (19 mm), while in the other months the changes are all less than 10 mm. Because the mean temperatures across the whole year are higher in the 2050s, more energy is available for driving evaporation or transpiration of soil water and intercepted water, which may cause the increases in ET in most months. Meanwhile, the reduced ET predicted for September and October, and the dramatic increase in ET from December to February for the 2050s are in general agreement with changes in precipitation for the 2050s, which indicates that precipitation has an impact on the availability of water for evapotranspiration.

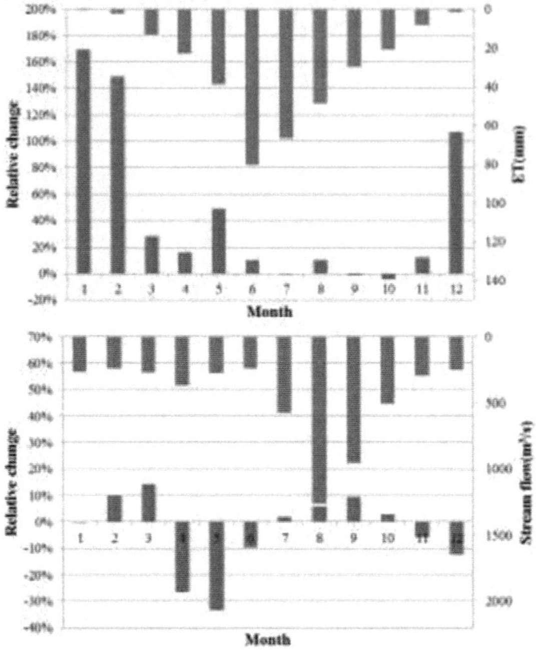

Figure 7. Averaged changes (red columns) of monthly mean ET of whole basin and stream flow at Fuyu station between baseline (blue columns) and future climate scenario in 2050s.

Stream flow is predicted to increase from July to October and from February to March in the 2050s. This is a consequence of increased rainfall in the same period. Due to the cumulative effects of increased rainfall in previous months and lower ET, an increase in stream flow was predicted for October (relative change of 3 %). The absolute increase in stream flow in these six months ranges from 11.4 mm to 89 mm (September). Although an obvious increase in precipitation (absolute increase from 0.7 mm-8.9 mm) was predicted for the five months from November to January and from April to June, the increase in ET was greater (absolute 0.9 mm-19 mm), with the result that absolute decreases in stream flow in these six months ranging from 17 mm to 96 mm (April) were predicted. Therefore, both precipitation and evapotranspiration have a significant impact on stream flow.

From the viewpoint of water resource availability, these results demonstrate that, especially in spring (particularly April and May), there will be a decrease in the available water resources under the 2050s A1B scenario, which may lead to a lack of irrigation water. As a consequence, agricultural drought may occur under the higher temperatures. Meanwhile, the increased stream flow will enhance the risk of floods in summer (especially August and September). The AR5 (Fifth Assessment Report) of IPCC Working Group II also pointed that flood hazards will increase over about half of the globe with variability at the catchment scale (medium agreement, limited evidence) [5]. The prediction in the Second Songhua River Basin is in accord with the result of Hirabayashi et al. [54] that Northeast China will face increased risk of flood hazard. Drought and flood due to climate change may challenge existing water management systems. In the Northeast China Plain, the staple food crops (e.g., wheat, maize, rice, and soybean) need water resources from April to October. This 2050s scenario of decreased stream flow in spring and increased stream flow in summer will have a great impact on the harvest.

5.3. Countermeasure for agriculture addressing climate change in this basin

As to the water resource management, IPCC-AR5 recommends that water management should be moved from the traditional "predict and provide" approach towards adaptive water management and the adoption of resilient approaches. Adaptive approaches to water management can successfully address the uncertainty due to climate change. There are two kind of framework about adaptive water management: the three-step framework which explicitly incorporates climate change and its uncertainty and threshold-scenario framework based on risk assessment. Adaptive techniques involve learning from experience and the development of flexible solutions that are resilient to uncertainty [6]. Therefore the integrated water resource management as a promising instrument for exploring adaptation to climate change should be carried out in this basin.

The changed temperature, ET, water and other agricultural climate resources change, sequentially lead to the changes of the crop yield, quality and planting area. In order to reduce these impacts and disasters caused by climate change, this paper put forward some adaptive measures. Making full use of the advantage of the increased accumulated temperature, improved variety with high photosynthesis ability and wide adaptability would be selected to promote in this basin. According to future climate resource redistribution and the new pattern of agricultural meteorological disasters, crop varieties layout should be improved. The rational use of water resources deserved more attention and needs to be strengthened. Irrigation and water conservancy construction, water-saving agriculture system and farmland shelterbelt are beneficial to the improvement of the agricultural ability to adapt to climate change. Agricultural disaster forecast and early warning research should be invested more. Strengthening the weather forecast system can timely avoid agriculture suffering huge losses.

6. Conclusion

Research about the availability of water resources under future climate change scenarios plays an important role for the future of agricultural production. To evaluate the influence of climate change on water resources in the Second Songhua River Basin and to predict the hydrological processes in this basin, we applied GCMs and the SWAT models. After demonstrating the reasonable of model in the basin, the GCMs meteorological variables were input to the SWAT model to predict hydrological processes in the 2050s.

As the main grain base, the staple food crops need large amount of water resource from April to October. However, the prediction of hydrological process in the 2050s showed that this basin may face water deficits in spring season, and that there would be enhanced risk of summer flood. In spring, the water resources may not meet the crop irrigation demands, which would impact on crop growth. Water resources must be regulated and saved to improve crop security. In summer, the more serious flood damage would result in soil erosion and would cause huge social and economic losses. Therefore, flood forecasting and warning, and agricultural water supply and drainage facilities deserve great attention to implement the integrated water resource management in this basin. The results of this study may be beneficial for developing practical water resource management policies in the Second Songhua River Basin.

As to the water resource management, this paper recommends that traditional water management should be moved towards adaptive water management. As to the comprehensive response of agriculture to the climate change, artificial weather control ability and emergency response capacity should be strengthened in this region in the future. To make the policy and decision making in a more targeted manner, an analysis about the amount of water needed is necessary. In order to actively confront the impact of climate change on agricultural production in the future, the future water resources prediction research needs aroused widespread attention. The country also should participate into the response of climate change, including the adjustment of agricultural construction of water conservancy facilities and the establishment of agricultural catastrophe risk fund, etc.

Acknowledgments
This study was supported by the Special Fund for Research on Public Interests, Ministry of Water Resources (No.: 201401036), the National Key Technology R&D Program (No.2016YFC0401308), the National Natural Science Foundation of China (Grant No.: 41201018, 91125015), and the Fundamental Research Funds for the Central Universities.

References

[1] Crossman J, Futter M N, Oni S K, Whitehead P G, Jin L, Butterfield D, Baulch H M and Dillon P J 2013 Impacts of climate change on hydrology and water quality: Future proofing management strategies in the Lake Simcoe watershed, Canada *J. Great Lakes Res.* **39** 19-32
[2] SWCS 2003 Conservation implications of climate change: Soil erosion and runoff from cropland A report from the Soil and Water Conservation Society
[3] Sun W, Wang J, Li Z, Yao X and Yu J 2014 Influence of climate change on water resources availability in Jinjiang Basin, China *The Scientific World J.* **2014** 908349
[4] Xu C, Widén E and Halldin S 2005 Modelling hydrological consequences of climate change-progress and challenges *Adv. Atmos. Sci.* **22** 789-97
[5] Jiang T, Chen Y D, Xu C, Chen X, Chen X and Singh V P 2007 Comparison of hydrological impacts of climate change simulated by six hydrological models in the Dongjiang Basin, South China *J. Hydrol.* **336** 316-33
[6] Alexander L, Allen S and Bindoff N L 2013 Working Group I contribution to the IPCC Fifth Assessment Report climate change: the physical science basis, summary for policymakers. Published online September 27
[7] Boyer C, Chaumont D, Chartier I and Roy A G 2010 Impact of climate change on the hydrology of St. Lawrence tributaries *J. Hydrol.* **384** 65-83
[8] Giorgi F and Francisco R 2000 Evaluating uncertainties in the prediction of regional climate change *Geophys. Res. Lett.* **27** 1295-8
[9] Tan G and Shibasaki R 2003 Global estimation of crop productivity and the impacts of global warming by GIS and EPIC integration *Ecol. Model.* **168** 357-70
[10] Schlenker W and Roberts M J 2009 Nonlinear temperature effects indicate severe damages to U.S. crop yields under climate change *Proceedings of the National Academy of Sciences* **106** 15594-8
[11] Brown R and Rosenberg N 1994 Sensitivity of crop yield and water use to change in a range of climatic factors and CO_2 concentrations: a simulation study applying EPIC to the central USA *Agr. Forest Meteorol.* **83** 171-203
[12] Deng X, Zhao C and Yan H 2013 Systematic modeling of impacts of land use and land cover changes on regional climate: a review *Adv. Meteorol.* **2013** 1375-83
[13] Araujo M B and New M 2007 Ensemble forecasting of species distributions *Trends in Ecology & Evolution* **22** 42-7
[14] Shi Q, Lin Y, Zhang E, Yan H and Zhan J 2013 Impacts of Cultivated Land Reclamation on the Climate and Grain Production in Northeast China in the Future 30 Years *Adv. Meteorol.* **2013** 153-6
[15] Bates B, Kundzewicz Z W, Wu S and Palutikof J 2008 Climate change and water *Intergovernmental Panel on Climate Change* (IPCC)
[16] Whitehead P, Wilby R L, Battarbee R W, Kernan M and Wade A J 2009 A review of the potential impacts of climate change on surface water quality *Hydrolog. Sci. J.* **1** 1-6
[17] Christensen J H and Christensen O B 2007 A summary of the PRUDENCE model projections of changes in European climate by the end of this century *Climatic Change* **81** 7-30
[18] Prudhomme C, Jakob D and Svensson C 2003 Uncertainty and climate change impact on the flood regime of small UK catchments *J. Hydrol.* **277** 1-23

[19] Fowler H J, Blenkinsop S and Tebaldi C 2007 Linking climate change modelling to impacts studies: recent advances in downscaling techniques for hydrological modelling *Int. J. Climatol.* **27** 1547-78

[20] Perks L A, Schulze R E, Kiker G A, Horan M J C and Maharaj M 2000 Preparation of climate data and information for application in impact studies of climate change over southern Africa Report to the South African Country Studies for Climate Change Programme

[21] Mo X, Guo R, Liu S, Lin Z and Hu S 2013 Impacts of climate change on crop evapotranspiration with ensemble GCM projections in the North China Plain *Climatic Change* **120** 299-312

[22] Zhang X, Liu W, Li Z and Chen J 2011 Trend and uncertainty analysis of simulated climate change impacts with multiple GCMs and emission scenarios *Agr. Forest Meteorol.* **151** 1297-304

[23] Ye D Z and Fu C B 1994 China's Global Change Research Advance (Part II) Scientia Atmospherica Sinica **18** 498-512 (In Chinese)

[24] Ding Y and Dai X 1995 Temperature variation in China during the last 100 years *Meteorological Monthly* **20** 19-26 (In Chinese)

[25] Zhao D, Zheng D, Wu S and Wu Z 2007 Climate changes in northeastern China during last four decades *Chinese Geogr. Sci.* **17** 317-24

[26] Sun F H, Ren G Y, Zhao C Y and Yang S Y 2005 An analysis of temperature abnormal change in Northeast China and type underlying surface *Scientia Geographica Sinica* **25** 167-71 (In Chinese)

[27] Sun F H, Wu Z J and Yang SY 2006 Temporal and spatial variations of extreme precipitation and dryness events in Northeast China in last 50 years *Chinese Journal of Ecology* **25** 779-84 (In Chinese)

[28] Yao X and Dong M 2000 Research on the features of summer rainfall in northeast China *Quarterly Journal of Applied Meteorology* **11** 297-303 (In Chinese)

[29] Yang S Y, Sun F H and Ma J Z 2008 Evolvement of Precipitation Extremes in Northeast China on the back-ground of climate warming *Scientia Geographica Sinica* **28** 224-8 (In Chinese)

[30] Kaiser D P 2000 Decreasing cloudiness over China. An updated analysis examining additional variables *Geophys. Res. Lett.* **27** 2193-6

[31] Liang L Q, Li L J and Liu Q 2011 Precipitation variability in Northeast China from 1961 to 2008 *J. Hydrol.* **404** 67-76

[32] Ma J, Xu Y and Pan J 2012 Analysis of Agro-meteorological Disaster Tendency Variation and the Impact on Grain Yield over Northeast China *Chinese Journal of Agrometeorology* **33** 283-8 (In Chinese)

[33] Zhao C, Wang Y, Zhou X, Yan C, Liu Y, Shi D, Yu H, Liu Y 2013 Changes in Climatic Factors and Extreme Climate Events in Northeast China during 1961-2010 *Advances in Climate Change Research* **4** 92-102 (In Chinese)

[34] Zhang H, Li J, Lv Z, Wang Y, Lin R and Dai X 2011 Quantitative evaluation on agrometeorological disasters in Northeast China *Journal of Meteorology and Environment* **27** 24-8 (In Chinese)

[35] Yang G, Han D and Cheng Y 2014 Studies on drought and flood transformation in northeast China from 1950 to 2010 *China Water Resources* **2014** 45-48 (In Chinese)

[36] Beven K J and Wiley C 2004 Rainfall-Runoff Modelling: The Primer, 2001 *Land Degradation and Development* **15** 449-50

[37] Neitsch S L, Arnold J G, Kiniry J R and Williams J R 2005 Soil and Water Assessment Tool theoretical documentation: Version 2009 *Comput. Speech Lang.* **24** 289-306

[38] Jayakrishnan R S 2005 Advances in the application of the SWAT model for water resources management *Hydrol. Process.* **19** 749-62

[39] Wu K and Johnston C A 2007 Hydrologic response to climatic variability in a Great Lakes Watershed: a case study with the SWAT model *J. Hydrol.* **1-2** 187-99

[40] Güngör Ö and Göncü S 2013 Application of the soil and water assessment tool model on the Lower Porsuk Stream Watershed *Hydrol. Process.* **27** 453-66
[41] Nash J and Sutcliffe J 1970 River flow forecasting through conceptual models: Part I -A discussion of principles *J. Hydrol.* **10** 282-90
[42] Fu G, Charles S P and Chiew F H 2007 A two-parameter climate elasticity of streamflow index to assess climate change effects on annual streamflow *Water Resour. Res.* **43** 2578-84
[43] Semmler T, Wang S, McGrath R and Nolan P 2006 Regional climate ensemble simulations for Ireland-impact of climate change on river flooding *Geography Compass* **4** 834-60
[44] Murphy C, Charlton R, Sweeney J and Fealy R 2006 Catering for uncertainty in a conceptual rainfall run-off model: Model preparation for climate change impact assessment and the application of GLUE using Latin Hypercube Sampling. In: Proceedings of the National Hydrology Seminar 2004, Tullamore
[45] Reichler T and Kim J 2008 Uncertainties in the climate mean state of global observations, reanalysis, and the GFDL climate model *J. Geophys. Res.* **113** 79-88
[46] [Matott L S, Babendreier J E and Purucker S T 2009 Evaluating uncertainty in integrated environmental models: A review of concepts and tools *Water Resour. Res.* **45** 735-42
[47] Ghaffari G, Keesstra S, Ghodousi J and Ahmadi H 2010 SWAT-simulated hydrological impact of land-use change in the Zanjanrood basin, Northwest Iran *Hydrol. Process.* **24** 892-903
[48] Sun W, Hiroshi I and Satish B 2012 Prospects for calibrating rainfall-runoff models using satellite observations of river hydraulic variables as surrogates for in situ river discharge measurements *Hydrol. Process.* **26** 872-82
[49] Williams J R 1995 the EPIC model In: V.P.Singh (Ed) *Computer models of watershed hydrology* Chapter 25 909-1000
[50] Saxton K and Rawls W 2006 Soil water characteristic estimates by texture and organic matter for hydrological solutions *Soil Sci. Soc. Am. J.* **70** 1569-78
[51] Abbaspour K C, Monireh F and Samaneh S C 2009 Assessing the impact of climate change on water re-sources in Iran *Water Resour. Res.* **45** W10434-5
[52] Wood A W, Leung L R, Sridhar V and Lettenmaier D P 2004 Hydrologic Implications of Dynamical and Statistical Approaches to Downscaling Climate Model Out-puts *Climatic Change* **62** 189-216
[53] Maurer E P, Brekke L, Pruitt T and Duffy P B 2007 Fine-resolution climate projections enhance regional climate change impact studies *Eos Trans. Amer. Geophys. Union* **88** 504
[54] Hirabayashi Y R, MahendranS, Koirala L, Konoshima D, Yamazaki, Watanabe S and Kanae S 2013 Global flood risk under climate change *Nat. Clim. Change* **3** 816-21

Spatiotemporal dynamics of LUCC from 2001 to 2010 in Yunnan Province, China

Z J Li[1], J S Yu[1, 3], X L Yao[1], X Chen[1] and Z L Li[2]

[1] College of Water Sciences, Beijing Normal University, Beijing 100875, China

[2] School of Water Resources and Environment, China University of Geosciences, Beijing 100083, China

E-mail: jingshan@bnu.edu.cn

Abstract. LUCC (Land use and land cover change) is increasingly regarded as an important component of global environmental change and sustainable development. In this study, regional land cover type maps were drawn using the MODIS products from 2001 and 2010 based on the modified classification scheme embodied by the characteristics of land cover in Yunnan. Dynamic change in each type of land cover was investigated by classification statistics, dynamic transfer matrices, and landscape pattern metrics. In addition, the driving factors of LUCC were discussed. The results showed that the land cover types of the Yunnan province, especially woodland (WL), cropland (CL) and grassland (GL), had experienced noticeable changes with an area of about 30% of land during the study period. And there was an obvious vertical distribution pattern for land cover types. The average altitude of different land cover types from the highest to the lowest were unused land (UUT), WL, GL, water (WT), urban and built-up areas (UB) and CL. The average slope for most of the land-cover types did not vary over the past 10 years. Stabilization and homogenization will be the direction of land cover in the future according to landscape metrics analysis. The regional differences of land use structure in the area are strongly influenced by such factors as the geographical position, level of economic development and land use policy. The new policy of land use, Construction of Mountainous Town, would be provided to achieve the economical and intensive utilization of land resources during the rapid development of urbanization and industrialization in Yunnan.

Key words: LUCC, landscape, MODIS, Mountainous Town, Yunnan

IOP Conf. Series: Earth and Environmental Science **41** (2016) 012033 doi:10.1088/1755-1315/41/1/012033

1. Introduction

Land surface conditions impact the regional and global climate through the water cycle and surface-energy budget [1, 2]. LUCC (Land use and land cover change) has become one of the focuses of global environmental change research, which is bound up with the environmental changes and socioeconomic development [3-5]. Regional land cover changes in the developed areas of Eastern China have been paid wide attention with the urbanization and industrialization development since a series of policy implementation in China [6-9]. While land cover changes in Western China have been drawn little concern, particularly with regard to the whole area of Yunnan with complex terrain.

The Yunnan province constitutes the majority of the Yunnan-Guizhou Plateau, accounting for 94% of the mountain and 6% of the plain, and the vertical zone from low to high elevation gradients can be divided into low heat, middle warm, high cold sectors [10]. As a result of typical climate changes and human activities, land cover in the Yunnan province differs from the other area of the same latitude [11]. Some researchers have discussed the land use change of cities, counties and some small watershed. Li et al. [12] studied LUCC from 1976 to 2003 in Xishuangbanna of Yunnan Province, clarified the impacts of land-use change on biomass carbon dynamics, and explored how the alternative land-use scenarios could influence future carbon budgets. Hu et al. [13] analyzed the changes in the ecosystem in relation to LUCC in the Menglun Township in Yunnan, using Landsat TM/ETM and Quickbird dataset from 1988 to 2006. They found that an abrupt variation in land use from ecologically important tropical forests and traditionally managed swidden fields to large scale rubber plantation brought about large losses of ecosystem services in this area. Xu et al. [14] investigated the impacts of local government-driven reforestation on LUCC, as well as its further impacts on livelihoods of upland farmers in Xizhuang watershed by aerial photographs and ASTER satellite imagery for a 15-year period (1987-2002). Fox and Vogler [15] summarized LUCC at eight sites in Montane Mainland Southeast Asiain, including three sites in Yunnan, over the last 50 years. They found that land use and land cover had maintained stable and the minor amount of land use had transfered from swidden to monocultural cash crops. They suggested that two forces, national land tenure policies and market pressures, would increasingly determine land use systems.

Many land cover change researches in Southwestern China have concentrated on counties or small watershed. Land cover change of the whole province on a large scale has not been reported yet. This deficiency is an obstruction of the development and utilization of land resources, ecological environment security, climatic change adaptation and water resource management of this area especially since the implementation of the western development strategy. To cover the shortage, we discuss the temporal and spatial variation of LUCC using datasets in Yunnan province between 2001 and 2010 and its driving forces in this study.

The goals of this paper were to study dynamic change in each type of land cover by classification statistics, dynamic transfer matrices and landscape pattern metrics, and to discuss the driving forces of this change. The study is organized as follows. "Data and methods" shows the study site, data collection and methods. "Results and discussion" displays the land cover change results using different methods, and discuss its driving forces. "Conclusions" draws conclusions according to the results.

2. Data and methods

2.1. The study site

Yunnan province is situated in the southwest of China between 97°31' and 106°11'E, and 21°8' and 29°15'N, covering an area of 394,100 km^2, with elevations ranging from no less than 100 m to 6740 m. Its topography is complex, with over 84% of the total land consisting of mountains, plateaus and hills accounting for 10% of the total land, and less than 6% of the total land composing of local plains (commonly known as "Bazi") and lakes. Three basins are covered in the Yunnan province, namely the Yangtze River Basin, the Pearl River Basin and the Southwest Region Basin. Among them, the Pearl River and the Red River originated from the region (Figure 1).

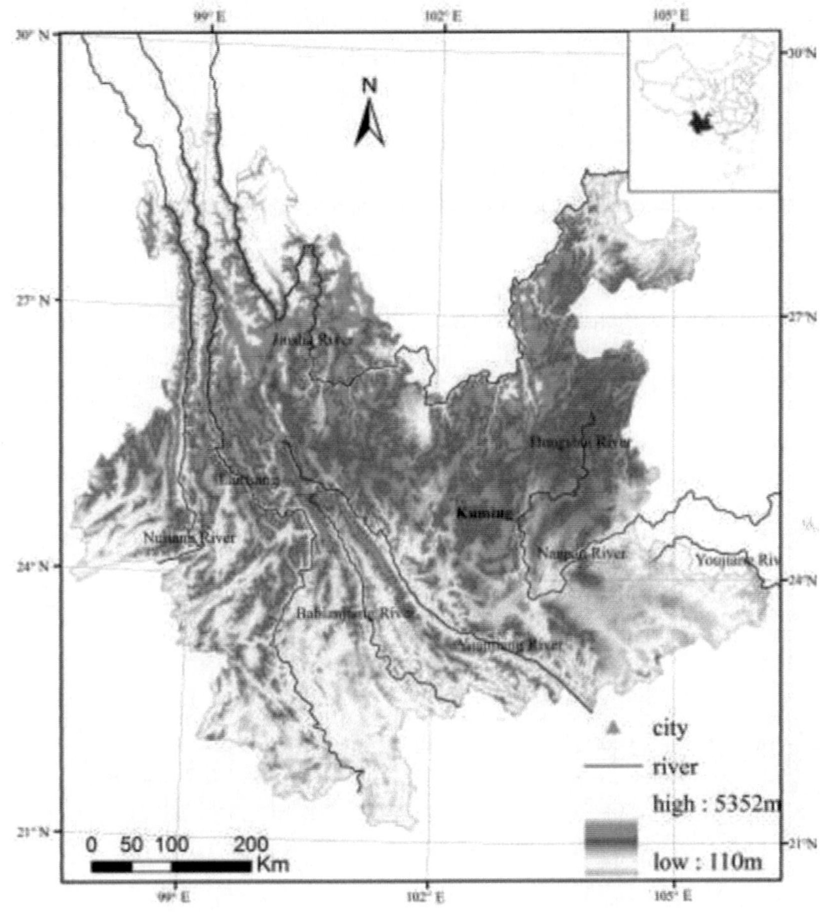

Figure 1. Location of the study area.

The climate of the Yunnan is affected by geographical position and altitude with tropical, temperate zones and highland climate zones, characterized by intense solar radiation and high daily variations in temperature, but low annual temperature fluctuations. The average maximum monthly (July) temperature is 22℃, and the average minimum (January) temperature is 6℃, with a mean annual temperature of 12℃. The mean annual precipitation changes from 300 mm in Benzilan located in the Jinsha river valley, where is a typical dry river valley climate style, to 4,000 mm in the western mountain with altitude more than 2000 m with most of the rainfall in June, July and August[16]. The vegetation mainly consists of tropical rain, seasonal rain, evergreen broad-leaf and cold temperate coniferous forests. The major soils include laterite, latosolic red, red, yellow, luvisols, and paddy soils. The province has a population of 45,966,000, and includes 25 minority nationalities, accounting for one third of the population.

The eco-environmental problems, such as ecological deterioration and soil degradation, have become more and more serious due to climate change and human activities in the recent years. These problems have been threatened the regional economic development and social well-being. This study intends to describe the temporal and spatial changes of LUCC in the Yunnan province for a 10-year period with the hope that the results can be useful for recognizing the driving forces and understanding the eco-environmental feedback mechanisms that may assist decision-makers to find sustainable solutions to optimize landscape patterns for maintaining the ecological security of this region.

2.2. Datasets

The land cover datasets in this study were determined using the MODIS (MOD12Q1-051) data from 2001 and 2010 (https://lpdaac.usgs.gov). The International Geosphere-Biosphere Programme (IGBP) global vegetation classification scheme was used to describe the land use/cover types of the Yunnan province by combining the 17 categories into six types, including cropland(CL), woodland(WL), grassland(GL), urban and built-up areas(UB), water(WT) and unused land(UUL) (Figure 2).

The Digital Elevation Model (DEM) dataset was from Computer Network Information Centre, Chinese Academy of Sciences (http://www.gscloud.cn), and was used to derive the mean elevation and slope of the region.

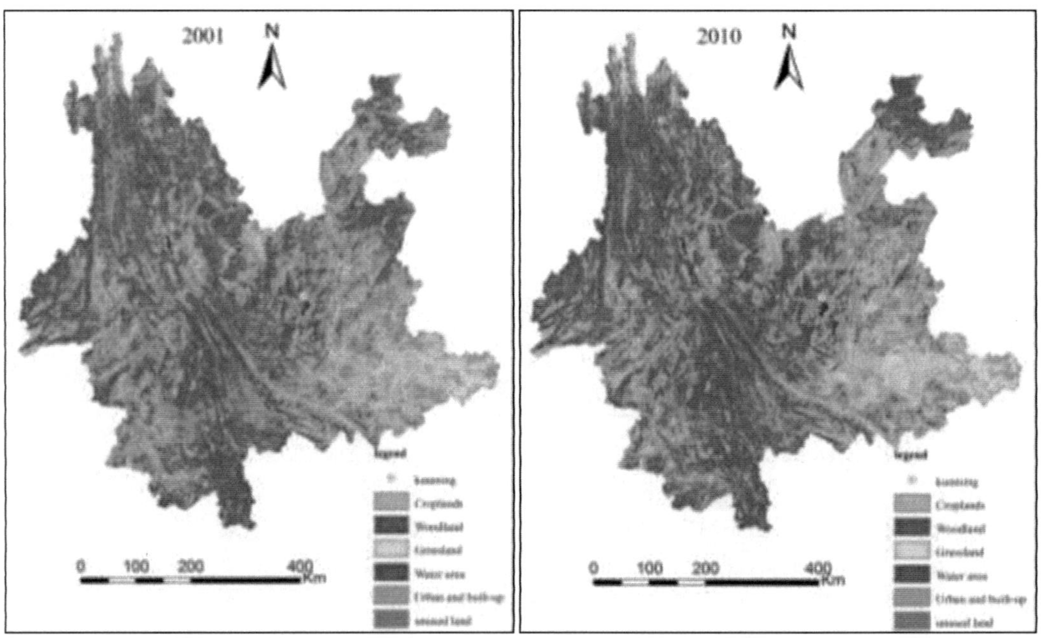

Figure 2. The land cover types in 2001 and 2010 in Yunnan province.

2.3. *Methods*

2.3.1 Methods for Land Use Change Analysis. Some existing models and indicators were used to calculate the extent and rate of LUCC [17, 18]. In this study, three models were selected to compute LUCC in the Yunnan province, a single dynamic index (k), a complex dynamic index (Lc) and a transition matrix [19, 20].

The single dynamic index (k) was used to compute the change of a specific type of land use. It was defined as

$$k = \frac{U_b - U_a}{U_a} \times \frac{1}{T} \times 100\% \tag{1}$$

where k was the dynamic indicator for a specific type of land use/cover. U_a and U_b were the areas of the specific land use/cover type at the start time and at the end time. And T was the time scale under consideration.

The complex dynamic index (Lc) was used to describe the regional differentiation of land use change rate, which was derived as

$$Lc = \left(\frac{\sum_{i=1}^{n} \Delta LU_{i-j}}{2\sum_{i=1}^{n} LU_i} \right) \times \frac{1}{T} \times 100\% \tag{2}$$

where LU_i was the area of land use/cover type i at the start date. ΔLU_{i-j} was the area turning from type i to type j in the computing period(i≠j). n was the number of the land use/cover types. and T indicated the monitoring period. If T was 1 year, Lc was the comprehensive land use/cover dynamic degree in the region.

The land use transition matrix came from the quantitative description of the status and status transition system during the system analysis [21, 22]. In this study, the transition matrix was adopted to conduct a trend analysis of land cover dynamics. The transition matrix and the cross-classification map were calculated by the Raster Calculator module of the software ArcGIS.

2.3.2 Landscape metrics. Landscape pattern analysis, which is based on geometrical features, can actually display the spatial pattern of LUCC and its impact on landscape ecology [23, 24]. Two groups of landscape metrics, including class level metrics and landscape level metrics, were computed using the FRAGSTATS software programmed by the Forest Science Department, Oregon State University, U.S.A. At the class level, three aspects were used to describe the characteristics of the landscape pattern, including area-edge metrics (LPI), shape metrics (PAFRAC) and aggregation metrics (NP, PD, IJI and CONHESION). At the landscape level, four aspects were used, including area-edge metrics (TA, LPI), shape metrics (Shape_AM), aggregation metrics (NP, PD, IJI, ENN_MN, CONTAG, CONHESION and SPLIT) and diversity metrics (SHDI). Although many indices have counterparts at the class and landscape levels, their interpretations may be somewhat different. Class metrics represent the spatial distribution and pattern within a landscape of a single patch type, while landscape metrics, considering all patch types simultaneously, represent the spatial pattern of the entire landscape mosaic. Most of the class indices can be observed as fragmentation indices because they measure the configuration of a particular patch type, whereas most of the landscape indices can be interpreted more broadly as landscape heterogeneity indices because they describe the overall landscape pattern [25].

3. Results and discussion

3.1. Analysis of LUCC

The main land cover types in the Yunnan province included WL, GL, and CL, probably 99% of the total land area as shown in Table 1. Among them, the WL area reached more than half of the total area. WT and UUL were the least common land types. In Figure 2, WL was mainly distributed in the middle and northern area of the Yunnan province, while GL was mostly focused in the southeast in 2010. CL was mainly located in the northeast, although it was also scattered in other places. The spatial distribution pattern was determined by the Yunnan region's unique geography and climate.

Table 1. The land cover types in Yunnan in 2001 and 2010.

land cover types	2001		2010		Change		Rate of Change	
	Area (km^2)	Percent (%)	Area (km^2)	Percent (%)	Area (km^2)	Percent (%)	K (%)	Lc (%)
CL	51576.75	13.42	59404.25	15.46	7827.5	15.18	1.686	0.39
WL	219757.25	57.18	224971.75	58.54	5214.5	2.37	0.264	2.3
GL	108574.75	28.25	95527.25	24.86	-13047.5	-12.02	-1.335	3.14
WT	1174.75	0.31	1248	0.32	73.25	6.24	0.693	5.16
UB	3017.25	0.79	3018	0.79	0.75	0.02	0.003	1.21
UUL	231.5	0.06	163	0.04	-68.5	-29.59	-3.288	1.02

The land cover statistics were used to describe the quantitative change of land cover. From Table 1, it was quite noticeable that the quantity of WL, CL and GL changed obviously during the past 10 years. The area of CL increased from 13.42% in 2001 to 15.46% in 2010 i.e., by 7828 km^2. The area of WL, the dominant land cover type, grew slightly from 57.18% in 2001 to 58.54% in 2010 i.e., by 5421 km^2. However, the area of GL greatly decreased from 28.25% in 2001 to 24.86% in 2010, representing a loss of approximately 13,047 km^2. The area of WT, UB had changed insignificantly.

The area of UUL had dramatically changed over the past 10 years, accounting for a smaller proportion of the total area.

To represent the change speed of land cover, we put to use formulas 1 and 2 to calculate k and Lc, and the results were presented in Table 1. According to k, the change speed of CL, WLL and UUL were fast. Among them, other types were transformed to CL, and the annual area variation for this area was 1.69% contrasted with a corresponding area in 2001. GL as well as UUL was converted into other types, with the annual rates of 1.34% and 3.29%, respectively. The complex dynamic index over the past 10 years was 1.74%. Overall, some changes took place in the regional land use.

For a better explanation of each type of land use transfer to other major types, transfer matrices were used (Table 2), including the area variation and percentage compared to the corresponding area in 2001. It could be observed that CL, WL and GL were largely transformed to other land cover types from 2001 to 2010. Among them, the transformation between WL and GL was the strongest. The area of GL transformed to other types was the largest and reached 53821.3 km^2. The percentage of UUL transformed to other types, which corresponded to the area in 2001, was the largest at 55.55 %. From 2001 to 2010, CL was especially transformed to WL and GL, of which the area added up to 24,477.6 km^2, about 47.46% compared to the CL area in 2001. WL was primarily occupied by CL and GL with an area up to 41,008.6 km^2 or approximately 20.08% of the total WL area in 2001. To sum up, major land cover type transformations (Table 2) indicated that nearly 30 % of land cover had changed during the study period.

Table 2. Major land cover changes in the Yunnan from 2001 to 2010

Land cover types (2001)	Land cover transformation (2010)	Change of the area (km^2)	Percentage of the changed area compared to corresponding area in 2001 (%)
CL	WL	13047.10	22.16
	GL	11430.50	25.30
WL	CL	14307.60	7.01
	GL	26701.00	13.07
GL	CL	18335.50	8.98
	WL	35485.80	17.38
WT	CL	22.09	1.88
	WL	220.65	18.78
UB	CL	139.99	4.64
	GL	37.48	1.24
UUL	WL	23.83	10.29
	GL	104.78	45.26

The spatial transformation distributions of the major land cover types, including CL, WL and GL in the Yunnan from 2001 to 2010, were presented in Figure3. The changes from CL to WL occurred in the northeast of the region and the transformation to other types was widely distributed throughout the region. The variation from WL to CL and GL mainly took place in the east. The conversion from GL to CL and WL mainly occurred in the west and east.

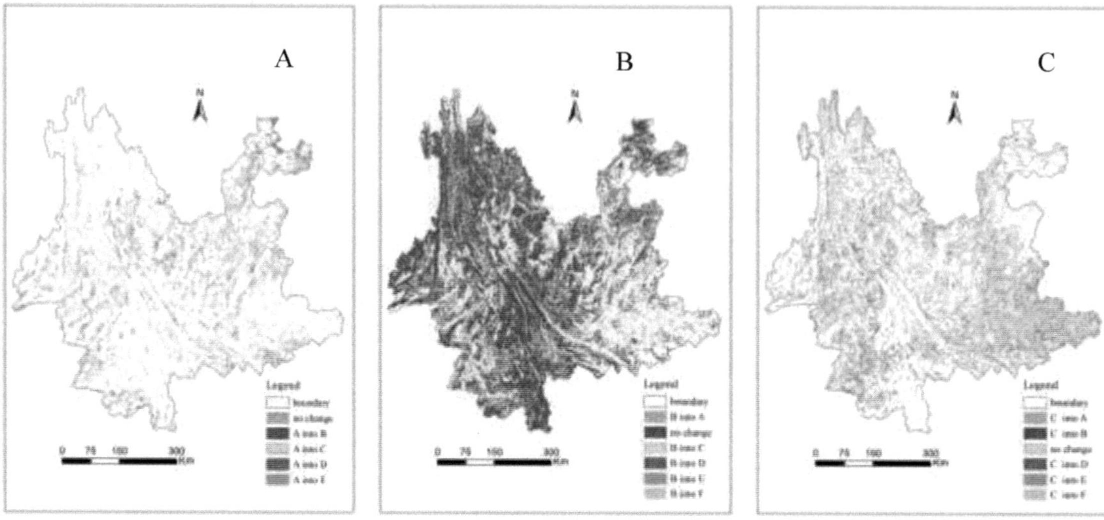

Figure.3 The major land cover types transition from 2001 to 2010.

(A:CL,B:WL,C:GL,D:WT,E:UB,F:UUL)

On the whole, the land cover types of the Yunnan during the study period had changed a lot. A major land cover variation happened when GL converted to WL, with an area variation up to 35485.8 km^2. Moreover, the percentage of UUL that transformed to GL is the largest, reaching 45.26 %.

Table 3. Elevation gradient changes of different land cover types in Yunnan.

land cover types	Average elevation gradient (m)			Area percentage (%)											
				Elevation gradient<2000 m			2000 m-3000 m			3000 m-4000 m			Elevation gradient>4000 m		
	2001	2010	Δ	2001	2010	Δ	2001	2010	Δ	2001	2010	Δ	2001	2010	Δ
CL	1655.30	1633.71	-21.59	77.79	76.37	-1.41	21.57	23.35	1.79	0.63	0.27	-0.36	0.02	0.00	-0.01
WL	1985.22	1973.19	-12.03	54.93	55.20	0.27	35.40	34.76	-0.63	8.70	8.96	0.26	0.97	1.07	0.10
GL	1779.71	1761.88	-17.83	71.60	72.60	1.00	19.63	18.64	-0.99	5.46	5.20	-0.27	3.31	3.56	0.25
WT	1736.34	1697.40	-38.94	93.31	92.44	-0.87	5.76	6.68	0.92	0.85	0.72	-0.12	0.08	0.16	0.08
UB	1705.76	1707.73	1.97	85.91	85.88	-0.02	13.69	13.71	0.02	0.41	0.41	0.00	0.00	0.00	0.00
UUL	4106.67	3747.77	-358.9	19.64	16.05	-3.59	1.34	0.00	-1.34	6.25	10.55	4.24	72.77	73.46	0.69

Δ: Difference

3.2. Land cover changes along with elevation and slope gradient

The distribution characteristics of land cover, which were closely related to the elevation gradient, were presented in Table 3. From the table, a prominent vertical distribution pattern in the Yunnan was summarized. The average elevation gradient of different land cover types from the highest to the lowest were UUL, WL, GL, WT, UB and CL.

CL, the lowest of land cover type in Yunnan with the average elevation gradient of 1600 m, was mostly distributed in the gentle-slope valleys below 1700 m with the comparatively lower elevation gradient and the relatively wetter and warmer climate, and this area was nearly 78% of the total CL area. The remaining area of CL was located between 2000 and 3000 m. About 55% of WL was distributed below 2,000 m, and 35% distributed between 2000 m and 3000 m. The average elevation gradient for WL was approximately 2000m. The GL was widely distributed in the east of the Yunnan, which had an average elevation gradient of almost 1800 m, with more than 70% at elevation gradients lower than 2000 m, approximately 20% between 2000 m and 3000 m and nearly 10% above 3000 m. WT and UB were located below elevation gradients of 2000 m, with the average elevation gradient up to 1700m. WT and UB, distributed in areas with gentle terrain and convenient locations, were the land cover types seriously impacted by human activities. UUL, the highest land cover type, was mainly spread above the elevation gradient of 4100 m. Only 17% of UUL was distributed below 4,000 m.

From 2001 to 2010, there were different decreases in the average elevation gradients of CL, WL and GL, the largest decline of average elevation gradients of land use type was UUL from 4100 m to 3750 m. However, there was a very small increase in the average elevation gradients of UB. Above 4000m, there were fewer changes to all land cover types because of few human activities in the high elevation gradient areas of the Yunnan.

The distribution of land cover types along with their slopes were summarized in Table 4. From the table, most of the land cover types under a certain slope had no significant changes with the passage of time. The average slope of the different land cover types from the lowest to the highest were WT, UB, CL, GL, WL and UUL. WT and UB were located in the zone with an average 2°, and only 13% of the area was located in a zone with a slope exceeding 6°. The average slopes of CL, WL and GL are approximately 6°. Almost 60% of CL as well as GL were distributed in zones with slopes lower than 6°, and 45% of WL was in zones with slopes below 6° as well as between 6° and 15°. UUL was the land cover type with the highest average slope, exceeding 11°, and approximately 39% and 36% of UUL in 2001 were located below 6° and 6°-15°, respectively.

Table 4. Slope gradient changes for different land cover types in Yunnan.

land cover	Average slope/			Area percentage/%											
				slope<6°			6-15°			15-25°			slope>25°		
	2001	2010	Δ	2001	2010	Δ	2001	2010	Δ	2001	2010	Δ	2001	2010	Δ
CL	5.76	5.27	-0.49	62.69	64.97	2.28	32.19	30.11	-2.08	4.92	4.75	-0.17	0.20	0.17	-0.03

WL	7.57	7.63	0.06	46.72	45.56	-1.17	43.96	44.94	0.98	8.72	8.90	0.18	0.60	0.61	0.00
GL	6.09	5.91	-0.18	59.64	61.75	2.11	33.78	32.07	-1.72	6.12	5.70	-0.42	0.46	0.48	0.03
WT	2.10	2.29	0.19	86.79	86.97	0.18	10.67	11.18	0.51	2.37	1.61	-0.76	0.17	0.24	0.07
UB	2.50	2.58	0.08	87.63	87.64	0.01	11.59	11.58	-0.01	0.74	0.74	0.00	0.03	0.03	0.00
UUL	11.15	9.80	-1.35	38.84	27.16	-11.7	35.71	38.27	2.56	18.30	27.8	9.47	7.14	6.79	-0.35

There were average slope decreases for most land cover types from 2001 to 2010, including CL, GL and UUL, and there were also small slope increases among the other types of land cover. The average slope of UUL dropped considerably, from 11.15° to 9.8°. Interestingly, the percentage of UUL in flatter areas (<6°) decrease and steep areas (15°-25°) increase. This finding implied that UUL was moving from flatter areas to steep areas.

3.3. Analysis of Landscape

Landscape configuration metrics were calculated using FRAGSTATS software based on the reclassified land-use map [23]. According to the characteristics of the landscape metrics, we found that the selected metrics had an ecological significance that described the characteristics of the landscape heterogeneity for operations. Results were shown in Table 5 and Table 6.

Table 5. The landscape patch types in Yunnan

	CA（ha）	PLAND	LPI(%)	PAFRAC	NP	PD	COHESION(%)
2001							
CL	5157675	13.4198	0.3812	1.5864	29253	0.0761	90.5247
WL	21975725	57.179	45.4179	1.5783	16727	0.0435	99.884
GL	10857475	28.2502	8.5471	1.6211	33267	0.0866	98.9083
WT	117475	0.3057	0.067	1.3844	758	0.002	85.4347
UB	301725	0.7851	0.0946	1.4667	1723	0.0045	81.2872
UUL	23150	0.0602	0.0104	1.4883	281	0.0007	69.4345
2010							
CL	5940425	15.4565	1.3859	1.5669	25679	0.0668	95.433
WL	22650375	58.9344	45.7657	1.5323	15193	0.0395	99.8725
GL	9399525	24.4568	7.8123	1.607	29044	0.0756	98.7974
WT	124800	0.3247	0.0729	1.4135	603	0.0016	87.8
UB	301800	0.7853	0.0947	1.4661	1723	0.0045	81.2238
UUL	16300	0.0424	0.0038	1.4694	167	0.0004	68.1645

CA: Class area. PLAND: Percentage of landscape. NP: Number of patches. PD: Patch density. LPI: Largest patch index. ED: Edge density. PAFRAC: Perimeter-area fractal dimension. COHESION: Patch cohesion index.

Table 6. Comparison of landscape metrics at the landscape level in Yunnan.

	TA (ha)	LPI	Area_ MN	SHAPE _AM	NP	PD	ENN_ MN	CON TAG	COHE SION	SPL IT	SHD I
2001	38433 225	45.4 2	468.65	68.14	820 09	0.2 1	1336.7 0	49.95	99.63	4.67	1.00 7
2010	38433 225	45.7 7	530.78	52.96	724 09	0.1 9	1377.8 8	51.80	99.59	4.61	1.00 5

TA: Total area. NP: Number of patches. Area_MN: Mean patch area. PD: Patch density. LPI: Largest patch index. SHAPE_AM: Area-weighted mean shape index. ENN_MN: Mean Euclidean nearest neighbor distance. CONTAG: Contagion. COHESION: Patch cohesion index. SPLIT: Splitting index. SHDI: Shannon's diversity index.

At the class level, the NP and PD of all land cover types had a decreasing trend over the past 10 years, which showed that the fragmentary of these types dropped. According to LPI, WL represented the highest concentration of land use types, with an LPI reaching 45%. GL are the second concentration, and the other four types were found in very low. The LPI of CL increased significantly from 0.381 in 2001 to 1.386 in 2010, which reflected the fact that the distribution of CL in the Yunnan was more fragmentary in the early of 2000s. It should be noted that all PAFRAC of land cover types were at approximately 1.5, which indicated that the patch shape of this region was complex, thus human activities were relatively obvious and we should reduce the impact of human activities throughout the entire region.

The COHESION can measure the physical connectedness of the corresponding patch type. Different types of ecological landscapes have quite different degrees of aggregation. The aggregation of CL, WL and GL were higher than 90%, with efficient material and smooth energy flow. Over the past 10 years, the degrees of aggregation were unapparent, within the fluctuation range of ±5%. In addition to CL, the fluctuation of aggregation in the rest of the landscape surface features was low. The COHESION of CL experienced growth from 90.5% in 2001 to 95.4% in 2010, indicating that the connection of CL improved.

At the landscape level, compared with 2001, NP in 2010 decreased from 82009 to 72409, and Area_MN expanded by approximately 62 ha, showed that many original patches joined together larger ones. The value of SHAPE_AM was declined from 68.14 in 2001 to 52.96 in 2010, which showed the shape of patch become more complex and irregular. SHDI changed from 1.007 in 2001 to 1.005 in 2010, which indicated that the landscape heterogeneity was decreasing slightly. During the 10 years, landscape fragmentation clearly dropped, the connection of landscape patches increased, and land cover gradually moved to the direction of stabilization and homogenization under the influence of human activities in the study area.

3.4. Analysis of driving factors

Land use change caused by human activities is the main cause of global environmental changes, and it is also the main factor of regional land use change. Differences of natural conditions determine the basic pattern of land use, while the conflicts between population and land accelerate the speed and extent of land use change. As well, traditional minority national customs, agricultural production mode and farming method can contribute to regional spatial heterogeneity of land use. Furthermore, under the impact of the market economy, seeking to maximize land use interest such as timber exploitation, agriculture planting, etc., is the direct cause of regional land use changes frequently. In addition, policy implementation has contributed to the change of land cover. The Yunnan is one of the provinces in the implementation of the Western Development Strategy that began in 2000, and its land cover has experienced changes influenced largely by the driving policy forces over the past 10 years. For example, the increase of WL areas indicates that the implementation of reforestation and afforestation of barren hills achieved some success as well as the promotion of sustainable economic development of Yunnan forestry.

Overall, its geographical features with more mountain less-flat plains determines the conflict between the growing demand for land and the limited supply that will exist for a long time with the development of urbanization and industrialization. To solve the land conflict of the Yunnan, the idea of constructing Mountainous Town was developed in 2011 and received a great deal of attention [10]. This idea suggested a slope-stratified and gradient -exploited land use pattern. The main areas of the Yunnan construction of Mountainous Town is located on the hilly slope land with slopes from 8° to 25° , with a total area of approximately 11570 km^2, especially on barren hills and slopes unable to be afforested. This area can provide a broad space for the construction of Mountainous Town in Yunnan. So the scientific implementation of Mountainous Town policy will lead to major changes in the land cover pattern in Yunnan.

4. Conclusions

The quantitative evaluation of the land use dynamics in the Yunnan were developed from MODIS dataset from 2001 to 2010 and coupled with GIS analyses. The results showed that the regional change of land use structure in the area took on a type of regular pattern. Mutual transformations between the three types of land cover, WL, CL and GL, changed relatively dramatically, total 30 % of land cover area changing during the study period. In quantity, the most obvious land use changes were the increases of CL and the decrease of GL. By percentage, the most rapidly changing area occurred in UUL with the annual variation of 3.29 %.

Different land use types indicated disparate distribution and variation characteristics related to elevations and slopes. The average elevation gradient of different land cover types from the highest to the lowest was UUL, WL, GL, WT, UB and CL. Bare land was distributed in the highest and steepest areas, and CL was distributed in the gentle sloping valleys below 1700 m of the region. There were small changes in the average elevation for CL, WL, GL, WT and UB from 2001 to 2010. The most obvious decline in elevation happened for UUL, particularly in the range between 3,000 and 4,000 m. With the passage of time, most of the land cover types under a certain slope had no significant changes. The average slope of dropped from 11.15° to 9.8°, which mostly occurred in the zone below 6°.

On the basis of landscape metrics analysis, the results show stabilization and homogenization will be the direction of land cover structure in the future in Yunnan region. The regional differences of land use structure in the area are strongly influenced by such factors as the geographical position, physiognomic situation, level of economic development and land use policy. The internal mechanisms of land use changes in Yunnan should be detected in the future research.

Acknowledgements
This study is supported by the National Key Technology R&D Program (No. 2016YFC0401308).

References

[1] Pielke R A, Marland G, Betts R A, Chase T N, Eastman J L, Niles J O and Running S W 2002 The influence of land-use change and landscape dynamics on the climate system: relevance to climate-change policy beyond the radiative effect of greenhouse gases *J. Mathematical, Physical and Engineering Sciences* **360** 1705-19

[2] Kalnay E and Cai M 2003 Impact of urbanization and land-use change on climate *J. Nature* **423** 528

[3] Pielke R A 2005 Land use and climate change *J. Science* **310** 1625

[4] Feng Y, Luo G, Lu L, Zhou D, Han Q, Xu W, Yin C, Zhu L, Dai L and Li Y 2011 Effects of land use change on landscape pattern of the Manas River watershed in Xinjiang, China *J. Environmental Earth Sciences* **64** 2067

[5] Turner B L, Skole D L, Sanderson S, Fischer G, Fresco L and Leemans R 1995 Land-use and land-cover change Science/Research Plan *Global Change Report*

[6] Seto K C and Fragkias M 2005 Quantifying spatiotemporal patterns of urban land-use change in four cities of China with time series landscape metrics *J. Landscape Ecol.* **20** 871

[7] Zhang X, Yu X, Wu S, Zhang M and Li J 2007 Response of land use/coverage change to hydrological dynamics at watershed scale in the Loess Plateau of China *J. Acta Ecologica Sinica* **27** 414

[8] Wang Y, Shao M A, Zhu Y and Liu Z 2011 Impacts of land use and plant characteristics on dried soil layers in different climatic regions on the Loess Plateau of China *J. Agr. Forest Meteorol.* **151** 437

[9] Yu W, Zang S, Wu C, Liu W and Na X 2011 Analyzing and modeling land use land cover change (LUCC) in the Daqing City, China *J. Applied Geography* **31** 600

[10] Ming Q, Wang J and Zhang W 2012 A geographical interpretation of mountain exploitation and construction of mountainous towns: a case study of Yunnan Journal of Yunnan Normal University (Humanities and Social Sciences) **4** 9 (in chinese)

[11] Tang C 2009 A comparative analysis on temporal spatial characteristics and dynamic mechanism of regional land use and land cover change-a case study on Huadu and Nanhai Districts *Journal of Natural Resources* **24** 414

[12] Li H, Ma Y, Aide T M and Liu W 2008 Past, present and future land-use in Xishuangbanna, China and the implications for carbon dynamics *J. Forest Ecol Manag* **255** 16

[13] Hu H, Liu W and Cao M 2008 Impact of land use and land cover changes on ecosystem services in Menglun, Xishuangbanna, Southwest China *J. Environ Monit Assess* **146** 147

[14] Jianchu X, Xihui A and Xiqing D 2005 Exploring the spatial and temporal dynamics of land use in Xizhuang watershed of Yunnan, southwest China International *Journal of Applied Earth*

Observation and Geoinformation **7** 299

[15] Fox J and Vogler J B 2005 Land-use and land-cover change in montane mainland Southeast Asia *J. Environ Manage* **36** 394

[16] Yungang L, Daming H, Jinming H and Jie C 2015 Variability of extreme precipitation over Yunnan Province, China 1960-2012 *Int. J. Climatol.* **35** 245

[17] Awasthi K D, Sitaula B K, Singh B R and Bajacharaya R M 2002 Land-use change in two Nepalese watersheds: GIS and geomorphometric analysis *J. Land Degrad Dev.* **13** 495

[18] Velázquez A, Durán E, Ramírez I, Mas J, Bocco G, Ramírez G and Palacio 2003 Land use-cover change processes in highly biodiverse areas: the case of Oaxaca, Mexico *J. Global Environmental Change* **13** 175

[19] Zhang F, Tiyip T, Feng Z D, Kung H T, Johnson V C, Ding J L, Tashpolat N, Sawut M and Gui D W 2015 Spatio-temporal patterns of land use/cover changes over the past 20 years in the middle reaches of the Tarim River, Xinjiang, China *J. Land Degrad Dev.* **26** 284

[20] Dong L, Wang W, Ma M, Kong J and Veroustraete F 2009 The change of land cover and land use and its impact factors in upriver key regions of the Yellow River *J. Int J Remote Sens.* **30** 1251

[21] Zhang K, Yu Z, Li X, Zhou W and Zhang D 2007 Land use change and land degradation in China from 1991 to 2001 *J. Land Degrad Dev.* **18** 209

[22] Jia B, Zhang Z, Ci L, Ren Y, Pan B and Zhang Z 2004 Oasis land-use dynamics and its influence on the oasis environment in Xinjiang, China *J. Arid Enviro.* **56** 11

[23] McGarigal K and Marks B J 1995 Spatial pattern analysis program for quantifying landscape structure. Gen. Tech. Rep. PNW-GTR-351 US Department of Agriculture, Forest Service, Pacific Northwest Research Station

[24] Haines Y R and Chopping M 1996 Quantifying landscape structure: a review of landscape indices and their application to forested landscapes *J. Prog. Phys. Geog.* **20** 418

[25] Raines G L 2002 Description and comparison of geologic maps with FRAGSTATS-a spatial statistics program *J. Comput Geosci-Uk* **28** 169

Institute of Physics
Dirac House, Temple Back
Bristol BS1 6BE UK

ISSN: 1755-1307
ISBN 978-1-5108-3051-6

CURRAN ASSOCIATES INC.
proceedings.com